SYSTEMATIC INSTRUCTION IN MATHEMATICS

SYSTEMATIC INSTRUCTION
IN MATHEMATICS
FOR THE MIDDLE AND
HIGH SCHOOL YEARS

MARGARET A. FARRELL
State University of New York at Albany

WALTER A. FARMER
State University of New York at Albany

 ADDISON-WESLEY PUBLISHING COMPANY

Reading, Massachusetts • Menlo Park, California
London • Amsterdam • Don Mills, Ontario • Sydney

ACKNOWLEDGMENTS

The authors wish to thank the copyright holders for permission to reprint in this text the following figures:

Figs. 9.4, 9.5, 9.6: From Book H of the *School Mathematics Project*, by J. Harris, D. A. Hobbs, K. Lewis, R. W. Strong, and T. Wilson (New York: Cambridge University Press, 1972), pp. 111, 112, and 114. Reproduced by permission of Cambridge University Press.

Fig. 9.7: From *Geoboard Geometry*, by Margaret Farrell (Palo Alto, Calif.: Creative Publications, Inc., 1971), p. 9. Reproduced by permission of Creative Publications, Inc.

Fig. 9.8: From *Nuffield Mathematics Project Problems*, Green Set. Copyright 1969 by John Wiley & Sons, Inc. Reprinted by permission of John Wiley & Sons, Inc.

Fig. 9.9: From *Mathematics One*, by Suppes *et al.* (New York: Random House, 1974), p. 409. Reproduced by permission of Random House, Inc.

Fig. 10.6: By permission of the editor, B. W. King, and the author, Constance C. Feldt, from "Ratios and Proportions," *New York State Mathematics Teachers' Journal*, Winter 1977.

Library of Congress Cataloging in Publication Data

Farrell, Margaret A
   Systematic instruction in mathematics for the
middle and high school years.

   Bibliography: p.
   Includes index.
   1. Mathematics--Study and teaching (Secondary)
2. Mathematics--Study and teaching (Higher)
I. Farmer, Walter A., joint author. II. Title.
QA11.F37 1979   510'.07'12   79-4250
ISBN 0-201-02436-5

Dedicated to Mary Farrell and Yvonne Farmer

PREFACE

This book evolved over a ten-year period as we collaborated to try to improve key parts of the undergraduate program for the preparation of mathematics and science teachers. Everyone we contacted indicated the need to develop a text significantly different from those currently available. Chapters in mimeographed form were tried out as each one was developed; the completed text was revised based on the classroom performance of the student teachers we instructed in both methodology and the related student-teaching experience. Extensive feedback from both students in training and cooperating school teachers supported the approaches we took. *Systematic Instruction in Mathematics for the Middle and High School Years* is a result of these efforts.

Our overall approach reflects the belief that teaching is much more than a simple stimulus-response activity. We view teaching as a problem-solving task involving knowledge, comprehension, application, analysis, synthesis, and evaluation. Thus we have employed a systems analysis approach that stresses the collection and analysis of feedback on the results of instruction. Emphasis is placed on the importance of considering both the intellectual development of students (Piagetian research) and the nature of the content to be taught (process and product aspects of the academic discipline) prior to selecting objectives. We then approach the design of instructional strategies (planning) using this base and applying the work of contemporary learning psychologists, such as Ausubel and Gagné. Both formative and summative aspects of testing/evaluation are treated as integral parts of the systematic approach. We also give attention to contemporary curricula options, resources for instruction, and student discipline, since these are of great practical importance to beginning teachers.

Each chapter begins by redirecting attention to the systems analysis model represented in schematic form; we ask you to rethink what has already been developed and to consider how the current topic fits into the total picture. Each chapter generally follows a sequence of: an advance organizer, an introductory activity, several topical subsections, cross-references to a sample Competency-Based Teacher Education (CBTE) evaluation instrument, summary and self-check, simulation/practice activities, and suggestions for further study. The CBTE instrument cross-referenced in the text is a field-tested evaluation tool whose clusters of competencies flow from the systems model. The instrument itself and a summary of the way it was developed and used is found in Appendix B. The annotated list of references at the end of each chapter was deliberately limited to a small selection of sources appropriate for the beginning student. Some of the activities and exercises are paper-and-pencil tasks, while others should be tried out in a classroom or micro-teaching situation. Throughout the text the reader is urged to avoid rote memorization and blind acceptance. Instead, the stress is on thinking things through by studying the printed materials and checking the ideas in the text against experiences of self, classmates, the instructor, supervisors of student teaching, secondary-school teachers, and administrators.

Since the systems analysis model is cyclic in nature there are several possible sequences for the presentation of chapters. We decided to use a psychological ordering, a choice that fits the logic of the model. We have found that students react well to starting with instructional modes because all students have a common experience base as consumers of instruction and because novices' thinking tends to focus on these obvious externals of the teaching act. There are several possible orderings of the text readings which instructors might use, depending on their objectives, the length of the course, and the background of their students. For example, an instructor might initi-ally assign only the section of Chapter 5 that deals with cognitive objectives and then ask students to study the section of Chapter 8 that deals with the writing of paper-and-pencil items. The sections on affective objectives in Chapter 5 and their assessment from Chapter 8 could be assigned later in the course. A similar choice holds for the psychomotor domain. Chapters 7 and 8 *each* deal with immediate, as well as long-range, concerns of the novice teacher. Some methods instructors will prefer to assign only those sections of Chapter 7 that treat the planning of single lessons. Similarly, the section on unit test construction, grading, and item analysis from Chapter 8 can be assigned late in the course, whereas the writing of items to match objectives might be profitably studied much earlier. Other sections in Chapters 7 and 8 could be treated as reference material to be used in more depth during student teaching. Chapters 9, 10, and 11 can be threaded throughout the course in diverse ways. Finally, if students have extensive background in areas such as psychology of learning, developmental psychology, or evaluation, some of the basic materials in Chapters 3, 6, and 8 can be quickly reviewed and more time can be spent on the novel applications presented.

This book was initially constructed to serve as a basic reading source for students both during coursework in methodology and throughout the student-teaching experience. Feedback from graduates indicates that many have found it to be a useful reference during their first year or so of contractual teaching, and several experienced teachers who borrowed copies told us they found much of it a valuable aid to analysis and improvement of their own classroom practices. Gratifying as this response has been, we would caution all who use this book that it is not a prescription for all the ills that beset education and should not be evaluated on that basis. Rather it should be considered as one tool in the total educational experience of mathematics teachers.

We acknowledge, with gratitude, the helpful suggestions of a former SUNYA colleague, Dr. Joseph Kelly, who read the early chapters of the manuscript. We owe a special debt to Dr. George Martin of the Mathematics Department, SUNYA, who critiqued several drafts of Chapter 4. His perceptive comments added much to that important chapter. We are also grateful to Mrs. Beverly Spiegel, who cheerfully and patiently cut red tape and eased countless minor burdens throughout the work on the manuscript. Finally, we thank our students, whose talents and questions challenged us to write this text.

Albany, New York
November 1979

M. A. F.
W. A. F.

TO THE STUDENT

So, you want to be a teacher! Are you *sure* that secondary-school mathematics teaching is for you and that you are for it? No expert can tell you the answer, in spite of thousands of studies designed to predict success prior to actual teaching experience. Nor is there any evidence to support the notion that teaching skills are passed on by heredity: the fact that Dad and Aunt Sally are teachers is largely in the realm of interesting but unrelated information. Neither will the mere reading of this book in and of itself provide you with the answer. What this book is designed to do is to promote an interaction of your talents with real-world classroom situations in a systematic way so that you will find an answer and so that you will know when you have found it.

The term *interaction* was deliberately chosen. Becoming a teacher is largely a do-it-yourself job. Of course a methodology instructor, a college supervisor of student teaching, and a school cooperating teacher can and will help in many important ways and at crucial points in the process. However, no one person or combination of these persons can do your part. What is your part? Our cumulative experience in working with over five hundred young men and women in methodology courses and student teaching points to several things *you* must bring to the situation.

Comprehension of subject matter is certainly one key ingredient. No one expects you to recall all the detailed information you have picked up in many semester hours of coursework in your major subject field. In fact, much of this information will find only the most limited use in teaching middle- and senior-high-school students. What is important is that your command of the fundamental concepts and basic principles enables you to apply those concepts to unique or novel situations and to treat them on various levels of sophistication while maintaining the integrity of the academic content.

Courage is also vital. Inexperienced teachers are generally "tested" by secondary-school students who want to determine what they can get away with in class. Novice teachers who cannot bring themselves to consistently insist on adherence to some simple rules of conduct—and thus risk temporary dislike by their pupils—don't last very long. Little things overlooked in the hopes that they will go away, soon grow into big things; confrontations follow, and growing disrespect for the teacher breeds overt hatred. When situations progress to this point the odds (25 students versus 1 teacher) are almost always insurmountable, and another teacher is lost to the profession.

You must genuinely like young people! Even the dullest of students can sense a teacher's dislike in spite of the teacher's best efforts to conceal the feeling. Even professional actors or actresses would be hard pressed to present a facade for five class periods a day, five days a week, throughout the entire school year. When students perceive a teacher's dislike of them as young people, problems are certain to develop. Then, too, consider how such teachers must feel having to work closely for extended periods of time each day with age groups they do not like. The start of each teaching day must seem like the beginning of a familiar bad dream.

Physical endurance is often overlooked. Beginning teachers are constantly amazed to discover how physically exhausted they are at the end of a teaching day. All the years you spent sitting as a student has done little to prepare your muscles for the amount of standing and walking required by teaching. On top of physical strain in the classroom, teachers have to face correction of papers and preparation of lesson materials after dinner. This proves too much for some to endure.

A realistic view of self is another prerequisite for success. All of us bring strengths and weaknesses to the teaching situation. You must learn how to capitalize on your strengths and be determined to shore up your weaknesses as rapidly as possible. This means that you must learn to be receptive to constructive criticism offered by supervisors and you must develop the skills of self-criticism. After all, shouldn't the main goal of teacher education be to help novices progress to the point where they can succeed in the classroom and continue to learn about instruction on their own throughout their teaching careers? Because we firmly believe that it is, the main purpose of this book is to assist you in becoming a life-long student of teaching.

If what we have said thus far makes sense to you and if you are determined to give teaching your best shot, you are our kind of person and this book was written with you in mind. Teaching can and should be an intellectually stimulating and ever-changing challenge. It should also be an endless source of satisfaction.

Note that the word *instruction* instead of the word *teaching* appears in the title of this book. Our choice of words indicates emphasis on producing learning according to specific intents of the teacher (objectives) by purposefully controlling those variables known to affect various types of learning. No doubt your students will learn in your classes things other than those you intended to convey. We pay some attention to these byproducts of instruction, but we decided to focus on a systematic approach to instruction. In particular, in this text we are referring to a systematic approach to instruction in mathematics.

Obviously there are many aspects of instruction that generalize across any subject matter. However, our experience indicates that novices learn best by starting to think about the specifics of their own discipline, to apply these specifics in particular instances, and to derive generalizations *after* a critical mass of first-hand experience has been accumulated. Further, each subject matter discipline has its own type of conceptual framework (structure of ideas) and its own kinds of processes for generating ways of finding out that system of ideas. Analyzing the discipline in terms of these process and product aspects reveals its potential as a source of objectives (specific goals to be attained by learners).

Does one teach *content* or *kids*? Think this over for a minute and see if you believe that question is really worth considering as an either/or proposition. Practitioners almost universally agree on a response such as "Both," or "One teaches subject matter to students," or "One teaches students the subject matter." As experienced secondary-school teachers, we long ago abandoned any inclination to consider either/or responses to this perennial question. This point of view should be obvious in the systems analysis model schema that precedes each chapter of the text.

Note that this schematic model incorporates both a planning and an implementation vector. Both the nature of the content to be learned and the intellectual development of students are treated as essential prerequisites to the specification of objectives. Once realistic and valid aims have been specified, knowledge about how humans learn various categories of content can be employed as a basis for designing instructional strategies. The strategies must include plans for *giving feedback to students* so they will know how they are progressing toward the objectives during the implementation stage. It is also vital to *plan for and collect feedback from students* so that planned strategies can be intelligently altered during the same stage. Similarly, the feedback loop that goes back to the specified objectives can function only if knowledge of student attainment of these objectives is obtained regularly and systematically.

Does this simple model of instruction make some sense to you? It should, because we have developed it by working with hundreds of prospective teachers. However, don't worry if all its ramifications are not clear at this point. That's what this book is all about. Read it, but don't stop there. Interact with the ideas, perform the activities, then apply the systematic approach with your students and *believe the data*.

CONTENTS

CHAPTER ONE

INSTRUCTIONAL MOVES
BY THE TEACHER
MODES OF INSTRUCTION

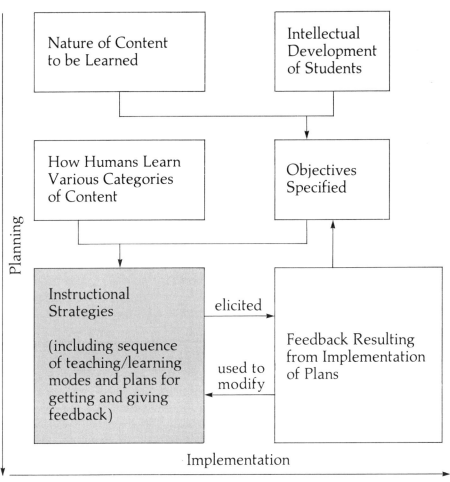

ADVANCE ORGANIZER

As explained in the introduction, we view teaching as a complex problem-solving task and are committed to a systems analysis approach. The schema of that systems analysis model appears on the title page of this chapter. Note the shaded box that illustrates the place where *Modes of Instruction* fit into the total picture. Why begin with components of instructional strategies? Certainly there are several logical alternatives. Why was the decision made to begin with modes?

We begin with modes of instruction because we believe it is the most psychologically sound beginning point for students who are preparing to teach. It is an area where you have a wealth of concrete experiences upon which to draw. Your teachers have used these modes since the day you entered kindergarten, so your experiences with these modes are readily available in your memory and will enable you to interact with the information and ideas to be presented in this section.

Before proceeding any further, let's be sure the phrase *Modes of Instruction* communicates the idea intended. Some prefer to call these behaviors *Teaching/ Learning Activities*. Either term means essentially the same thing. Just ask yourself the question, "What kinds of things do teachers do in an effort to help students learn the content of subject matter?" If you reflect upon your own experiences as a student, certain things are sure to come to mind. Most students recall lecture, discussion, homework, question/answer, audiovisual aids, demonstration, and laboratory work. If you came up with an assortment similar to these, we are off to a good start. Now consider whether you, each of your classmates, and your instructor all mean exactly the same thing when using each of these labels or terms. Our experience tells us no. This is not at all surprising, since we tend to use several of these terms very loosely in our speech and since most textbooks do little better on this point. If we are to communicate ideas effectively, we must first get operational definitions of the key terms to be used.

Let's begin with an operational definition of *operational*. A definition is of the *operational* type if *it refers to key observable traits and, wherever possible and appropriate, how these are measured.* Unlike most dictionary definitions, our definition does *not* describe the new entity in terms of abstractions (non-observables). Do you have the idea? The only way to tell for sure is to try to use that idea in appropriate ways.

You will get feedback as you interact with the main part of the next section, where we consider the most commonly employed modes, their operational definitions, their uses in instruction, and their potential for getting and giving feedback (knowledge of results). We begin with the lecture mode because it is the one that college students have experienced most frequently. The sequence of others that follow was selected according to a preconceived pattern of verbal interaction. See if you can detect this pattern as you read. Above all else, be sure to interact with the material, as rote memorization will net you nothing for your effort when the time comes for you to step into the role of teacher.

1.1 OPERATIONAL DEFINITIONS OF COMMON INSTRUCTIONAL MODES

Lecture During most of the time you spent in college classrooms you were expected to sit quietly and listen to the professor talk about the subject matter. Often you were expected to take notes, and sometimes your teacher used the chalkboard. Frequently at the end of such a class period the teacher asked, "Any questions?" Questions from students may not have been either expected or desired, and such questions were seldom forthcoming. What, then, is a brief operational definition for *lecture*? *The teacher talks, perhaps with some use of the chalkboard, while the students listen quietly and sometimes take notes.* Check this operational definition to be sure that it communicates meaning in terms of observables. Can you tell one when you see one? If so, this is a bona fide operational definition. Note that no time requirement is made in the definition, so this mode may occur for either an entire class period (as is frequently the case in college classes) or for smaller segments of a class session (as is more frequently the case in secondary schools).

Question/Answer This is one of the most common modes employed at the secondary-school level. Novice observers are usually amazed at the results of keeping tally of the number of questions asked by the teacher during a 40-minute class period. The total often exceeds one hundred and not infrequently approaches two hundred! The pattern typically begins with the teacher asking a question and then recognizing one student, who answers. Next the teacher reacts verbally in some way to the student's response and asks a question of another student, who then responds. Again the teacher handles the student's response and poses another question, which in turn is answered by a third student . . . and so the pattern continues. As in the case with the lecture, the teacher may write something on the chalkboard and sometimes the students take notes. Thus the operational definition for *question/answer* is: *The teacher asks a question; one student answers; the teacher reacts and asks another question which is responded to by a second student, and so forth.* Again note that the definition does *not* specify a time limit, so the mode may continue for all or any part of a class period.

Discussion This term is used very loosely by many teachers, students, and textbook writers. We have all observed instructors who began class with "Today we are going to discuss the very important applications of ———" and who then followed with 40 minutes of lecture on the topic. Textbook writers are also prone to make the same error, as evidenced by a host of chapters that begin with something akin to "The following topics will be *discussed*."

Throughout this book we use the term *discuss* to refer to *student-to-student talk with occasional verbal intervention by the teacher*. Think back to the last time you were involved in a buzz-group session. Your most recent participation was probably in the dorm rather than in a college classroom, for this mode is rarely employed by college teachers. If your last experience was an actual discussion, most or all of the participants contributed information and ideas *without* filtering them through the leader of the group every time. Have you caught the essential difference that discriminates this mode from the question/answer mode? If so, you are doing well. If you are not so sure, talk it out with a few classmates and refer back to the operational definition above. Note that no time limit is implied, nor is the size of the group narrowly prescribed. The entire class need not be (and seldom is) involved as a single discussion group. On the other hand, one person can hardly constitute a discussion group. A group consisting of five to seven participants is the size recommended as optimal for effective discussion; we'll outline more about size later. Have you caught on to the pattern we had in mind as we sequenced the presentation of the three modes defined (not *discussed*) thus far? If so, try to verify your conjecture as you read about the next three modes.

Demonstration The literal meaning of this word is "show" and that meaning pinpoints the operational definition of the mode. Typically, but not always, *the teacher shows something, such as a specimen or a model, while students watch.* In some instances, one student may do the showing while other students watch. You may be thinking at this point that usually some talking (such as lecture or question/answer) is used with a demonstration. Some talking usually is involved, but silent demonstrations can be used very effectively. What do we call a situation where demonstration, short lecture, and question/answer are used in an integrated fashion? This is one example of an *instructional strategy*.

We will consider the design of strategies in Chapter 7. If you reflect on your own experiences as a secondary-school student, you are probably about to conclude that, with rare exceptions, modes are usually combined in these strategies. This is good thinking. For now, however, let's focus attention on the individual modes, since this background will serve us well when all other prerequisites to strategy design have been mastered.

Laboratory This mode does not refer to a place nor a special class period in the weekly schedule, but to an activity. This activity may occur in a regular classroom, a specially equipped room, at home, or outdoors. The key idea, of course, is that *students manipulate concrete objects, specimens, or equipment under the direction of the teacher.* This is in clear contrast to demonstrations in which only one individual does the manipulation and in which all others watch. (The word *manipulation* is frequently used by mathematics teachers to mean "calculation." Note that this is not the sense in which we refer to manipulative tasks.)

Individual student projects At first glance, this may look like the laboratory mode. The critical differences are that here the *students are all doing different manipulative activities or varied library research, or different problem-solving tasks on an individual basis.* Notice that all students are *not* doing the same things in the same place at the same time as so often is the case in laboratory. Some student choice of activity is also implied by this mode.

By now you have probably discerned the pattern of sequencing of modes presented so far. We began with the "talking" modes, placing the most teacher-dominated mode first (lecture) and moving through to the least teacher-dominated mode (discussion). Next we considered the "showing" and "doing" modes, again moving from the most to the least teacher-dominated mode (see Fig. 1.1).

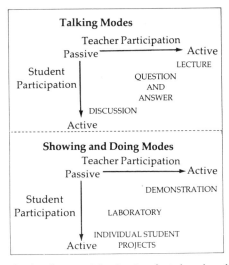

Fig. 1.1 Continua of teacher/student participation in selected modes of instruction.

We are indebted to Dr. Joseph R. Kelly, a former colleague, for devising this schema to illustrate our verbal description.

Audiovisual aids (AVAs) Actually all modes involve hearing, seeing, or both. However, this mode typically means that *students are viewing and hearing a motion picture or television, seeing slides, a filmstrip, or a loop film, watching an overhead transparency, or listening to a recording of some type.*

Supervised practice This mode involves having *students try to perform some task at their seats or at the chalkboard while the teacher observes their progress and gives help as needed.* Like all other modes, this one has an important role to play in effective and efficient strategy design. However, novices often either overlook its importance or carry out this mode in an ineffective manner. We will outline more about this later, but think about the relationship of this mode to the mode that follows. Perhaps you can get the jump on us.

Homework "Boo, Hiss!" is the typical student reaction to homework. Could this be due in part to the narrow definition given to the mode by typical teacher use? Let's not limit the possibilities to the odd-numbered problems for tonight. (Guess what is coming for tomorrow night?) We prefer to define it as *any activity relevant to the achievement of objectives that students perform outside of the scheduled class sessions and without the supervision of the teacher.* This definition encompasses such potentially meaningful activities as spending 15 minutes in the supermarket after school to get data from labels on competing brands of the same product, getting data for growth curves on plants grown over a two-week period, and interviewing a random sample of people to elicit their beliefs on a mathematics-related topic. It also leaves room for doing the even-numbered problems from a textbook when this makes pedagogical sense.

Although we might try to identify and operationally define a few more modes, the nine defined thus far are the common ones and will suffice for our purposes. By now if you have been interacting with the text, you should be anxious to get on to meatier questions of what each mode has to offer as input into strategy design and what the potential of each mode is for getting feedback (finding out how well each student is doing) and giving feedback (letting students know how they are progressing).

1.2 INTRODUCTORY ACTIVITY

There's no observation without
an observer.

A. With at least one other classmate, observe a junior/senior-high school class in your principal teaching field. As the class progresses, record data following the format shown in Fig. 1.2.

Time	Subject/Grade: **Math 11**	Topic: **Logarithms**
	Names of modes observed	Samples of feedback observed during the use of each mode
8:10 am to 8:15 am	1. T lectures on logs— history	1. 3 Ss have heads on desk, all others look front, are quiet, and take notes.
8:15 am to 8:20 am	2. T: Q/A on content of lecture	2. 10-12 Ss volunteer and T calls on 4 in front, praises ans., and clarifies.

Figure 1.2

Record instances of feedback the teacher got and any reactions by the teacher to that feedback. Be specific—for instance, list the approximate number of hands raised.

Use of the following shorthand will save time:

T: teacher
S (Ss): student (students)
Q/A: question/answer
Demo: demonstration
HWPM: Homework post-mortem ("going over" last night's homework)

B. Meet with one or more of your classmates who observed the same lesson. Compare the data collected and check for agreements/disagreements as to the modes observed and the related feedback. Account for any discrepancies in the data that might have resulted from lack of consistency in use of the operational definitions of the modes (consult Section 1.1).

C. Consult with classmates who observed other lessons and look for any common patterns, such as number of changes of modes employed within a single class period and amount and kinds of feedback given/received by the teacher. (Note: The above activity can occur in a live classroom or while viewing a videotape of a class.)

1.3 USES OF MODES

Lecture This mode provides opportunity for the teacher to provide verbal input that is not available to students in other forms (such as textbooks) or at an appropriate level (such as vocabulary or conceptual sophistication). It is assumed that the words of the lecturer convey the same meaning to all of the students. It is also assumed that the students are listening with the intent to comprehend and remember, not just displaying the "lecture syndrome" (eye movements following the teacher, smiles on faces when the teacher tells a joke, and so on). Since it is difficult for most secondary-school students to remain passive listeners for long periods of time, the lecturer should either keep it short or be a dynamic lecturer, or both. Few novices, if our experience is any guide, are dynamic lecturers. What has been your personal experience with lectures that extend for long periods of time? How long can you pay rapt attention? Does the lecturer seem to know when you are confused on a point, or does he or she continue to retread the same ground while you are anxious to get on to the next idea? What feedback is available to the lecturer? How can he or she interpret the feedback signs, particularly if the group is sophisticated in playing the game of school? Further, how do students find out if they are following the thinking of the lecturer? Obviously the feedback-giving and -getting potential of this mode is minimal and, along with the limited attention span problem, it is a strong argument for using lecture sparingly and for very brief time spans.

Question/answer Many novices, and some veterans, believe question/answer is a mode that enables the teacher to find out who knows what. Is it?

Only one student responds at a time, a sample of only one out of twenty-five or thirty! So much for feedback potential—better than lecture, but still extremely limited. Furthermore, many a novice has been misled by apparently positive feedback from one segment of a diverse group of students in a question/ answer sequence focused on the solution of a complicated problem. Quiz scores subsequently reveal far less success with a similar problem. Consider the following illustration: Jack answered question 1; Mary, question 2; Sam, question 3; and so on. That's right! Mary and Sam may have been cued in by the earlier response, but may be unable to *begin* the problem. Other students may be unable to *complete* the solution. What, then, is the *forte* of question/ answer?

The question/answer mode is extremely valuable as a way to guide developmental thinking, to stimulate creative problem solving, to initiate discussions, and to stimulate quick recall of prerequisites needed for the day's lesson. Indeed, the question/answer mode can be used effectively in combination with every other mode. The kind of question posed, the preamble to the question posed, and the variety of ways used to encourage and accept responses are all skills that make the difference between thoughtful interaction and humdrum, dull sequences.

There are three major components of the question/answer mode that need special attention: the questions themselves, the ways in which student responses can be obtained, and the ways in which student responses can be handled. Beginners can get disastrous results from trying to use the question/ answer mode when a complex question is dropped on the class like a bomb. Lack of student response may also be a function of the teacher's inability to handle earlier responses. Miss Goldberg complains that only Bill and Sara ever respond to her questions, but in Mr. Boswell's class, most or all members of the class mumble something whenever he asks a question. What can be done to help either of these teachers? A lot! But let's start at the beginning.

The following do's and don'ts are intended for the novice teacher whose difficulty seems to be related to the nature of the questions posed.

The question—do's and don'ts

1. Write down the major question in a developmental sequence and analyze the possible responses ahead of time.

2. Precede a question/answer sequence by a brief lecture or demonstration designed to set the stage for the sequence.

3. Do not ask frequent yes-no questions or fill-in-the-blank questions, such as "Does anyone know the answer to number five?" "No."

4. Increase the number of questions requiring a phrase or a sentence in response.

5. Do not try to elicit developmental thinking by the all-encompassing "What about" question, such as "What about the circle?"

6. Use a variety of opening question-phrases, such as "How? Who knows? When is . . . true? What is . . .? What seems to be . . . ? Why?"

Notice that suggestion 1 in the previous set refers to your planning, as does suggestion 2; but all the other suggestions require some on-the-spot analysis. An observer can help gather appropriate data; so can a tape recorder. In either case, a teacher can begin to alter his or her questioning skills over time. Why is suggestion 2 made? As you read the section on the discussion mode, look for a parallel guideline. Even the newest student of teaching realizes that if wide interaction does not occur, the teacher has no way of immediately assessing the extent to which the question/answer mode guided developmental thinking. Suggestion 3 now takes on more significance, since the constant violation of this guideline usually results in shouted (or mumbled) one-word responses from a small sample of those present.

As in the first set of suggestions, there are diverse ways of *getting* responses from students (the second critical ingredient in an effective question/answer mode); none should be used exclusively.

Getting responses from students—do's and don'ts

1. Pose the question *before* you call on someone.

2. Do not call on students in only one area of the room for all answers.

3. Ask shyer and slower students low-level questions.

4. Save high-level questions for brighter students.

5. Do not direct a series of quick questions to students row by row (or in any clear pattern).

6. Do not call *only* on students who volunteer.

7. Wait at least five seconds prior to accepting responses to high-level questions. Inform the students that you are going to do this.

8. Tell the students that there is no penalty for incorrect or partially correct answers. Tell them it is not a quiz, but a learning experience.

Since the teacher wants to optimize both the number and the quality of responses from the students, many of the above suggestions must be used in combination with one another. Occasionally, to achieve the same result the teacher should violate one of the previous guidelines. For example, Mrs. Frost cannot implement suggestion 1 from the responses list. As soon as the end of her question is heard and before she is able to call on a student, her seventh-grade class shouts out responses. What should Mrs. Frost do? See if the next set of suggestions helps.

These suggestions deal with the third component of the question/answer mode—handling student responses.

Handling student responses—do's and don'ts

1. Ask another student to agree or disagree and give his or her reasons for doing so.

2. Take a "straw vote" ("Let me see the hands of those who agree with Jack, those who disagree, those who aren't sure."), and follow up with a request for justification.

3. Frown a bit and ask, "Are you sure?"

4. Ask other students to add to the answer of the first student.

5. Ask the student to explain how he or she arrived at the solution.

6. Ask if there is another way to solve the problem.

7. Do not accept *mixed* chorus responses.

8. If a student cannot answer a difficult question, ask a contingency backup question on a lower level.

9. Refuse to accept responses that are not audible to all students.

10. Give praise for partially correct responses to complicated questions.

Mrs. Frost, who has to deal with a "too eager" class, may now use one of the above suggestions to direct attention to one student. One clear way to convince a class of the limitations of the mixed chorus response is to use suggestion 1 on a student. That student's confusion as to what he or she is asked to defend is more convincing than a hundred reminders from the teacher. These suggestions must be used with some care and in combination with those listed earlier.

In addition to all the concerns noted so far, the physical location of the teacher can greatly influence the nature of the verbal interaction. Look at Fig. 1.3 and analyze the flow of verbal interaction. Each number represents either a question or an answer, and it is clear that a large number of these occurred, but where? Why? Where was the teacher? (*T* designates teacher location.) Now analyze Fig. 1.4. There is still only one teacher in the classroom, but that teacher has moved to four different locations during the sequence. Why? Compare the sequence of numerals in the box designated as teacher responses in each case. A different pattern of verbal interaction is occurring in these two classrooms.

Chalkboard
T

1, 3, 5, 7, 10, 13, 15, 17, 19, 21, 23, 25, 27, 29, 31, 33, 35, 38, 41, 43, 45, 47, 49, 51, 53, 55, 57, 59, 61, 63, 65, 67, 69, 71, 73, 76, 79

	66	4, 64, 68	18, 34, 52, 54, 56	14, 32, 42, 44, 46, 48, 50
		24, 28	8, 11, 58, 70	9, 12, 26, 37, 40
2		30	36, 39, 74, 78	16, 60, 62, 75, 77
	6		72	20, 22

20 Students

Fig. 1.3   Flow of verbal interaction.

Chalkboard
T

1, 3, 5, 7, 9, 11, 14, 17, 19, 21, 25, 27, 29, 33, 35, 37, 39, 41, 43, 46, 48, 51, 53, 57, 59, 61, 63, 65, 67, 70, 72, 75, 77, 79

T

	22, 24	2	12, 15	10, 13, 16
20, 23	30, 56	76, 78	4, 6	55, 64, 66
18	26	36, 38	8, 69, 71, 73	68, 74
28, 31	32	52	42, 54	62
34	44, 49	40	45, 47, 50	58, 60

T

25 Students
T

Fig. 1.4   Flow of verbal interaction.

To what extent can the teacher control this pattern so that the broadest possible feedback sampling can occur? The teacher who stands next to a timid, soft-spoken student can encourage a louder voice by moving away from the speaker. The right-handed teacher may find that he or she frequently calls on students to the left of the teacher's desk. Your position is important and must be consciously varied to achieve optimum results.

Questioning skills are so crucial to instruction that books, tapes, and slide presentations have been constructed to help the teacher. Some sources you may find useful have been listed at the end of this chapter.

Discussion As noted earlier, most activities labeled "discussion" are really lecture or question/answer. In a discussion, as defined here, the teacher usually initiates the interaction, but is only sporadically heard from after that. What, then, is the first essential ingredient for a discussion? There must be a topic—a question, a problem, or a situation in which the students can share ideas and compare or contrast views. What are some suitable topics or situations that have been used effectively by mathematics teachers in initiating discussion?

	Class	Topic	Procedure/Purposes
1.	Geometry	Original proofs	Small groups seek varied methods of proof, contrast efficiency, difficulty level of each, advantages of one over the other.
2.	General math	Traffic pattern in school corridor	Groups to discuss ways of reducing congestion in school corridors by establishing new traffic pattern; must design plan to get all needed data.
3.	Any subject level	Test post-mortem when results range from very good to poor	Each group to be arranged by the teacher to reflect the range. Group to discuss all problems on which any member needs help, consider reasons behind errors, pinpoint strategies for correct solutions.
4.	Math lab	Results of individual or group lab work	Groups to compare results, conjecture possible generalizations based on lab work, pose questions, analyze reasons for varied results.

Obviously, a second major ingredient for a successful discussion is appropriate student prerequisites. In the first illustration, students who know only enough geometry to obtain one solution to the proofs posed will have nothing to compare. If all or most students in the class are at this level, the "discussion" will degenerate into a practice session. For a similar reason the teacher must structure heterogeneous groups in the third illustration above. If all the failing students are in the same group, you have the blind leading the blind. And, of course, a teacher who attempts a discussion on traffic congestion models before the students have learned simple sampling procedures and ways to collect and relate data on number of students and size of corridor is asking for a shared ignorance pool on the topic. You may have experienced such ignorance pools in many out-of-school discussions. A surprising number of adults seem willing to share views on topics about which they know little.

The necessary ingredients for the effective use of the discussion mode, then, are the suitability of the topic and the students' grasp of needed prerequisites. But even with the presence of both of these, many discussions slump because the teacher ignores a simple structural ingredient. Compare Figs. 1.3 and 1.4 with Fig. 1.5. Students who see only their nearest neighbors and only the backs of heads are not likely to engage in useful interaction. Seating arrangement becomes of prime importance. Figure 1.5 offers two possible seating arrangements for the discussion mode. A teacher who makes considerable use of discussion may want to explore alternative seating arrangements *within* subgroups. Some guidelines that make a lot of sense have been verified by research in group interaction:

1. Communication tends to flow *across* a circle, not around it. Remember that a massive amount of communication is nonverbal and we can see more clearly the facial expressions of the person across the circle or table.

2. A corollary to the above guideline is the principle that communication is maximized between students who sit opposite each other and lessened between those who sit side by side.

Although it is probably impractical to assign each student to a seat, you can use the above guidelines to encourage a shy student or to discourage an overtalkative student from monopolizing the group.

Finally, where should you be during the discussion? To find the answer to that question, you must first answer several more important questions. What can a teacher find out *during* a discussion? How can a teacher contribute without altering the mode to Q/A?

If you've prepared the class for the discussion, carefully selected a suitable topic, structured the groups and seating arrangements thoughtfully, and

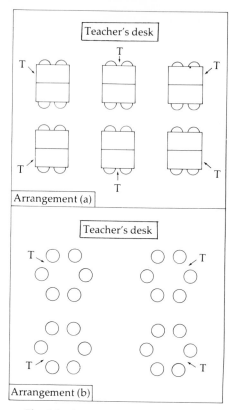

Fig. 1.5 Arrangements for discussion.

clearly given all instructions *before* the students move into their groups, then your role in the first few minutes is that of a helpful overseer. You are making sure that all students follow those initial instructions by making a quick tour around the perimeter of the groups.

What next? If you're only interested in the end product of the discussion, you can hire a babysitter and take some time off. Be careful! You may scoff at that idea, and yet some teachers reward only the end product and neither attend to nor reward the steps students take to get to that end. You want to get feedback on each group's work and on as many individuals' work as possible. Who is contributing? How? Who agrees, disagrees? You must systematically circulate and look and listen carefully to get this kind of feedback. And if students are to believe that their efforts in the discussion are worthwhile, you must give verbal and nonverbal praise to individuals and groups for such things as creativity, thoughtfulness, intelligent skepticism, and efficiency. However, you must be cautious that your praise isn't misconstrued as a closing off of discussion, because your other major role is to sustain the discussion by cuing students as to potentially useful avenues, by answering some questions, and by posing other questions. Take a long careful look at the various positions of the teacher in Fig. 1.5 and notice the various directions the teacher faces. Why are these directions and positions recommended? In what areas is the teacher never found in these diagrams?

When does a discussion end? When the attention of subgroup members starts wandering, the discussion mode has already continued too long. You may call a halt when all groups have made some inroads on the topic and when sharing of ideas seems desirable. After all the students have responded to the signal to end the discussion, each spokesperson can report on that group's findings. Then students from other groups can be asked to indicate agreements or disagreements. Finally, you can help all to summarize class results.

Notice that the feedback getting and giving of this mode has good potential. You may sample everyone a number of times under optimum conditions.

Demonstration Used in combination with lecture, question/answer, or laboratory, this mode can be remarkably effective. The following examples illustrate just a few of the ways in which demonstrations can be used to enhance mathematics instruction. (See Chapters 10 and 11 for many more examples.)

Examples of effective demonstrations

	Class (topic)	*Demo*	*Use*
1.	Math 7 (Symmetry in Living Things)	T cuts grapefruit in half	To show model of radial symmetry
2.	Math 8 (Units of Measure)	T holds up can of soup, box of cereal	To get attention on practical problems
3.	Geometry (Locus)	T stretches and shrinks shape on large geoboard	To illustrate the paths of moving points
4.	Algebra 2 (Conic Sections)	T produces shadow images of a metal ring by tilting the ring between the wall and a powerful bulb	To illustrate the conic sections using a "cone" of light rays

To focus student attention on a demonstration, you must first be sure that the object can be seen. The specimen must be large enough to be visible from the rear, or the teacher must move about the room with a smaller object. White cardboard sheets or the overhead screen can provide an effective backdrop. A model of the hyperboloid or a dissectable cone can be highlighted by placing them on the stage of the overhead; the overhead light in a darkened room shines on the object as in a theater. What are the students to see? Do they know where to look and why? If not, the effectiveness of the demonstration will suffer.

Although demonstration is often used simultaneously with other modes, a silent demonstration can be very effective. Such a demonstration might be used either to set the stage for, or to further develop, a topic.

What is the potential of this mode for feedback getting? Almost zero. Oh, experienced teachers learn to "read" the nonverbal signs—the alert gazes, the "aha" expressions, the scowl, or the worried look. Unfortunately, experienced students have also learned to play the game of school and many manage to put on the face the teacher wants to see.

Laboratory Since the students are manipulating equipment or material in this mode, they are doing so to collect data. Therefore, you might employ this mode to help students reach a generalization, test a conjecture, observe the application of a rule, or learn and practice a psychomotor skill. Some examples follow.

Examples of effective laboratory exercises

	Class	Materials	Activity	Purpose
1.	Math 7	String, meter stick, cans of different shapes and sizes	Students meaure circumference and diameter, record data, and calculate C/d	To obtain the generalization that C/d is a constant
2.	Algebra 2	Wax paper with a fixed line drawn and a point placed above it	Students fold the paper so that the point coincides with the line; they obtain a large number of folds	To observe the evolving parabola
3.	Geometry	Geoboards, rubber bands	Students place 2 different trapezoids with diagonals on the geoboard and find the areas of the 8 triangles formed in each case; results are compared among students	To find generalizations as to equal areas and to use geoboard relationships to verify the conjectures (a geoboard *proof* can follow if geoboard axioms are identified)

| 4. | Math 7, 8 | Poker chips | Students move chips to form rectangular and, where possible, square shapes containing 1, 2, 3, 4, 5, 6,..., 24 chips; they record numbers which can be represented by squares, by only a row of chips, or by one or more rectangular shapes | To form a concept of prime and composite numbers, of perfect squares, and so on |

How does a teacher implement the laboratory mode? Perhaps more than any other mode, this one requires careful preparation of materials, instruction in safety precautions, organization of equipment, and possible reorganization of the usual physical facilities in the classroom. Directions need to be pre-planned. Some teachers write directions on an overhead projection sheet so that all students can refer to them repeatedly. If the laboratory activity involves small groups, the teacher must decide on the nature of the groups. Again, it may save time to list group names and locations on an acetate sheet or the chalkboard. Don't forget the need to survive! If Tony and Sam fight continually, you're asking for trouble if you place them in the same group.

Is grouping a necessary part of the laboratory mode? In example 1 above, group work may be essential, with members of the group alternating jobs—measuring, checking, recording, calculating—while each student needs to work independently on the paper-folding task in example 2 and there is no real advantage in grouping. However, beware of being hypnotized by eager, active students who never attain the purpose of the lab because you forgot the most important ingredient of all: you must tell the students where they're headed, emphasize the need to look for patterns (as in example 3), insist on careful recording of data, and structure the follow-up to the lab so that the different observations can be shared. Here is a place where developmental questioning combined with lab can be used with profit.

What feedback potential is possible with this mode? Little or none if you are ill-prepared, must repeat directions often, or must continue to carry materials to different students. However, if you have done your job well, the students should be busy with the materials in a matter of minutes. Now you can move systematically to each individual or group, listening, asking questions, and observing the way in which the students are proceeding.

In addition, you have unlimited opportunities to praise, to cue, and to correct intermediate errors.

Why use the lab mode with bright, older students who are well motivated? Isn't it a waste of time? Recall your own experience in recent college classes when a new concept was defined for you. Did you always grasp the concept? When you had difficulty learning the concept, what seemed to lead to the difficulty? These and related questions occur again in Chapter 3. They are crucial questions for the teacher.

Some references to help the mathematics teacher design interesting laboratory activities will be found in Chapter 11, and the current curricular emphasis on the use of laboratory mode is described in Chapter 9.

Individual student project Projects can be student initiated or teacher initiated. They may be used to explore some topic in depth, to introduce relevant applications of the subject being studied, to investigate historical background of the subject, and so on. Project work may occur during the class period, but most of the project work is usually carried on outside of class time. Student reports may be shared with the class, and student productions can be displayed.

Here are two examples of teacher-initiated projects:

1. *Math 10 Project*   *Date:* February 8
    Choose one of the following topics:

    a) Linkages

    b) Locus

    c) Curve stitching

    Design an original model. Check out the mathematics of your model with the teacher. Then build a durable, attractive model and be prepared to explain its properties to the class.

    Submit your plan on a five-by eight-inch index card by February 15. Completed projects are due February 22, and class time will be reserved for reports the week of February 25.

2. *Math 12 Project*   *Date:* October 1
    Choose *one* of the following topics:
    Algebraic Balancing of Chemical Equations
    Algebraic Solution to Electrical Circuits
    Minimal Surface Area Experiments with Soap Films
    Numerology
    Ciphers, Codes, and the Way They Are Broken
    Mathematical Aspects of Population Growth
    Statistical Study of Finger Length Variation in Adolescent Hands

Mathematics in Sound and Music
Function Before Fashion—Mathematical Designs
Timetables, Calendars, and Clocks
Analysis of Hurricane Paths

Research the topic and prepare an 8- to 10-page typed paper on the results of your findings. Include a bibliography of at least four references. You may *not* include an encyclopedia as one of the four sources! Be sure to cite these references in the format given in class.

Submit a one-page outline by October 15.
Completed papers are due November 1.

If students complain of the extra work involved in a project, then the teacher has goofed somewhere along the line. The project should be threaded into the coursework; it is up to the teacher to design assignments and daily work so that students see this mode as an alternative but different way of learning relevant aspects of the subject matter. The students soon learn that the teacher doesn't mean what is said if no class time is reserved for reports, or if the end of the project means little more than a grade in the teacher's record book.

Because much of the project work is completed outside of class time, the feedback potential prior to project completion is limited. However, the teacher is able to gather feedback on a host of student capabilities upon completion of the project, and he or she can give feedback in the form of written or oral comments. Project work affords an excellent opportunity to encourage students to give feedback to each other—formally during oral reports, and informally as they inspect posted projects.

Project work also poses a problem for the novice. You've probably already identified it. Your 24 students all handed in reports, models, or posters of varying quality sometimes due to the presence or absence of artistic talent. "This isn't art class!" "How did you judge the projects?" "What's my grade?" These are not easy questions to answer, but you must answer them if you assign projects. See if the material on all aspects of evaluation found later in this book meshes with your present response.

Audiovisual aids (AVAs) Lights out! Is it a signal for paper-wad battles or siesta time? All too often, even in an interested class, AVA time means neither the battle nor the nap but an extraneous short feature that is clever but hasn't much to do with the topic. Again teacher preparation is critical if this mode is to help students learn.

There are four major areas of preparation in the use of AVAs: (1) use of the equipment; (2) quality and suitability of the tape, slide, film, filmstrip, or transparency; (3) preparation of the students; and (4) follow-up and integration of the AVA into other modes.

The suggestions listed below may seem to include the obvious, but our experience in the classrooms of novice teachers convinces us that the obvious is often overlooked.

AVAs—do's and don'ts

1. Use of the equipment:
    a) Check out your school's rules and regulations and follow them. (Some schools have student operators of equipment.)
    b) Dry-run equipment you will operate *in the classroom in which it will be used*. Check visibility and /or audibility from various areas of the room.

2. Quality and suitability of the materials:
    a) Review all material you hope to use and *do not* use material of poor quality or limited applicability, even if last year's teacher did order it.
    b) Be a ruler of, not a slave to, equipment and materials! If parts of a film are excellent for your purposes, show only those parts. If you wonder how the students would have considered a question on a topic before they heard the answer, stop the film, engage in Q/A, then start the film and let it provide the reinforcement. The possibilities are unlimited.

3. Preparation of the students:
    a) Provide an introduction and an overview. Let the students know where and how this AVA contributes to the subject being studied.
    b) Explain to the students their responsibilities. Will you ask questions during and/or after the viewing/listening? Should they take notes? If so, make sure they can see. (Obvious, isn't it? But how many pitch-black classes have you sat in and been expected to take notes?)

4. Follow-up and integration of AVA mode:
    a) Plan alternative ways of combining this mode with others to get feedback, to give feedback, and to keep attention on the task. One route, the interspersing of Q/A with the AVA, was alluded to in 2b. Another effective route is to prepare an outline with questions or problems to be responded to either during the viewing/listening or after, or both. Hand this to the students prior to the viewing/listening.
    b) Keep the particular AVA material alive after the lights go on. Conduct Q/A or small-group discussions if suitable. Future lessons, assignments, tests, and quizzes should refer to examples from this AVA whenever possible.

It is clear that all this time and effort should not be wasted on worthless material. But the old saying, "A picture is worth a thousand words," has been proven true too often for you to ignore the immense potential of the AVA mode. However, suppose you have equipment available for either a lab or a demo and have a film that portrays the same lesson. On what basis would you make a choice of mode? How would you rate this mode as to feedback potential? Why? Check out your conjectures against the ideas presented in Chapter 2.

Supervised practice Practice makes perfect—or does it? Have you ever practiced a golf swing with little improvement in your drive? It's probable that you learned an error to perfection. That's right. The students must be involved in the practice of a learned rule; practice must follow relevant instruction. You must know how to get feedback on the results of that instruction before and during the supervised practice. In fact, the word *supervised* suggests the intimate tie between this mode and feedback. In Chapter 2, several suggestions aimed at feedback getting and giving are detailed.

Here are a few suggestions aimed at helping you implement supervised practice:

1. Use several short periods rather than one long, marathon session at the end of class. (Of course, this will keep students busy; but there is a more important reason. It has to do with feedback and learning. Chapters 2, 6, and 7 have more on this.)

2. Sequence and cluster your practice examples so that all or almost all students can begin the work and so that no one is finished before you have some time to make a tour of most of the room. Tour of the room? That's right. "Supervise" does not mean that you monitor the group as if you were the guard in the tower of a prison yard.

3. Follow the arrows in Figure 1.6 *after* all the students are at work. The arrows illustrate the direction you should face. Why is it a good move to turn the body as diagrammed here? Remember, there are 25 students in those seats, and they are not all like thee or me. The tour should be quick and complete. So try not to interrupt the tour to spend time with a single student. If you haven't guessed why, the next chapter offers some reasons.

4. Systematically move into the room to check the work of individual students. But five hands are up! Use verbal and nonverbal signals to let students know you see their hands and will be with them in order. In the meantime, they could try another problem or be directed to share their question with another student.

5. Be aware of the entire class even while helping one student. Turn to face most of the class. Bend at the knees, rather than the waist. (It's good for

posture, too!) Give a cue, but insist on student responsibility to try. Students soon learn that some teachers will do the work for them if asked.

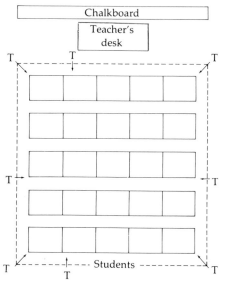

Key: - - - = route for quick tour (inital 1-2 minute maximum)
T→ = direction T faces when helping individuals

Fig. 1.6   Positioning of teacher during supervised practice.

Supervised practice is such a critical mode that more attention is paid to it in later chapters. But for now, we leave you with questions. When should a teacher end a supervised practice session? Do honor-level twelfth-graders need supervised practice?

Homework Three examples of nonroutine homework assignments were described in Section 1.1 Refer back to these examples and consider how the teacher would make use of the products of the students' efforts during the subsequent day's lesson. You should be able to think of several good possibilities. However, actually making use of the homework is based upon the assumption that nearly all students have done the assigned task. How can the teacher ensure that such will be the case?

One often-overlooked technique is to inform students how data from an assignment is made. This technique encourages students to view homework as an important part of the course rather than simply as "busy work." However, in spite of your best efforts to head the problem off, you must be prepared to cope with the reality that one or two students will come to class unprepared. What alternatives exist to handle this kind of problem? Shouting, screaming, and idle threats will do nothing to salvage the situation nor will they prevent its recurrence. Positive action, which demonstrates the consequences of students' failure to prepare, is needed. For example, if the plan calls for students to form groups to share and analyze data, the offenders can be directed to read relevant material in other texts instead of participating with their classmates. However, if your plan calls for a question/answer session based on the collected data, then you should refuse to allow unprepared students to participate in that portion of the lesson and emphasize to them the importance of the assignment in a private teacher/student conference after class.

Does the typical paper-and-pencil homework have a place in instruction? Yes. In fact, the most obvious use of this mode is to provide *relevant* practice on skills introduced in the day's lesson. Notice the emphasis on the word *relevant*. Recall the statement in Section 1.1 on the danger inherent in routine selection of the even-numbered problems that leads students to perceive homework as mindless practice. But caution is advised! There is considerable evidence that practice has no effect—or even negative effects—after a certain point. When students perceive homework as a way to keep them busy, they tune out. The key to this is to err on the side of too little, rather than too much, of this kind of homework. Choose only a few exercises or activities, and make sure the ones you choose make sense to the students.

Another sure way to turn off the class is to use homework as a punishment. And if your class has already learned to associate homework with punishment, you will need to be particularly thoughtful about your use of this mode. Resource books are filled with appropriate possibilities: open-ended questions, practice in coded or puzzle form, take-home labs. See Chapter 11 for specific references.

A less often used, but effective, approach to get the most out of relevant paper-and-pencil exercises is to build the foundation of tomorrow's lesson into today's homework. For example, a carefully structured sequence of quadratic equations could be assigned to give practice in use of the quadratic formula. Then if the roots range from real, rational roots to imaginary roots, the students can be asked to look for a pattern relating the kind of roots to the value of the expression $b^2 - 4ac$. The teacher now has a ready made entree to the topic: the nature of the discriminant. A teacher who comprehends and anaylyzes subject matter is at a distinct advantage in designing such assignments.

The "what" of homework is crucial, the "how" is equally important. When possible, let students begin work on exercises in class, and *don't call it homework!* The alert student often has his or her own definition of homework, and will tell you that it is to be done at home. There are a variety of reasons for allowing students to start such work in class. One of these has to do with feedback. By now you probably have some idea what that is. Think of some other advantages of an early start.

"How" also refers to the post-mortem procedure (the day after). Will you collect the homework? Why? When? Will you take class time to check on the results of the class work? How? When? The answers depend to an extent on your need to get and give immediate feedback to the class. There are some specific post-mortem procedures suggested in Chapter 2, but two rules of thumb are basic commandments. First, don't do the homework for the students—"Do problems 7, 10, 12, 15, ..." Second, don't ignore the home-work. If it can be ignored, it wasn't worth doing.

You've heard the expression, "Different strokes for different folks." Can you think of any implications of this saying for use of the homework mode? If not, you are typical of most people with long careers as students who have seldom, if ever, experienced differentiated assignments. Do those students who whiz through five typical application problems at the end of today's lesson need to do more of the same for homework? Then, too, consider that small group of students who could solve none of the problems by the end of the lesson. Diverse feedback of this kind is a clue that different homework assignments need to be designed and matched to various student needs. Think of other situations where it makes maximum instructional sense to ask that different students pursue varied tasks as homework.

As is the case with all instructional modes, effective and efficient use of homework requires careful consideration of alternatives and thoughtful planning by the teacher. *Effective* and *efficient* are key ideas you will find emphasized throughout this text. One of the greatest rewards of teaching is evidence of student learning as a result of teacher attention to these ideas.

1.4 CROSS-REFERENCES TO A SAMPLE CBTE EVALUATION INSTRUMENT

The instrument referred to throughout this section and all comparable sections in subsequent chapters is the one we developed and field-tested in the Albany Mathematics Science Teaching Project. The total instrument and a description of its development can be found in Appendix B.

In the long run, the precision with which you can define modes, describe their uses, and list their advantages and limitations may be interesting, but merely academic, knowledge. We believe the final critical test of teaching ability must occur in the classroom. Thus, we refer you to our CBTE (Competency-Based Teacher Education) instrument.

If performance alone counts, why are we working through this text? Good question! One answer is that knowledge and the related ability to analyze instructional modes and their usefulness are prerequisites to more effective classroom teaching for most, if not all, novice teachers. Knowledge is not sufficient (doesn't guarantee) future success in the classroom, but it is necessary (needed) to ensure success.

Category 2.3 (on page 1 of the instrument) includes most of the modes we've defined. Where is lecture? Item 2.32 paraphrases the lecture mode definition and emphasizes that duration of teacher talk is not a criterion for judging the existence of the lecture mode. In some cases more than one item refers to a mode. See items 2.33, 2.34, and 2.35—all of which relate various aspects of the question/answer mode. The other multiple-item set (2.310 and 2.311) refers to two different aspects of homework. Where do we talk about independent student projects? Item 2.65. Why was this item placed here rather than with all the other modes? That's a choice we made for a reason. Continually check up on the choices of sequence (order) and content throughout this text, and ask why we made those choices. That's the beginning of thinking about teaching.

Now you've had a glimpse into the future. You will be the teacher being observed by your college supervisor or cooperating teacher. Will your selection of modes be appropriate to the content and the students? These are competencies we'll be working towards throughout the rest of this text.

1.5 SUMMARY AND SELF-CHECK

Where have we been and where are we going? We began with consideration of a model of a systematic approach to instruction and devoted our initial attention to that part to which novices typically bring the greatest range of personal experience. Yes, we asked you to step into the systems analysis model box labeled "Instructional Strategies" and to begin identifying with the way teachers promote student learning—modes. These are important parts but *only parts* of what is meant by the term *strategy* (overall plans targeting on the achievement of an objective). Nine typical modes were identified and defined operationally. Common uses and abuses of each were explained, and you were introduced to the feedback-getting and-giving potential of each of these modes. More questions were raised than were fully answered; and that was by design rather than by accident. Some of these answers will have to wait exploration of other components of the model in chapters to follow.

Right now, however, you should be able to:

1. Name and operationally define nine common instructional modes.

2. Identify modes being employed during either a "live" or "canned " lesson.

3. List at least two uses and two abuses of each teaching mode and provide reasoned justification for your choices.

The exercises in the next section are designed to test your understanding of some of these modes as they relate to your role as a teacher of mathematics. Your understanding of the content of your subject matter discipline will be put to the test, as well as your ability to move from the role of a student to that of a teacher. These situations are based on actual classroom happenings, and the content analysis exercises have been used by classroom teachers to improve instruction. Give all of them your best efforts. Then go back and check your responses against your experience and your learning as you progress through other sections of the text.

1.6 SIMULATION/PRACTICE ACTIVITIES

A. Each of the following questions is defective in its potential for eliciting unambiguous evidence of student learning. Identify the defect and construct a more effective question.

    1. How about the circle?

    2. Is 22/7 equal to π?

    3. Can you tell me how to differentiate $f(x) = x^2 + 3$?

    4. Does everybody understand the law of sines?

    5. "Where do you want me to put the decimal point?" (Hint: While we were observing a class, a student teacher asked this question. A response by students, inaudible to the student teacher, was rather specific in terms of an anatomical reference.)

B. As soon as the teacher started a demonstration, several students began to stretch to the right and the left in their seats; three students in the last row stood up. Which fundamental rule for effective use of the demonstration mode was probably violated? Suggest two or three ways of overcoming the problem.

C. Select one of the following topics: probability, volume, congruence, ratio, fraction. Describe a demonstration that would help teach the central idea. Be specific about the material(s) you would use, how you would manipulate them, and other modes you would use in connection with the demonstration. Limit this teaching segment to a five- to ten-minute portion of the class period.

D. During successive periods you observed the same laboratory taught to comparable groups by two different teachers. Teacher A's class paired off quietly, got their materials quickly, returned to their desks, and immediately begin to work along the lines directed by the teacher. In Teacher B's class, students jammed up at the front of the room and began jostling and making extraneous noise. The students returned to their desks with needed materials only after a period of several minutes, which was interspersed with loud reprimands by the teacher. Next there was a rapid series of requests for teacher's help from various parts of the room and sporadic verbal commands by the teacher to "follow directions," "get to work," "stop wasting time."

    List at least three variables in teacher behavior that could account for the differences in the students' performance in the classes. Defend your selections.

E. Choose any topic from the content of secondary-school mathematics and design a homework assignment to serve two purposes: (1) It should provide relevant practice of some skill. Identify the skill. (2) It should be structured to allow you to elicit a subsequent idea or skill.

    Provide answers to the exercises, and write a brief justification of the ways in which your assignment meets both of the above purposes.

SUGGESTIONS FOR FURTHER STUDY

Carin, A., and R. Sund. *Developing questioning techniques: a concept approach.* Columbus, Ohio: Charles E. Merrill, 1971.

The aim of this book is to help classroom teachers select and use questioning techniques not only to promote cognitive development, but also as a way of building student self-esteem. This goal is realized to a high degree. The taxonomies of the cognitive and affective domains (Bloom and Krathwohl) are used throughout, and numerous examples are provided to illustrate the ideas presented. The relevance of the work of well-known cognitive psychologist (Bruner, Gagné and Piaget) is shown and Rowe's research on the effects of "wait-time" is used effectively. The chapters on creativity and the construction of guided discovery lessons are highlights of this book.

Hunkins, F. P. *Questioning strategies and techniques.* Boston: Allyn and Bacon, 1972.

This very useful paperback assumes no previous knowledge of techniques and attempts to deal with all aspects of the subject in a clear but concise way. For the most part, the author has accomplished his goals quite well. The entire presentation is made within the frame of reference of contemporary curriculum study groups in mathematics. Sufficient examples are included to illustrate the major points, and frequent ties are made to Bloom's Taxonomy. Novice teachers would do well to skip over the detailed treatment of various categorizing schemes (such as those of Taba, Suchman, and Hunkins) and to concentrate their attention on the other parts of the book.

Sanders, M. *Classroom questions: what kind?* New York: Harper and Row, 1966.

This brief paperback has a number of features helpful to teachers seeking to improve their questioning skills. Technical terms are kept to a minimum and those used are defined in simple words and clearly illustrated. Each chapter ends with a set of questions on its contents, and an answer key is provided to further help readers check their understanding. Bloom's Taxonomy is used throughout, and the emphasis is on questions above the level of recall.

Useful as it is, this book is not without its weaknesses. The illustrative questions are almost entirely in the social studies area. Readers should also be wary of the author's heavy dependence upon true/false questions, which thoughtful readers will recognize as answerable by students on a guessing basis. Fortunately, almost all of these could be made to achieve their avowed purposes by adding "Justify your answer."

CHAPTER TWO

THE HEART OF SYSTEMATIC INSTRUCTION
FEEDBACK

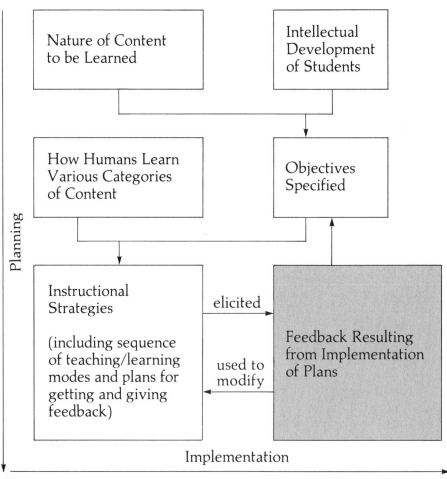

ADVANCE ORGANIZER

Many teachers assess their effectiveness in the classroom in terms of student reaction. Some teachers do this sporadically; some, naively; a few, regularly and systematically. Yet this kind of assessment, the *getting of feedback,* is the basis of all learning theories and the core of all attempts to communicate with others.

Some college instructors are notorious ignorers of feedback procedures, according to students who sit through lectures where no questions are asked or answered and no recognition of student attention or confusion occurs. In these cases, students learn to psyche out the teacher on the basis of feedback unknowingly given them. However, when instructors consciously and systematically provide feedback, students learn when their performance is correct, nearly correct, or totally incorrect both from the feedback provided by the instructor and from the built-in feedback components of the content.

In nonschool areas of life, people get and give feedback regularly. The mechanic who diagnoses a car's malfunction does so by testing and then observing a reaction, subsequently modifying his test and performance on that basis, and so on. The doctor, the lawyer, and the golfer—all with varying degrees of success—use feedback to modify their performance so that they may reach some goal. The systematic approach exemplified by computer work and characteristic of all kinds of systems analysis requires the feedback loop. Notice the position of the feedback box in the systems analysis model on the chapter opening page. Follow the arrowheads, and analyze the reason for the multiple connections of this component to others in the model. Your reading in Chapter 1 has already introduced you to the connecting link between modes and feedback. The following reading and activities spell out those connections and engage you in the problem-solving task of interpreting and analyzing feedback.

2.1 OPERATIONAL DEFINITIONS

Feedback The student gets feedback when the teacher says "incorrect" in response to a student's answer; the teacher gets feedback when he or she notes that a student has written the correct solution to a problem posed by the textbook; a student gets feedback while checking the results of a solution to a problem. Therefore, an operational definition of *feedback* is *any information, communicated to teacher or student, verbal or nonverbal, on the results of instruction.*

Notice that the definition of feedback makes it clear that both the teacher and the student need to *get* feedback if instruction is to be both efficient and effective. From the teacher's point of view, plans must be made to *get* feedback and to *give* feedback. Now you've guessed it; we've spiraled back to modes. *When the teacher employs a sequence of two or more modes with the planned intention of getting and/or giving feedback, that type of sequence is referred to as a feedback strategy.*

Feedback often appears in subtle forms and can be easily overlooked. In the following activity you will be introduced to one such aspect of feedback.

2.2 INTRODUCTORY ACTIVITY

A two-person team is needed for this activity. One person, P_1, rests one arm on a table so that the heel of the hand is on the edge of the table. P_1 should turn that hand sideways and prepare to catch an object between the thumb and the index finger. The other partner, P_2, holds the object, a 30-cm ruler, at the 30-cm mark between the thumb and index finger of one hand. P_2 then positions the ruler so that it is almost entirely above P_1's hand—actually with the 1-cm mark at the level of P_1's thumb (see Fig. 2.1). When P_2 drops the ruler, P_1 tries to catch it. The measurement at which P_1 caught it is recorded.

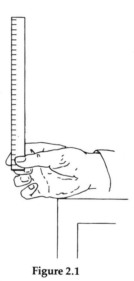

Figure 2.1

A. Repeat the procedure ten times, and record P_1's results in a table such as the one shown in Fig. 2.2.

Results for P_1

Trial #	1	2	3	4	5	6	7	8	9	10
Ruler mark										

Figure 2.2

B. Switch roles and record P_2's results in a similar table.

C. Identify the various kinds of feedback that were observed throughout the entire experiment. Describe the changing nature of the feedback. Who got feedback? Who (what) gave feedback? Account for any improved performance on the basis of feedback received by the "catcher." What other variables may have interfered with an improvement in performance?

Activities such as the ruler activity are specifically designed to teach students the nature of feedback in tasks involving motor reflexes, such as braking a car suddenly. See the description of the interdisciplinary materials entitled *The Man Made World* (Suggestions for Further Study section) for other everyday examples.

2.3 FEEDBACK: GETTING, USING, AND GIVING

Assumptions The major assumption that forms the theme of this chapter is that feedback is a key part of all instruction.

The byproducts of this general assumption are four specific assumptions:

1. Feedback should be given to and elicited from the widest possible sampling of students.

2. Feedback strategies should be implemented frequently and at key points during *each* instructional period.

3. Feedback should be interpreted and used to pace instruction and/or to alter the sequence of instruction.

4. Feedback should also be utilized to assess the feasibility of the stated objectives.

Getting feedback Throughout the previous chapter, you were urged to consider each mode's feedback potential. Check your understanding of this against Fig. 2.3. If you do not agree, it may be due to the fact that you are thinking about various forms of nonverbal feedback while we are considering only verbal feedback at the moment. On the other hand, disagreement might indicate a need to review operational definitions of the various instructional modes. In any case, try to resolve this matter in your mind before reading any further.

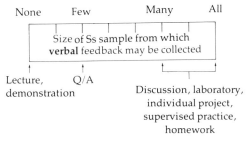

Fig. 2.3 Verbal feedback potential of various instructional modes.

Since each mode does have unique purposes in instruction, modes with zero verbal feedback-getting potential need to be followed by and interspersed with positively valued feedback strategies in order to assess the effects of instruction and modify it when necessary. For example, a short lecture can be followed by questions specifically designed to assess the listener's grasp of the lecture. Questions can range all the way from the "What did I say" type to the more meaningful "How" or "Why" types. At all costs the lecturer should avoid the "Any questions?" habit unless the typical silence from the class response is quickly followed by a specific question from the teacher. Most students have learned to shrug off "Any questions?" as merely rhetorical. Further, sometimes the student who is confused and knows that the teacher is willing to field questions just can't phrase a sensible question and doesn't want to admit it! Such a student profits from class reactions to thoughtful teacher-constructed questions on the content of the lecture. The teacher might ask students to write the answers to appropriate questions then and there. Feedback from many students can then be obtained as the teacher quickly tours the room. Such a procedure might be followed by a *straw vote* (poll).

The straw vote is one of the most useful techniques for efficiently obtaining feedback. The teacher asks to see the hands of those who agree, who disagree, who aren't sure (or any analogous classification). The procedure takes less than a minute and provides information to the teacher on the diversity of class views. But the teacher who immediately begins to reteach the topic or to quickly move on to the next topic on the sole basis of a straw vote may be in for a shock. Why? Suppose that one-third of the class disagreed with the expected answer, but that all of these students had simply misinterpreted one minor aspect of the question!

You can probably imagine many other possible reasons for diverse feedback. Therefore, straw votes should frequently be followed by a more sophisticated kind of feedback strategy—for example, a series of questions to members of each faction probing the reason(s) behind their response, or a quick tour around the room sampling papers of members of each group, or a student demonstration in defense of a response. The alternatives are many, but the teacher must plan these alternatives based on the likely stumbling blocks in the lesson and the teacher's knowledge of the needed prerequisites for subsequent phases of the lesson.

All of the preceding suggestions could be adapted easily for use during or after the other modes that have no feedback potential. Moreover, it is equally important to intersperse higher-rated modes to get feedback *after* a sequence of developmental questions has been used. There is a serious caution to be observed here. Suppose your class was being led through a careful sequence of developmental questions that was interspersed with straw votes at every third question. You've got it! Many students would stop voting and would probably stop thinking—about the lesson. Developmental questions must flow as smoothly as possible with members of the class learning from one another's answers as well as from the teacher's cues. With the modes rated high in feedback potential, the teacher can get a broad sampling of specific feedback if the modes are being used in appropriate ways.

Two of these higher-rated modes, laboratory and supervised practice, are critical in feedback getting. However, most novices and some experienced teachers frequently overlook the enormous potential these modes possess for improving instruction. The two illustrations that follow include elements of feedback getting, using, and giving.

Sample 1: Supervised practice

T has just completed a lecture and Q/A sequence on writing the logarithm of any three digit numeral with the tables given. A straw vote on a single practice example showed most Ss with the correct answer. Therefore, T directs the Ss to begin working on examples 4, 8, 9, 15, and 24. Before T moves from the front, T checks to make sure that all Ss have started, reminds two Ss of the location of the table in the text, tells a S whose hand is raised for help to reread notes until T has had a chance to check on others' progress. All others are now at work. T tours the room (as suggested in Chapter 1) and sees that only a few Ss are having difficulty with examples 4 and 8. T praises individual Ss quietly, indicates in a clear voice that many Ss are making good progress, quietly gives a cue to one S in difficulty, then moves to a second S and notices that some Ss have almost finished all five examples. T sends selected Ss to the board, asks one S to serve as tutor for any S needing help, assists another S, and checks progress of class. (T has to decide if enough Ss are far enough along to stop the practice and look at the models of correct performance on the board. T must also decide if progress was sporadic or sub-stantial. Do the Ss need additional practice on the same kind of examples or on a more complex set, or should T move into another mode to introduce a new rule?) T calls the class to attention, tells them to compare their work with the board work and to be prepared to identify areas of difficulty. T stands in the rear quietly while this occurs and then takes straw votes on the number of Ss who agree, disagree with board work and seeks corrections where needed. Ss are frequently asked to respond to Q from other Ss.

Notice that the teacher refused to immediately move to the first student who raised a hand. Why? Suppose the quick tour had uncovered a far different story? If most students were blithely working away on the basis of a misconception, the delay caused by the teacher's conversation with an individual could lead to, at least, wasted time and, at worst, chaos. The same kind of rationale can be advanced for not interrupting the tour. (Remember the question asked of you in Chapter 1?) This teacher was also alert to differential progress and tried to use positive feedback in more than one way. Students were used as individual tutors and were asked to display their work on the board so that all other students could get feedback on the nature of the completed work. The other instances of feedback getting and giving in this sample are numerous. After you read the next section on feedback giving, check this illustration again for specific examples.

Sample 2: Laboratory

T begins with a ten-minute lecture with demo related to finding the empirical probability of an event. A 3-minute Q/A session, which sampled eight students, gave positive feedback on Ss understanding of (1) the empirical techniques used to obtain number of favorable ways the event could happen, (2) the "sample space" method used to record data, (3) the location of the various games of chance and problem cards, and (4) the roles of each lab partner.

T then gives the signal to begin work and takes a position in the room from which the Ss collection of materials and equipment can be observed. All Ss gather materials quickly and quietly, then return to their assigned laboratory places. At this point, T begins to move quickly and systematically from group to group. T notes that the first four groups visited are proceeding exactly as directed but just as T reached the fifth group, loud voices were heard coming from the opposite corner of the room.

T moves quickly to the loud group, touches one S on the shoulder, and uses the finger-across-the-lips signal when the S looks up. Both partners quiet immediately and T asks if they are having a problem with the exercise. Both partners start to talk at once, each complaining that the other is misinterpreting the card problem which seeks the probability that two jacks will be drawn from a standard deck of 52 (when two cards are drawn without replacement). "Herman is drawing the cards one at a time," says Yvonne. T holds up one hand in a "stop" gesture, the Ss quiet down, and T uses Q/A with each in turn to clarify the procedure. The partners return to work; T observes they are now working amiably, and T moves on to other groups that are all proceeding in an orderly fashion.

As T continues to move from group to group, observations indicate that all Ss are now interpreting the directions correctly in the case of the coin, dice, card, marbles, and the roulette wheel events. However, the first three groups to progress as far as the event utilizing the roulette wheel and the spinner have difficulty agreeing on the nature of the events that are to occur. T turns off the lights for three seconds, all the Ss quiet, stop work, and look at T, who then turns the lights back on and gives a lecture-demo on these particular events. All groups then return to work at T's signal, and T's subsequent observations while touring groups indicate all are now obtaining appropriate data.

With about five minutes of work time left in the class period, T overheard the following conversation among one pair of partners:

Saul: "We calculated the probability of a black number showing on the roulette wheel as 42/100 and of a 2 occurring on a spin as 3/100. If we want the probability of both happening at the same time, we would have a combined probability equal to the product of 42/100 and 3/100."

Alice: *"No, we wouldn't! We would just be adding one event to another, so the total probability would be the sum of 42/100 and 3/100."*

Saul: *"Look, think of it this way. The chance that both events will occur at the same time is less likely than for just one to occur. A sum would signify just the opposite."*

T praises Saul and Alice privately for thinking ahead and gives each credit for having a certain amount of logic. A short Q/A sequence by T gets Saul and Alice to compare this problem with the card problem where two jacks had to be drawn. T asks both to rethink the definition of probability based on sample space usage. Then T goes to the chalkboard, writes a brief statement of the problem posed by Saul and Alice, and for the evening's assignment asks each student to write out a sample space using ordered pairs to solve this problem. Then each student is to arrive at a generalization for the $P(A \cap B)$ and to obtain the probability of the first event or the second happening, given the same data. A brief flickering of the lights signals "clean-up," which progresses smoothly and quickly. T calls attention to the assignment, publicly gives credit to Saul and Alice for thinking ahead, and praises all for their very good work during the period. The bell rings, the Ss leave, and T notes that all is in readiness for the next class, which is to begin in four minutes.

Note the highly positive feedback the teacher received on the effectiveness of the introduction to this laboratory exercise. Identify all the things the teacher did to ensure that all students would begin the laboratory work effectively and efficiently. Also note how the teacher found out that only one group was encountering difficulty with the card problem and how this problem was handled without interrupting the other students who were proceeding without problems. Nonverbal feedback giving, as well as feedback getting, is highly effective for all who learn how to use it.

Apparently the teacher was quick to read the negative feedback collected on the roulette wheel/spinner problem and made use of it to reorient the entire class. If the teacher had failed to interpret this as a widespread problem, no doubt it would have been necessary to treat the same problem over and over via successive individual visits to nearly all the groups.

The student-to-student discussion between Alice and Saul not only provided the teacher with valuable feedback, but also presented a golden opportunity to reinforce prior learning on the nature of the mathematical enterprise and to provide for an in-context extension of that learning via a relevant homework assignment.

Most of us are quick to criticize and slow to praise. Note that the teacher in this instance did find numerous opportunities and ways to give well-deserved praise to individuals as well as to the entire class. In the long run, this approach yields big dividends for all and helps put in proper perspective those occasions when a teacher must offer constructive criticism as part of instruction. It will be well worth the effort to reread the entire description of this laboratory class and recount the occasions and techniques of praise-giving that occurred.

In addition to the specific modes referred to in this section thus far, there are two other common instructional situations that are frequently mismanaged because the teacher ignores available feedback. The two situations are labeled homework post-mortem (HWPM) and test or quiz post-mortem (T/QPM). In the former case, post-mortem refers to what the teacher plans to do about yesterday's homework. In the latter case, post-mortem refers to the return of a test or quiz and the teacher's plans for "going over it." Unfortunately, "going over it" is literally what happens in many classrooms. In the case of a test or quiz that has been corrected and analyzed, there is no excuse for boring the class by requiring everyone to sit through a rehash of each answer regardless of class performance. The teacher already has specific feedback from every student. Feedback so general that the teacher is still not sure where the students' learning problems are signifies a defect in test construction (more on that in Chapter 8).

Homework post-mortems are a somewhat different kind of beast. The teacher who let the students start the homework in yesterday's class may have some feedback on the initial success of the students, but many teachers are misled by this and assume far too much. At other times, teachers can receive useful feedback from students who come in for extra help after class. But once again, it would be a mistake not to check out the generalizability of the data before acting on it. So Mary, Jim, and Abby had trouble with example 5! So what?! Why should the teacher assume that the other 23 students need to sit through a lecture on the ins and outs of this question? What can you do to avoid these pitfalls and others like these?

Several illustrations of feedback strategies used in both homework post-mortem and test or quiz post-mortem are detailed below. Identify the ways in which the teacher gets, gives, and uses feedback not only at the start of the post-mortem, but during it.

Homework post-mortem. If a teacher wishes to ascertain the extent of student mastery of homework tasks, both of the following illustrations have proved useful:

1. In a class of 24, T sends 6 Ss to the board (at or prior to the opening bell) to put answers to 6 questions on the board *without* taking homework papers with them. T instructs Ss at seats to check their solutions against work as it is being put on board. T then tells the seated Ss that they will be required to talk through one of the examples, agree and tell why, correct errors and give reasons, or explain alternative correct solutions as soon as Ss at the board finish. Meanwhile, T walks around, makes a quick check of homework papers, identifies (a) those papers that are complete, (b) those on which various problems have been left out, and (c) types of

errors made. Then T sends 6 *other* Ss to the board to critique the earlier work. Next, a third set of 6 students may be used to settle disagreements among the previous 2 Ss who worked at a problem.

or

2. T distributes ditto with answers, or solutions and answers (overhead may be substituted for ditto) *as Ss enter* the room. Ss are directed to check their work against the standard displayed by ditto or overhead while T walks around to check as in example 1 above. Then T selects Ss to explain typical errors T has detected from papers.

If the teacher is satisfied that most students have achieved mastery of the task but wishes to efficiently provide feedback on details of the procedure, the teacher may:

1. Check papers as Ss walk in, identify correct answers, and instruct Ss with correct models to put work on board. T directs Ss to stay there to explain and to answer questions from other Ss.

or

2. Give selected Ss acetate sheets to take home. T tells them to do the assigned problem on acetate for projection the next day. Also, T informs them that they will be expected to explain and answer the questions of other Ss.

Test or quiz post-mortem. After a full-period test has been given and graded, the teacher does an item analysis (see Chapter 8) *prior* to handing back papers and uses the results to:

1. Identify those problems that will be treated in class. Procedures similar to those detailed in the section on homework post-mortem might be followed.

or

2. Make an instructional decision to completely reteach selected parts of the content.

or

3. Divide class into groups of five Ss whose grades were distributed over the range. T instructs Ss to work cooperatively until all students in the group can do each problem. T walks around and acts as a resource person.

After giving a short quiz and collecting papers, the teacher may display answers on overhead immediately if:

1. major disagreement is not expected;

2.   materials tested are prerequisite to next topic; and/or

3.   little time remains in the instructional period.

Perhaps at no other time in the classroom is the teacher's identification of positive and negative feedback and use of both so crucial to attention keeping. In the interest of calling on everyone, teachers sometimes ask students to read answers to homework or test questions. If this can be done quickly and efficiently, the practice has some merit. But read the conversational sequence that follows.

> *T:* "Question 5. Rachel."
> *Rachel:* "I didn't get it."
> *T:* "Why?"
> *Rachel:* "I just didn't try."
> *T:* "Jack? 5?"
> *Jack:* "3057?"
> *T:* "Is that what all of you got?"
> *Ss* (mixed chorus of shouts): "No," "Yes," "I don't know," "Who cares?"
> ...

Variations of the above conversation have occurred with disturbing frequency in classes we've observed. However, effective teachers try to avoid this kind of time wasting and provide the feedback needed by the students who tried to do the homework in one or more of the following ways:

1.   T reads all answers and tells Ss to record them and check their responses.

2.   T projects a prepared acetate with answers and proceeds as in 1 above.

3.   T has answers posted on the bulletin board for Ss to check during supervised practice time.

As soon as the class has seen or heard the correct answers, it then makes sense for the teacher to take a straw vote on the total number of correct answers or on examples most often missed. Here you must be alert to both positive and negative feedback.

When should the class wait for one or two students to be helped? If all but three students got examples 5 and 7 correct, how can the teacher differentiate subsequent instruction? If many students missed all of the examples, what factors might cause the teacher to decide to rehash only one example? These questions and others like these are representative of the need for intelligent decision making by the teacher during the ongoing instruction.

Take another look at the systems analysis model that appears at the beginning of this and every other chapter. Notice the arrows leading *from* the Instructional Strategies box *to* the Feedback box. As a result of planned feedback strategies, the teacher gets feedback on the effectiveness of the instruction thus far. Now the teacher must use that feedback and may need to alter the planned instructional strategies—the arrows leading *to* the Instructional Strategies box *from* the Feedback box. The loop continues as long as instruction continues and will be utilized effectively to the extent that the teacher gets feedback, can interpret feedback, and can draw on background knowledge to generate immediate modifications of planned strategies.

These are some aspects of feedback use that depend on an analysis of the subject being taught, or the intended objectives, or the intent of the planned strategies. You should be prepared to make more sophisticated decisions on these aspects after working through Chapters 4, 5, 6, and 7. But even at this point in the text, you have read enough about modes and feedback to make some tentative hypotheses about next steps. See if your hunches are included below.

Using feedback Sometimes the best immediate use of feedback is to get more feedback. More feedback may imply a larger or broader sampling of students, or it may refer to the nature of the responses and the need to obtain more specific or more extensive responses. More extensive feedback getting is essential if instructional strategies are to be effectively implemented.

Some of the more common uses of feedback are listed below.

1. After obtaining feedback on the number of students who possess the prerequisite learning for the next topic, T may:

    a) Reteach the prerequisite material if all or most Ss have given negative feedback to T. If T has received positive feedback from a few Ss, they may be encouraged to help T in the instruction by demonstrating, justifying correct answers, analyzing the reasons behind common misconceptions. In any case, T should reteach the topic in some novel way. The original set of strategies obviously failed!

    b) Ask Ss to try a practice example (while T makes a quick tour of the room) if there seems to be ambiguous feedback.

    c) Use appropriate recall Q/A and lecture to weave specific instances of the prerequisites into the ongoing lesson if only a few Ss give negative feedback on the prerequisites. Tell the Ss that you will be doing this, and reinforce the "old" when it occurs.

2. When beginning HWPM, T displays or reads the answers and then gets feedback (in one of the ways suggested earlier) on the extent of the students' problems with homework.

    a) If only a few Ss give T negative feedback, T may:

        (1) Use Q/A with members of the rest of the class assisting *if* the rehash can be accomplished in a short period of time and *if* it can be made meaningful to the rest of the class.

(2)   Work with the small group of Ss while putting the larger group to work on other materials.

(3)   Tell the small group of Ss that they will be helped later in the class when supervised practice occurs, or after class in a remedial period.

b)   If a diverse number of examples were missed by varying numbers of Ss, T may:

(1)   Tell Ss only example 5 will be treated at this time, since all other examples are simply variations of example 5. T can then follow the rehash of example 5 by supervised practice on one or more of the other examples.

(2)   Group Ss by cluster of examples missed *if* there are some Ss who got most or all of the examples correct. These latter Ss will serve as teachers of the small groups.

(3)   Reteach the basic idea underlying the homework *if* the nature of the feedback suggests serious learning problems. T may use one of the homework examples in this set of teaching strategies and should include a supervised practice session on one or more of the remaining examples.

(4)   Tell Ss to hand in the papers so that T can locate the sources of the errors *if* T has noticed that diverse, careless errors seem to be common. T should return the papers the next day and may add some general oral comments to the individual written comments.

Many of the suggestions given on the preceding pages could be used in other instructional situations—after a sequence of developmental questions, during a lab, or during the use of an AVA, for example. Notice the recurrence of "ifs" in almost every statement. There is no recipe for perfect teaching. In each case you must consider the consequences of your actions and make those moves that seem most promising in relation to your objectives.

One of the important characteristics of any of the preceding suggestions is letting students know what to expect. Suggestion 2a(3) spells this out, but you should add similar notes to each of the suggestions. Students can get the impression that their successes and failures are being ignored or dismissed as unimportant if you don't *give* explicit feedback on your expectations. You have noticed many examples of feedback giving in earlier sections. In the following paragraphs, specific attention is paid to feedback giving.

Giving feedback If the reading thus far has seemed to de-emphasize non-verbal feedback, you might ask yourself why. Are there modes where non-verbal feedback may be especially misleading and others where it provides information not otherwise available to the teacher? Check out your ideas on this with a classmate, then reread the Advance Organizer section of this chapter and the sample practice and lab situations in Section 2.3.

But if nonverbal feedback is sometimes misleading when the teacher gets it and tries to interpret it, it is often doubly misleading when the teacher *unconsciously* gives such feedback. For this reason, it is often asserted that we teach students many things in addition to those we intend. Teachers give students feedback on the importance they attach to a topic by verbal emphasis, projected seriousness of purpose, and time spent in instruction or in correcting students' misconceptions. Teachers convey to students that they are progressing toward the intended objectives by smiling, giving a "thumbs up" signal, or patting a student on the back during supervised practice. It is important that the nonverbal feedback given by the teacher parallels the verbal feedback. How would you feel if an employer verbally praised your work but *looked* disgusted at the same time? Imagine the confusion of a toddler who has just said a four-letter word and been scolded verbally by chuckling parents. No wonder the tot repeats the word for more laughs!

Feedback giving needs to be incorporated in all instructional strategies. If we follow the events of instruction in a typical lesson, we can pinpoint places where the teacher must *plan* to give feedback.

The events of instruction* and feedback giving

1. *Gaining and controlling attention.* T gives Ss feedback on what to observe, its relevance to past work, and so on.

2. *Informing the student of expected outcomes.* T tells Ss whether they will be expected to reproduce a derivation, apply a rule to typical problems, or write down observations made during a film.

3. *Stimulating recall of prerequisites.* T may tell Ss the relevant aspects of prerequisites needed today.

4. *Presenting the new material.*

5. *Guiding the new material.*

6. *Providing feedback.* T praises correct responses, modifies partially correct ones, and clarifies in case of errors. T has the Ss copy a sample problem with its solution.

7. *Appraising performance.* As the Ss check out their learning in supervised practice, Q/A, or discussion, T gives feedback as in event 6.

Two categories of feedback giving that are especially important are: (1) providing a model of correct performance and (2) praise. As noted in the

---

*The "events of instruction" used here are an adaptation of Gagné's (1970, p. 304) description of the events of instruction.

events of instruction, one time when the teacher must provide a model of correct performance is in the teaching of type problems (see event 6). The teacher does this when he or she puts type problems on the board (overhead), shows correct procedure step-by-step *after* appropriate rule teaching has occurred, and emphasizes format or labeling that is acceptable. Now the students have a model against which they can compare their efforts on similar problems. The teacher should give praise to individual students for correct alternative solutions, for fast as well as correct work, for good questions about procedures, or for improvement toward a solution even if the solution has not yet been obtained. Some teachers have learned to effectively use nonverbal as well as verbal signals to give feedback to students on their efforts. Feedback often needs to be provided by immediate reinforcement for correct verbal and nonverbal responses from an individual student, from small groups of students, or from the entire class. This kind of positive feedback (or praise) should occur:

1. during supervised practice,

2. when Ss ask good questions,

3. for S responses that show connections,

4. after quizzes or tests, and

5. during homework review.

There are also times when the teacher must give feedback in response to an incorrect answer. A mistaken interpretation of discovery teaching leaves some teachers with the impression that they should never give students this kind of information. As a result, students try to guess what is in the instructor's mind. Experienced teachers learn to give this kind of feedback by means of contingency questions or by using analogy, a physical model, or a numerical illustration. In these ways students "read" the indirect feedback and conclude for themselves that their original response was in error. A similar kind of feedback is inherent in a design or puzzle based on the coding of the correct answers to a set of exercises (see Fig. 2.4). The students note the existence of an error when the design is lopsided and most, if not all, students will then attempt to eliminate the error. Such built-in feedback systems are a tremendous aid to both the teacher and the student. You will find some sources of these built-in feedback systems in Chapter 11, and many more in the experienced teacher's files. Finally, teachers provide feedback on written work by writing comments on test papers, homework papers, projects, and so on.

From our point of view, it would be hard to overemphasize the importance of feedback. We're willing to bet that you, as we, have sat in classes where the ineffective and the inefficient use of feedback was a central

THE PROFESSOR

Directions: Using the rules for signed numbers, find the answers to examples 1-18. Then, in order, connect the dots that appear with the answers you obtained for examples 1-14. Use a straightedge. Next, connect the dots that appear with the answers you obtained for examples 15-18.

1. $7 - 4 = ?$ 7. $19 - (-4) = ?$ 13. $(-11) - (-15) = ?$

2. $(-18) + 2 = ?$ 8. $56 \div (-7) = ?$ 14. $(-27) + 30 = ?$

3. $9 + (-4) = ?$ 9. $(-12)(3) = ?$ 15. $14 + (-40) = ?$

4. $(-13) - 5 = ?$ 10. $(-24) + (-6) = ?$ 16. $(-4)(-17) = ?$

5. $42 \div 6 = ?$ 11. $(-50) \div (-2) = ?$ 17. $(-3)(-5) = ?$

6. $(-9)(-8) = ?$ 12. $7 \times 7 = ?$ 18. $-28 + 2 = ?$

−1	32	3	−55	−13	−16	20
13	−32	55	68	−11	2	−5
49	4	−26	15	16	−20	11
−51	25	33	51	1	5	−45
−2	−30	−36	−14	−10	−3	45
−44	−8	72	50	−33	14	44
10	23	7	−18	37	−50	−37

Fig. 2.4 A sample exercise with built-in feedback.

problem. Without the collection, interpretation, and use of feedback, the assessment of classroom activities must fall back on unreliable factors such as the desires of a supervisor, the gut feelings of the teacher next door, or the words of wisdom of some quasi expert enshrined in a textbook.

2.4 CROSS-REFERENCES TO A SAMPLE CBTE EVALUATION INSTRUMENT

It doesn't take very long to find feedback mentioned on the CBTE instrument. Check each of the competencies in Section 2.5, and be sure you would "know one when you see it." Notice the modifiers: *appropriate, immediate, frequently,* and *broad.* Each was referred to in some illustration, hint, or caution in Section

2.3. If you were using this instrument to assess your own progress, you would try to exhibit all of these competencies consistently while teaching your classes.

Check out the other categories on the instrument to find additional places where feedback giving, getting, or using is implied. Without stretching that notion too far, you would agree with our classification if you chose 2.61, 3.6, 5.4, 5.5, 5.6, 6.5, and 7.5. If you added other items, you are probably sensing the critical role of feedback and the complex nature of the knowledge the teacher must input into feedback plans. You'll have an opportunity to assess your own intellectual understanding of the problem in the next two sections of this chapter. Your ability to improve that understanding will be tested again and again in this text. Translation into instructional strategies in the classroom is the crucial test. Whenever you have an opportunity during student teaching, collect evidence on your own skills. Audiotape yourself. Ask another student teacher to observe you and to record data on feedback that you missed. Videotape yourself. You'll need practice in observing and a willingness to observe objectively. After you've identified a problem with which you need help, seek the advice of experienced teachers, especially your college supervisor and your cooperating teacher. If you want to be an effective teacher, you must work at it!

2.5 SUMMARY AND SELF-CHECK

The present chapter elaborated on the concept of feedback and considered specific strategies for getting, using, and giving both verbal and nonverbal feedback. Several illustrations of instructional situations were described and analyzed in detail to focus attention on the crucial importance of feedback. A host of specific recommendations for getting, using, and giving feedback were incorporated throughout. Novices would be well advised to heed these carefully, since many of our former students have confided that they reread this material several times with increasing profit as they progressed through student teaching. We consider this to be excellent evidence that feedback is indeed the heart of instruction.

Right now you should be able to:

1. Operationally define feedback, feedback getting, feedback using, and feedback giving and describe several classroom illustrations of each.

2. Identify instances of effective versus ineffective feedback getting, using, and giving occurring during either a "live" or a "canned" lesson and state reasons for your judgments.

3. Design effective feedback strategies that can be incorporated into actual lessons you will present to students.

The exercises that follow are designed to test your comprehension of feedback as it relates to your role as a teacher of mathematics. These situations are based on actual secondary-school classroom situations typical of those we regularly observe. Give them your best efforts. Then compare your proposed solutions with those of your colleagues, instructor, college supervisor of student teaching, cooperating public school teacher, and/or other experienced secondary-school teachers. Discuss all differences in proposed solutions thoroughly, and reconsider and amend your responses where this seems appropriate. However, defend your ideas vigorously wherever evidence and reason support your views and resist the temptation to react only on the basis of gut feelings and the basis of established tradition.

2.6 SIMULATION/PRACTICE ACTIVITIES

A. T follows homework strategy of checking all papers at seats while Ss work on a review problem. T collects the following feedback:

1. Four Ss have perfect papers in every respect.

2. Six Ss either have no papers or papers with little work done on the assignment. (Two of these Ss rarely bother to do homework.)

3. Ten Ss have most of the assignment correct. There is no common pattern in types of errors made and no common pattern of missed problems.

   Design a promising strategy that makes use of the positive and negative feedback described.

B. T has just taught a rule and received some feedback that a variety of Ss could apply the formula in stereotype situations. Now T gives Ss three or four examples to do in an eight-minute supervised practice session. T begins the supervised practice in the fashion outlined in the preceding chapters. When T collects initial feedback, T finds that 8 Ss can't even begin the work while the other 17 Ss seem to be progressing very well.

   Design a promising strategy for using this positive and negative feedback.

C. T administered a ten-minute surprise quiz at the beginning of the period. T observed from the rear of the classroom and noticed that several Ss just sat and stared for the last three minutes while others seemed to finish during the last minute. T collected papers systematically. Some Ss complained that T had not taught some of the material, that the material was not in the book, or that T had not told Ss that they were responsible for this material. Other Ss complained that the quiz was too long. Some asked: "How much is this going to count?" and "Can I take a makeup?"

   Interpret the feedback and outline the T's immediate strategy for handling the situation.

D. T planned to introduce a complex rule by a carefully designed sequence of developmental questions. To T's astonishment some Ss moved quickly from correct responses to the early easy questions to a correct generalization before T had asked all the planned intermediate questions. T asked a sample of other Ss questions designed to diagnose whether they also accepted the generalization. This feedback and a straw vote gave T feedback that the class had moved faster than expected.

The next step in T's plan for the day called for several short practice sessions with problem sets of gradually increasing difficulty. Use the feedback described above as the basis for suggesting an appropriate change in the spacing of T's planned practice sessions.

SUGGESTIONS FOR FURTHER STUDY

Engineering Concepts Curriculum Project. *The man made world.* New York: McGraw-Hill, 1971.

Chapter 7 of this unique interdisciplinary textbook is devoted entirely to the subject of feedback. Many specific examples are included to illustrate the clear verbal descriptions and the thorough explanations provided in the text. Both the nature of feedback and its central role in living and nonliving systems receive thorough treatment, and readers are bound to gain increased understanding of the total concept.

Gagné, R. M. *Essentials of learning for instruction.* Hinsdale, Ill: Dryden Press, 1975.

This brief paperback incorporates a presentation of Gagné's views on the place and importance of feedback in human learning. His model of the act of learning phases depicts feedback (reinforcement) as the final phase in any act of learning, and the text explains its interrelationship with other components of the model. The need to involve a large sampling of students in both feedback getting and giving is stressed, and some specific techniques for achieving this end are included.

Garrison, K. C., and R. A. Magoon. *Educational psychology: an integration of psychology and educational practices.* Columbus, Ohio: Charles E. Merrill, 1972.

The subject index of this book contains numerous references to portions of the text that deal with various aspects of the topic of feedback. Novices would be well advised to begin by reading pages 288-89, where the place of feedback in cognitive learning is described and where a schematic model is presented to clarify the connections between cybernetics and human behavior. The role of feedback in academic gaming and directed discussion (pp. 315-16) and the use of testing information as feedback to both teacher and students (p. 548) can then be read with increased understanding. Results of research studies supporting the importance of feedback in motivation, learning, and transfer are also reported (p. 235).

CHAPTER THREE

ADOLESCENT VERSUS ADULT THINKING
THE INTELLECTUAL DEVELOPMENT OF STUDENTS

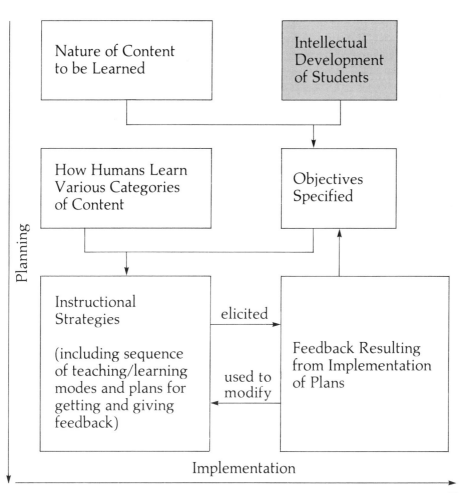

ADVANCE ORGANIZER

At the turn of the century, magazine pictures of children depicted them as shrunken adults. Some psychologists and teachers believed that children think in exactly the same way as adults think, but that they simply need to be fed smaller doses of ideas. Some teachers still behave this way. Others apply the theme "more of the same" for the honors class. If you had such an experience, you know that this approach placed a new burden on the memory bank but didn't necessarily help you understand.

How do students think? Do they learn some subjects differently from the way they learn others? The most promising source of help on all of these questions, in our experience, is the work of Swiss psychologist Jean Piaget. His massive collection of data puts to eternal rest the myth that children are intellectually shrunken adults. But even more exciting is Piaget's explanation of the changes in intellectual development from infant to adult. A four-year-old boy asks, "Why do the clouds move?" or "Why is the grass green?" A tenth-grader cannot estimate the capacity of an irregular container. How is each of them thinking? What factors will alter the four-year-old's approach to his world? What obstacle is confusing the tenth-grader? Piaget directs our attention to the *active* nature of the human intellect. The key to both cognitive development and to apparent stumbling blocks seems to be in the nature of the *interaction* between the individual and the environment. Much more on that later. For now, know that another ghost has been laid to rest—the assumption that the mind is a blank slate upon which the teacher writes!

Check the shaded box in the systems analysis model on the opening page of this chapter. The subjects of the first two chapters were modes and feedback, concepts found in the lowest layer of the systems analysis model and concepts tied together by a pair of vectors in that model. Now we've jumped up to one of the boxes at the top of the model—a box apparently not tied directly to either feedback or modes. Why did we make this kind of sequencing decision? We did have one other possible choice: the nature of the content. (A major problem for the teacher is encapsulated in the question just posed. What kind of sequencing has potential, and for whom? Look for help on this question in this chapter and also in Chapters 6 and 7.) However, our experience with over five hundred student teachers attests to a great chasm in their knowledge of the students they are about to teach and immediate concern for help in this area. These are two of the factors that led us to our sequencing decision.

3.1 JEAN PIAGET: THE MAN AND HIS WORK

Jean Piaget (born in 1896) was a precocious adolescent who, at 15, published a research article on molluscs. This early interest in the biology of shell creatures later changed to an interest in the development of knowledge in humans. In 1922, armed with a doctorate in biological science from the University of

Lausanne, Piaget began working in Alfred Binet's laboratory school in Paris. While doing some routine intelligence testing, Piaget became interested in what lay behind children's answers, particularly their incorrect answers. He began to utilize the semiclinical interview procedure he had learned while serving a brief internship at Bleuler's psychiatric clinic in Zurich. The semi-clinical interview, later augmented by tasks involving manipulation, has become the trademark of Piagetian research.

Piaget in his early books (such as *The Origins of Intelligence in Children*, *The Reconstruction of Reality in the Child*, and *Play, Dreams and Imitation in Childhood*) reported the results of systematic study of his own children's reactions throughout their early development. During the 20 years between the two world wars, Piaget's colleagues, along with psychologists in France, Russia, and Great Britain, replicated and extended Piaget's research efforts with thousands of children from birth to ages 16 or 17. Since the early 1950s, American psychologists have begun to give serious consideration to Piaget's work. His more than 30 books and hundreds of articles (some published over 50 years ago) have been resurrected and dissected. His studies have now been replicated throughout the world, and Piaget has been hailed as the contemporary "giant of developmental psychology," one destined to rank with Freud or Einstein for his long-range contribution to psychology, education, and related fields.

To appreciate both Piaget's findings and the reaction to those findings, it is necessary to understand his position on knowledge. To Piaget, knowledge is the *transformation of experience* by the individual, not just the accumulation of pieces of information. With that view of knowledge, Piaget sought evidence of the nature of that transformation in the behavior of infants when a toy is hidden from them and in the behavior of adolescents faced with a complex solutions task. For perhaps the first time, a psychologist seriously classified the development of motor reflexes in infants as components of cognitive behavior—the earliest manifestation of the development of intelligence.

As the titles of some early Piaget books suggest—for example, *The Child's Conception of Number*—he applied this view of intelligence to an investigation of the ways in which children's view of specific areas (such as number) changes from a distorted notion to a sophisticated concept. His data resulted in the formulation of a two-pronged theory—both aspects of which continue to be tested. One aspect of his theory, designated as the stage-dependent theory, resulted from the appearance of *qualitatively* different intellectual abilities in a *sequence* of stages related to age. Characteristically, for example, the ability to engage in logical classification was observed in the performance of children beginning at age seven or older. Piaget's analysis of a host of responses to tasks resulted in the identification of four major *stages* (age ranges characterized by distinct intellectual abilities), which form an *invariant sequence*. These four stages, in order of appearance, are the sensori-motor, the pre-operational, the concrete operational, and the formal operational stages.

The second aspect of Piaget's theory, designated the stage-independent theory, includes Piaget's explanation of the development of intellectual structures and his view of the nature of knowledge and the nature of knowing. This aspect of Piaget's work is of such vital concern to the teacher that a presentation of its chief characteristics and their implications for instruction is given extensive treatment in Section 3.3, which is devoted entirely to that topic.

The rest of this section presents an overview of Piaget's stage-dependent theory, beginning with a description of the four major stages. The *sensorimotor stage* (usually birth to two years) is characterized by the gradual refinement of motor reflexes as the child begins to "know" the world by using the five senses. An infant touches, tastes, smells, listens to, and looks at everything within range. In addition, older babies begin to show evidence of a developing awareness that a hidden object still exists, whereas younger children behave as if "out of sight, out of mind." This latter characteristic, a kind of conservation of objects, involves reasoning on the basis of mental images. These abilities, plus the ability to coordinate hand-eye movements and other motor complexes, typify the stage Piaget calls sensori-motor.

The second stage, the *pre-operational stage* (usually two to seven years) is the period when language development increases the child's ability to symbolize—that is, to represent things in a variety of ways. Pots and pans become cars and trucks, and the child may become upset if one of the "cars" is put on the stove. Labels, to this child, are what they name. Perception develops, and with perception the child refines notions of large, small, wide, narrow, and so on. But it is perception that sometimes confuses the child into decisions that seem unreasonable to the adult. If five-year-old Mary is shown two jars of identical capacity and shape (Fig. 3.1) and allowed to fill each with limeade, she will readily agree that both jars contain the same amount of limeade. However, if the limeade from container B is poured into container C—a taller, thinner jar that has the same capacity as the other two jars—Mary no longer agrees that the amount of limeade has been *conserved* (remains the same), and usually says jar C has more limeade because it's taller. Mary's inability to divorce her thinking from the perceptual change, to consider the two compensating perceptual changes simultaneously, or to mentally reverse the action of pouring limeade from one jar to another are all characteristics peculiar to the pre-operational stage. On the other hand, *operational* implies a mental ability that is reversible as well as an action that can be internalized.

The third and fourth stages, the *concrete operational stage* (usually 7 to 12 years) and the *formal operational stage* (usually 12 to 16 years), are characterized by operational thinking and are differentiated from one another both by the nature of the reality about which children are able to reason as well as the form of their reasoning. Since these two stages include the preadolescent and adolescent, both stages are given extensive treatment in Section 3.4.

All of the preceding remarks on the Piagetian stages are a capsule treatment of the stage-dependent theory of Piaget. Beyond identifying and ana-

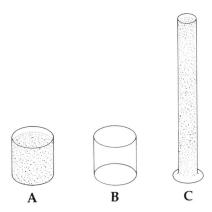

Fig. 3.1 Conservation of liquid capacity.

lyzing clusters of intellectual development that appear in an invariant sequence, Piaget has reinforced the concept of prerequisites and has radically altered preconceived notions of what these prerequisites might be. For example, his data suggest that children are able to handle direct variation tasks but not inverse variation tasks at the concrete operational stage. A table of selected tasks and the stage or ages at which they have been solved is included in Section 3.4. But the *what* of prerequisites has often been overemphasized while the *need* for prerequisites has been ignored. Throughout all of Piaget's work, the theme of developmental stages implies that early intellectual abilities are not eradicated by later ones, but are modified and extended. They, at one and the same time, are the cornerstone from which the structure is built, and yet they remain viable structures in their own right. The schema that follows is an attempt to illustrate this dual concept (see Fig. 3.2).

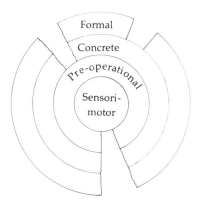

Fig. 3.2 The nested stages model.

The stages are nested in an invariant order. The *nesting* illustrates the formal operational student's ability to use abilities of an earlier stage when needed, while the crevices in the schema illustrate that same student's inability to reason formally about tasks for which he or she lacks prerequisite experiences. When you took calculus, you may have had the experience of trying to learn to differentiate without any tie being made to the concept of slope. If so, you joined the army of calculus students who then were unable to decide *when* differentiation would be helpful in problem solving.

Perhaps the most widely misunderstood aspect of the stage-dependent theory is that of age. The age ranges are meant to identify the *earliest* age at which characteristics of a stage begin to be exhibited by at least 75 percent of the children tested. The oldest age in a range identifies the age at which all characteristics of that stage have been demonstrated to be stable intellectual characteristics by at least 75 percent of the children tested. Further, the children originally tested were European children.

There have been some striking variations in age ranges in tests with children from other cultures. In America, Elkind (1961) has placed the age range for attainment of some characteristics of the early formal stage as a year or more later than those of European children. Experiments by Bruner and his colleagues (1967) in the subcontinent of Africa have demonstrated the invariance of the sequence of *stages* while indicating great differences in the *age* ranges. Indeed, in this case, the first three stages were attained at much later ages and the formal operational stage characteristics were not found to be stabilized in the oldest sample tested (16- and 17-year-olds).

What are the implications of all this diversity?

1. Children do not move from one stage to another overnight or on a birthday.

2. Children of the same age may exhibit widely differing intellectual characteristics.

3. As a group, American adolescents may safely be considered to be a year or more *behind* their European counterparts.

4. The most useful age to serve as an indicator of certain intellectual abilities is the *later* age in an age range.

These four summarizing statements are practical conclusions from which you will be asked to draw some inferences about instruction in later sections and chapters of this text.

Believe it or not, we've barely scratched the surface of Piaget's work and what it can mean to the teacher. But past experience has convinced us that learning about Piaget, as about anything else, requires interaction with something more concrete than reading material. You're ready, we think, for an introductory activity.

3.2 INTRODUCTORY ACTIVITY

I hear and I forget; I see and I remember;
I do and I understand.

A. The success of this activity depends on locating one pair of junior-high students between the ages of 11 and 13, one pair of senior-high students between the ages of 15 and 17, and one pair of college juniors or seniors who are not science majors. (Try to avoid senior-high students who are taking or have just taken a physics course.)

You will need to have on hand string or twine, scissors, one watch with a second hand, weights (such as croquet balls, large nuts, or any other compact but heavy weight), a ruler, tape, and a book.

B. Each of the three pairs of subjects will be asked to solve the following problem:

Use the materials to construct an inexpensive timing device (pendulum) which will take precisely one second for each complete swing. (See Fig. 3.3.)

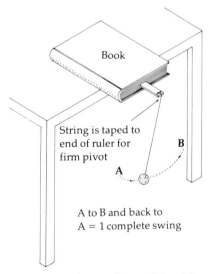

Fig. 3.3 Sample setup for pendulum lab.

One member of each pair should start the pendulum swinging and should count the number of complete swings silently during a 60-second interval. It is the other member's job to observe the second hand of the watch or clock and to tell the partner when to start the pendulum and when to stop counting. Adjust any aspect of the pendulum mechanism until 60 complete swings take exactly 60 seconds.

C. Your job is to systematically observe the approaches taken by the subjects in solving the preceding problem. Record at least the following kinds of observations:

   1. Do they manipulate only one, or more than one, variable at a time? (For example, do they try changing the length of the string *and* the distance through which they swing the bob during a single trial?)

2. Do they seem to progress with the task in a systematic or nonsystematic fashion? (Is their response to errors random?)

3. Is there any attempt to keep written records of the results achieved by changing the pendulum in different ways?

4. When and if the subjects solve the problem, question them in such a fashion as to convince yourself that the solution was not obtained by luck, but rather by isolation of the one variable (length of string) that affects the period of the pendulum.

D. Compare the data obtained for each of the three pairs. Compare your data with that of as many of your classmates as possible. What similarities (differences) were found between the junior-high and senior-high subjects? Between either of these and the college students? Explain your results on the basis of your understanding of Piagetian theory. What differences would you expect between college students who are science majors and those who are not? Test your conjecture by trying this activity with pairs of college students who are science majors. (Don't forget the need for a representative sample).

The above-described activity was engaged in by 14 pairs of college seniors at the State University of New York at Albany in the fall semester of 1975. Each pair consisted of one mathematics major and one physics, chemistry, or biology major. (There were exactly 1 physics, 2 chemistry, and 11 biology majors.) The authors observed the experiment and collected the following data:

Behavior	Number of pairs		
	At start	By end	Never
1. Manipulated exactly one variable at a time	1	11	2
2. Systematically made changes as a result of feedback from trials	1	11	2
3. Kept written records of results	1	7	6
4. Able to justify success	1	9	4

The single pair who are recorded as reacting positively to each of the four behaviors at the start of the experiment was the pair that included the single physics major. However, it should be noted that several of the mathematics majors were physics minors and that all the science majors had taken high school physics. Some who were able to recall the principle governing the pendulum were still observed changing variables other than the length of the string!

There are several possible explanations for this apparently contradictory behavior. First, it is clear that some students had learned the "statement of the rule" of the pendulum but they had never really learned the "rule." (We will look into this problem of rule learning again in Chapters 6 and 7). Why does this happen? Perhaps these students were never given a relevant concrete situation that would help them make sense of the rule statement. But aren't intellectually sophisticated college students capable of reacting in a systematic way even to a novel situation, especially when the situation is as simple as this one? Surely these college students are at the formal operational stage of intellectual development. Yet according to the data, numbers of them behaved as if they were *not* formal operational. What explanation can *you* give for such behavior? Hint: Consider the cracks in the nested stages model. Then rethink this entire activity after you read the following sections—especially the Implications for Instructions section.

3.3 THE STAGE-INDEPENDENT THEORY OF PIAGET

The activity of constructing a clock outlined in the previous section and the limeade in different-shaped jars task described in Section 3.1 are representative of the tasks through which Piaget evolved his developmental theory. Nowhere is the child required merely to listen or look and to subsequently repeat what he or she hears. In particular, the tasks (in contrast to some of his earlier ones that were criticized for emphasizing vocabulary) generally require some kind of manipulation of materials. The nature of the possible manipulation always embodies a cognitive activity. In any case, all tasks necessitate active mental participation on the part of the subject with or without manipulation of concrete materials. This characteristic of the experimental tasks is a salient feature of Piaget's theory—a feature he calls *interaction*.

Interaction Piaget holds that individual intelligence develops through the person's *interaction* with his or her environment. As a biologist, Piaget accepts evolutionary theory and conceives of intelligence as plastic and durable, for while each human being possesses a unique genetic makeup, all people share a common genetic base. The former provides the human learner with inborn structures—for example, a nervous system capable of a limited number of perceptions; the latter provides an operational mode common to all living organisms in their interaction with reality.

Piaget insists that knowledge is active. For him, to *know* an idea or an object requires that the student manipulate it physically or mentally and thereby transform it. At different developmental stages, this activity of transforming knowledge takes on different forms. Nine-month-old Marcy shakes a toy, tastes it, looks at it, and listens to it. In her own way, through her senses she transforms the object and so "knows" it to the extent that these actions transform it for her. At the other end of the spectrum Mr. Spotle, a banker faced with a business problem, may "turn it over in his mind," may doodle, may prepare charts, and may confer with colleagues. He transforms the set of ideas in a combination of symbolic and concrete ways and so "knows" the problem.

This construct of interaction may be the most important of Piaget's contributions. According to this concept, when you want to solve a problem about hobbies or finance, in the home or at the garage, on the ski slope or at church, you will spontaneously and actively interact with those characteristics of the real situation that you perceive as relevant to your problem. An observer might easily identify the external characteristics of your interaction if he or she saw you systematically trying to weight and unweight your skis across a steep slope in response to slides and skids. However, only you might be able to attest to the interaction occurring during prayer at church. What happens in school? Think through your own school instruction. Which modes were utilized often, seldom, or not at all? The answer often is that listening to lecture was emphasized; small-group discussions were minimized; and laboratory, outside of science class, was never used after elementary school. Might there be some connection between the mode used and the probable lack of meaningful interaction? If your results on the pendulum experiment paralleled ours, then you know that somewhere the appropriate interaction had not occurred. By the way, you also know that those who succeeded with the activity, and didn't just make a lucky guess, now *know* the principle of the pendulum. That's pretty good evidence of the kind of interaction that works.

Are there times when listening is an appropriate way to introduce ideas? Are there times when reading can be said to encourage interaction? You can answer that last question now. Do you recall reading an article throughout which you were struggling with the ideas being communicated? You may have even felt constrained to argue with the author. Have you also had the experience of reading an article and not being able to recall its major theme a few hours later? What are the characteristics of each article, the differences in the knowledge you brought to the task, the differences in your desire to "know"?

In many of your college classes, listening is expected to be the student's major role. What do you find are essential prerequisites for you, the listener, if listening is to result in knowledge, not memorized bits and pieces? Suppose you, the listener, have all the needed prerequisites. What is required of the lecturer? Must the lecturer also write on the board? Have you been in a college class where no board notes accompanied the lecture? What were the results of the occurrence, or the lack of occurrence? If you haven't already done so, compare your answers to the above questions with those of two other readers of this section. Try to come to some consensus on major points. In a later section, we'll reconsider the questions raised here.

Adaptation Piaget provides us with other constructs that explain the functioning of interaction and suggest the parameters of teaching behaviors most likely to result in interaction. Just consider your own mental activity as you read these paragraphs. Each of you comes to the reading with different experiences, differing degrees of understanding of Piaget, and different retention of past school learning. You read the same words. Some of the words are familiar, but may be used in a slightly novel way. Some of the ideas may be, for you, old friends; for another reader, the same ideas might present a totally new notion. If you were to talk out your mental activity as you read for understanding (We all know that it's possible to read just to finish an assignment and then wind up with a cognitive residue of zero!), you might describe it as a *matching-patching* task. You may try to *match* the inferred ideas to ones you possess that seem to be similar, or you may *patch* previous cognitive knowledge on the basis of a new twist. It is important to realize that 30 readers may conclude the same reading with 30 different shades of meaning. It must be obvious that if you possess no previous conceptual glue with which to do the matching-patching, then, no meaningful knowledge will become part of your cognitive structure.

Piaget has specific labels for the match-patch function described above. He labels the process where an individual interacts with an experience (a real object; a situation; inferred ideas through reading, listening, or seeing) *adaptation*. He describes adaptation in terms of two concurrent functions: *assimilation* and *accommodation*. In the illustration above, the reader *assimilates* (matches) the ideas inferred from the paragraphs as he or she simultaneously *accommodates* (patches) prior cognitive structures to the new input. Notice that when adaptation (matching-patching) occurs, the individual is changed (in a cognitive sense) while he or she changes the experience.

What does all this have to do with you, the future teacher? How can you use these ideas to help your students learn? In order to answer these questions, we need to consider in detail those two stages related to adolescent intellectual development.

3.4 THE CONCRETE OPERATIONAL VERSUS THE FORMAL OPERATIONAL STAGE

The last two of the Piagetian stages are designated as *concrete operational* and *formal operational*. These are the stages that span the age range from 7(8) to 15(16)—an age range that includes your students.

Characteristics of the concrete operational stage Operations, you will recall, are mental actions that are reversible. The limeade jar example that confused the pre-operational child would be easily solved by the older student who is not misled by perceptual changes. Indeed, the first indicator of this stage is the ability of the child to conserve number—that is, to realize that the number of objects is unchanged regardless of their arrangement. However, the same child may assume that the length of a pipe cleaner is changed when it is bent or that the amount of clay in a round ball is altered if the clay is molded into a long, thin pipe. In other words, ability to conserve number, length, mass, volume, and so on, are not all achieved at the same time even though all those schemas require the ability to consider several perceptions simultaneously and the ability to reverse a mental action.

By the end of the concrete operational stage the preadolescent is able to conserve all of the preceding relationships. Table 3.1 illustrates the diverse ages at which various conservation tasks have been accomplished by subjects

Table 3.1

Approximate ages of attainment of selected conservation tasks

Task	Approximate Age
Conservation of number	7-8
Conservation of weight	9-10
Conservation of length	7½-8½
Conservation of area	7½-8½
Determination of area by use of length and width multiplication	11-12
Conservation of volume	11-12
Determination of volume by use of length, width, and height multiplication	12-13

tested by Piaget and his colleagues. The ability to apply some of these concepts to standard measurement tasks occurs much later.

If you are a physics minor, you will realize that volume can be a fairly complex concept, because it includes the ideas of interior volume, liquid volume, and displacement volume. Even the age range of 11-12 may be a lower limit for stable grasp of the conservation of all facets of volume. With this in mind, look at the specific ages in Table 3.1 as a guideline only.

Underlying all the above major advances of this stage are the ability of the child to classify and the ability of the child to work with relationships where order makes a difference. What a surprise to adults who have heard four-year-old Jack recite the first ten numbers when they realize that he does not understand the relationship between 5, 6, and 7 and they realize that he cannot correctly assert that 6 toys, 6 books, and 6 miscellaneous objects all have the common characteristic of sixness. The ability to classify—for example, to distinguish the cowbird from the starling and the grosbeak from the finch—is perhaps the most powerful of our thinking tools. In and out of school, we are asked to learn hierarchies of classification systems. In addition, much learning depends upon the ability to perform multiple classifications. That is the task we give Kenneth when we ask him to select an isosceles right triangle from a collection of shapes.

Piaget's data show that these abilities are developed gradually throughout the concrete operational stage as a result of interaction with sufficient experiences that require multiple classification. Ordering relationships are an inseparable part of learning to classify. For instance, Mary learns that she has a father and a brother but so does her friend Bart and her cousin Paul—the beginnings of an understanding of the class of "fathers" or of "brothers." However, Mary gets confused when she hears that cousin Paul's father is her father's brother. The problem of order relationships is involved, and it is not until the concrete operational stage that most children begin to use these correctly. The full meaning of statements such as "5 is greater than 4 which is greater than 3" is not grasped by the pre-concrete operational youngster. But the concrete operational child correctly concludes that "5 is, therefore, greater than 3" and, later, is able to reverse these relationships and handle "less than" statements.

However, all of the above mental abilities are developed during this stage by interaction with "concrete" content. What does "concrete" signify? Must the child manipulate physical models of the ideas to be learned? Usually, but not necessarily! What is essential is that the experiences be real to the child and that those experiences reflect in as tangible a way as possible the concept or rule being developed. This characteristic has been the most troublesome one when trying to use Piaget's findings in the classroom. For, as we all know, much out-of-school learning occurs because you and I are motivated to get an answer or result or to resolve a conflict. An enormous concrete base of activities, verbal responses, visual responses, and visual experiences was required in the initial learning of an idea. However, adults tend to forget about the various concrete phases that underlie their present state of learning.

Which instructional modes offer the best potential for providing the necessary "concrete" basis for learning? Look back at the Continua of Teacher/Student Participation in Chaper 1 and the scale illustrating verbal feedback potential in Chapter 2. These should provide a basis for a third continuum—one you should construct to depict potential for interaction with concrete experiences. Is laboratory at the top of your scale for providing a concrete basis? It should be. Laboratory was the key mode used in the pendulum activity, and the data from that experiment support the need to use concrete experiences in instruction even when the students are assumed to be formal operational. Try comparing the results of that activity with the one described in the next section—one that Piaget's colleagues have used to identify the characteristics of the formal operational stage.

Characteristics of the formal operational stage Since this stage, like the one before it, is labeled *operational,* you should be alerted to the nested stages idea once again. In other words, the mental structures that develop by the *end* of the concrete operational stage are still available for use by the adolescent or preadolescent. In fact, they *must* be used to solve many real-life problems. In some cultures these concrete operational structures are the most complex that can be identified. This suggests that more sophisticated intellectual structures do not develop in the absence of a need for them.

What's new about this stage? The new aspect is cued by the label *formal,* which suggests that the adolescent can deal with the "form" of the situation and need not resort entirely to the concrete aspects of the problem. But when does this happen in a problem-solving situation, and how does one recognize that "form" is being considered? Piaget has identified *four* characteristics of the formal operational stage—all of which depend upon one another. They are

(1) the treatment of the real as a subset of the possible, (2) hypothetico-deductive reasoning, (3) combinatorial analysis, and (4) propositional thinking. All four are described in the context of problem situations in the following paragraphs and frequently are characterized by citing differences in the behavior of concrete operational and formal operational children.

For example, the concrete operational child draws hypotheses only from observed or experienced reality and the potential is seen only as an extension of the real. On the contrary, at the formal level reality is considered as a subset of the possible with the result that hypotheses may proceed from nonobserved and nonexperienced phenomena. This characteristic of the formal stage—the ability to imagine the possible as containing the real—frees the formal thinker from the restrictions of his or her senses. The real-possible relations at the two levels can be exemplified by responses to a problem that requires looking at a picture of an auto accident and suggesting the circumstances surrounding the accident. Concrete operational children may suggest a flat tire or a heart attack—realities derived from the picture or experience—while formal operational children might propose that the car was designed to move forward when the driver put it in reverse—possibilities not observed in the picture and not experienced by the respondents.

Inhelder and Piaget (1958) describe responses to experiments that illustrate this characteristic repeatedly. An excellent illustration comes from an experiment on the law of floating bodies. The experiment involves the classification of objects as floating or nonfloating and is directed toward the discovery of the law on which this classification depends. After some preliminary tests, young children incorrectly predict that a coin will float because it is "little" or that a wooden plank will sink because it is "heavy." Even the concrete operational child is limited to predictions based on observable elements such as the level of the water, the volume of the object, and the weight of the object. It is not until formal operations have stabilized that the student hypothesizes about a situation that is not directly observable, such as the comparison of the weight of the object with the weight of an equal volume of water. For although the entire volume of water in the container is observed, the volume of water equal to that of the object in question has no observable shape, and it requires that an abstraction be conceptualized. In addition, the solution implies the understanding of density or specific gravity and so presupposes an understanding of weight and volume, the latter not being fully realized before the early formal stage. In this particular case, therefore, the solution demands formal operations for several reasons, whereas in the accident example the solution might have been among those obtained without recourse to these second-degree operations. This difference in the nature of problems is reflected in different approaches to the problems at each stage, which, in turn, exemplify the nested quality of cognitive structures and the possibility of thinking according to the mode of a prior stage. In other words, the formal operational child *has the capacity to use formal operations but is not compelled to do so.* He or she may revert to any of the earlier modes of thinking as they now issue from the transformed cognitive structures, for earlier stages are not eradicated but integrated into later stages.

This first characteristic of formal operations, the changed relation of the real to the possible, is dependent for existence upon the presence of another characteristic, the potential for combinatorial analysis. In the concrete stage, the child faced with a multiple variable situation usually is limited to trying one-many correspondences or to testing unsystematically other possible correspondences, but the adolescent able to employ combinatorial analysis can consider all possible combinations of variables in a systematic manner. This ability is a *necessary* condition for generating all possibilities and so determines the shift in the orientation toward the real and the possible.

Another example from Inhelder and Piaget's experiments points out behavior specified by this characteristic. In a solutions experiment, the elements are four similar flasks filled with perceptually, but not chemically, identical liquids; a fifth flask with a dropper is added to the array. The experimenter shows the child two glasses of apparently similar liquid that has been taken from one or more of the four original flasks (without being seen by the child). Into each glass is placed a few drops of liquid from the fifth flask. In one case the solution turns yellow; in the other case no change is observed. Then the child is asked to reproduce the color yellow, using the liquids from the four flasks and the one with the dropper, in any combination. Jane, a concrete operational child, may try the fifth flask with flasks one, two, three, and four separately, a one-many correspondence represented by $(5 \times 1) + (5 \times 2) + (5 \times 3) + (5 \times 4)$; and she may even attempt $[5 \times (1 \times 2)]$, but she soon gets lost and either forgets earlier trials or neglects possible ones. In contrast, the older child, Mark, seems to realize the totality of possibilities from the start and he seems to proceed systematically in the testing of these possibilities. His language, "If 4 were water, then . . . " and his actions support the assertion that he possesses and is using combinatorial analysis. This assertion does not mean that he actually tests each and every possible combination; in fact, he rejects those not necessary to a final determination of the solution.

It is often puzzling to comprehend this last assertion. On what basis does the older child reject solutions without an empirical test? He does, in fact, use the results of early trials to make conjectures, and these conjectures reduce the total number of trials. In this case one of the chemicals is a bleaching agent and one is neutral. (The four liquids are water which is neutral, thiosulphate which is the bleaching agent, diluted sulphuric acid, and oxygenated water. The fifth bottle contains potassium iodide. The experiment is based on the fact that oxygenated water oxidizes potassium iodide in an acid medium. Thus the mixture of these last three liquids will produce a yellow color.) The first time the formal operational youngster adds the bleaching agent to another combination and sees the effect, he or she may conjecture that this agent will bleach other combinations, may test it one more time, and may then eliminate it from consideration.

It is this kind of reasoning from the effects of early trials that allows the intellectually mature adolescent to identify combinations "not necessary" to a final solution. However, the potential for combinatorial analysis does not imply that the youngster thinks about the system itself. Rather it implies the kind of awareness in which he or she displays a motivated attempt to consider all possible combinations of a set of elements *before* experimentation has occurred. Moreover, it is customary to see this adolescent recording the effects of trials in a systematic way. On the contrary, while some concrete operational children may begin to keep a record, there seems to be no attempt to list possibilities in a patterned way so that no trial will be overlooked. There are some adults who display similarly unsystematic approaches to a multifaceted problem. Thus, possession of this ability does not guarantee that it will be used.

Both of these properties of formal operations, the real-possible relation and the use of combinatorial analysis, are apparent in the actualization of a third characteristic, hypothetico-deductive reasoning. The formal thinker's reasoning is less "This is true, therefore . . ." and more "If this were true, then. . . ." This kind of reasoning is essential if the possible is to include the real in the set of hypotheses. It also follows hand-in-hand with the ability to systematically check all possible combinations. It is not true that the use of "If . . . , then . . ." statements is *necessarily* evidence of hypothetico-deductive strategies. The younger child makes use of "If . . . , then . . . " sentences but is restricted to statements deduced from experience. An example of such a statement is "If it rains, then the ground will be wet." This kind of statement is not what Piaget has in mind when he speaks of hypothetico-deductive reasoning, which seeks both a general hypothesis and a necessary one. According to Inhelder and Piaget (1958), this quality of necessity is a distinguishing mark of formal thinking. In their experiments with children of various ages, they found that the formal operational adolescent searched for necessary causes and was not content with sufficient ones. Thus, the concrete operational child makes deductions from the observed or experienced situation while the older child's thought results from a union of possibility, hypothesis, and deductive reasoning. The concrete operational child's inability to go beyond experience and observation leads that child to think in a disconnected manner in which each link in the chain is seen as independent of the others. In contradistinction to this mode of thinking is that of formal operations where the adolescent considers each link in relation to the others.

Finally, as implied in the illustrations so far, formal operations are characterized by propositional thinking. The elements manipulated by the formal thinker are logical propositions, statements containing raw data, rather than just the raw data itself. In other words, the older youngster may utilize concrete operations of the earlier stage by organizing reality into classes, ordering them, and so on; but then that adolescent proceeds to form propositions using these results and to operate on the propositions via conjunction, disjunction, implication, negation, and equivalence. This type of thinking is what Piaget calls second-degree thinking, operations that result in statements about statements.

In the solution experiment described earlier, the formal operational youngster obtained the set of all possibilities by operating on propositions about the effect of the four variables when combined with drops from the fifth container, one at a time, two at a time, three at a time, and four at a time. A hypothesis framed like this one, "If $[5 \times (1 \times 2)]$ tests as yellow, is it true that 3 and 4 must be absent for yellow to occur?" represents a statement about statements. The reader who has studied symbolic logic will recognize the construct "proposition" in Piaget's construct "propositional thinking." An engineer searching for a solution to a complex problem will undoubtedly begin to symbolize the factors involved and will logically consider the multiple ways these factors might be combined (for example, both together, one and not the other, one or the other or both, and so on). If this explanation is beginning to sound very much like your interpretation of combinatorial analysis, then you have identified a fundamental tie among these and the other two characteristics of formal operations. Although they can be discussed separately, they occur simultaneously and are more like varied aspects of the same complex action than mutually exclusive notions.

The four characteristics of formal operations outline the manner in which the intellectually mature adolescent thinks. Presented with a new situation, that adolescent begins by classifying and ordering the concrete elements of the situation. The results of these concrete operations are divested of their intimate ties with reality and become simply propositions that the adolescent may combine in various ways. Using combinatorial analysis, the student regards the totality of combinations as hypotheses that need to be verified and rejected or accepted.

However, these powerful tools of formal operations were not initially employed by the formal thinker in the solution experiment. It is significant that all students *first* made some tests and so obtained concrete data. Then the formal thinker began to hypothesize before continuing the tests while the concrete operational child simply tested more cases engaging only in reasoning from the data, not in predicting what might occur next. Therefore, it seems clear that in the case of material that is completely new to the child and for which he or she has no previous concrete base, that the process of transforming reality into concrete relations and those into propositions must occur *before* propositions are available for formal thought. Important distinctions between the thinking patterns of the concrete operational student and the formal operational student are summarized in Table 3.2.

Now perhaps you have some answers to the questions posed earlier—

Table 3.2
Thinking patterns

Concrete	Formal
Needs reference to concrete experiences, familiar objects or events.	Can reason with concepts and indirect relationships, assumptions, and theories.
Is restricted to reasoning inductively (generalizes on the basis of data).	Can also use hypothetico-deductive reasoning.
Needs step-by-step instructions in a lengthy procedure.	Can plan a lengthy procedure and tends to use systematic approaches in recording data and in deciding next steps.
Does not attempt to check conclusions; seems unaware of inconsistencies in own reasoning.	Is aware of own reasoning and tries to resolve inconsistencies; seeks necessary, as well as sufficient, conditions for a conclusion.

questions regarding the interpretation of the data from the pendulum activity or the nature of that particular task. In the next section we'll be making some initial connections among the theory, the data, and classroom instruction.

3.5 IMPLICATIONS FOR MATHEMATICS INSTRUCTION

To study Piaget's data and to accept his view of knowing cannot help but change a teacher's approach to students. Errors become sources of new approaches to instruction rather than evidence of failure.

From the stage-independent theory Regardless of the age of the students, the stage-independent constructs are a gold mine of instructional ideas for the teacher. There are some clear signals to the teacher in the functioning of adaptation. First, if assimilation-accommodation is to occur, the gap between the new experience and past knowledge cannot be too large. How large? There is no easy answer, but there are hints. If you analyze the nature of the content and search for prerequisites, some obvious prior needs will be identified. Then you can informally diagnose through homework assignments or a short question/answer period or any number of other modes.

Warning: what students were exposed to is not necessarily what they've learned! A tenth-grade geometry teacher preparing for the unit on similarity might be impressed and misled by a review of the seventh- and eighth-grade syllabi and texts. The junior-high geometry unit may list similar triangles and their properties, proportions, and their application in shadow problems and more. The pleased tenth-grade teacher might thus assume that the sophomores in the class have "had" all the necessary concrete experiences and visual interpretations they need to begin the axiomatic study of similar figures profitably. You've already identified some of the flaws in that assumption!

Yes, what's between the covers of a syllabus or text may or may not have been included in instruction. Even if it were, you now know enough about the variable potential on the receiving end of instruction to question the probable retention of that intuitive base. What are we saying here? Assume ignorance? Reteach everything needed? If you do, you'll bore a great many capable students. We're back to the necessity of getting feedback on the nature and extent of prerequisite knowledge. Diagnosis!

That brings us to the second signal. If adaptation is such an individual matter and the resulting knowledge is more heterogeneous than homogeneous, then diagnosis must be able to identify differences in individual understanding. The ideal is to optimize individual learning, but we work toward the ideal in small steps. Diagnostic questions or tasks can be constructed to assess recall, comprehension, or the ability to use concepts in a novel way. Isn't there some commonality about that which we intend students to learn? Very definitely. There are standard definitions, standard rules, and standard ways of using rules; teachers want a uniform kind of behavior to eventually occur with those who start from different points but are expected to reach the same finish line. To some we give enriched or remedial instruction; to others we give additional time (self-paced instruction). However, we must get feedback on their different needs *before* treating them differently.

The major implication of the constructs of interaction and adaptation seems to be the need to encourage an *active* learner and the need to individualize. As a careful reader you will have noticed that the concept of individualizing is tied to diagnosis, Piagetian stage descriptions, and the need to help the student transform reality. Nowhere was it suggested that "individualize" equals "learn alone." In fact, Piagetian theory presents some strong evidence *against* the efficacy of "working alone" as the most profitable way of interacting with reality. Keep this in mind when you read the "Differentiating Instruction" section of Chapter 11.

Now we're ready to respond to the earlier questions as to the roles of the lecturer and the listener in an effective college class. The lecturer has to know his or her "stuff"; equally important, though, the listener has to have enough background knowledge so that the spoken words trigger meaningful associations. The lecturer also has to capture the listener's attention at the start and keep it throughout. Is the listener expected to recall data and to dig into the lecture for deeper meanings? Then the listener will probably need to take notes and the lecturer may need to facilitate the note taking with a written aid. Notice how much more is demanded of the lecturer if listening is to be active. And regardless of the lecturer's oratorical charms, how difficult it is to keep the atmosphere right for interaction on the part of the listener! Perhaps this is why most college students take notes during a lecture. That physical activity may be a stimulus to mental activity, although sometimes note taking can be done automatically with the more complex areas of the brain tuned out.

While Piaget's stage-independent theory sets out routes that lead us and our students further in knowing, his stage-dependent theory gives teachers milestones by which they can understand students better and plan more effective instruction.

From the stage-dependent theory The section on formal operations emphasizes the meaning of *formal* versus *concrete*. The reader should notice that the Piagetian tasks were carefully constructed to elicit distinctions that would highlight the sophisticated cognitive structure of the formal operational child. Too often, however, teachers of superior students use the Piagetian descriptions as justification for a highly verbal, highly symbolic treatment of content. There are some messages they've missed. The fact that cognitive structures are nested and that one never loses the mental abilities of earlier stages is significant; if a task is truly one that requires formal operations, the student cannot operate on it meaningfully unless he or she has had sufficient concrete experience on which to draw. Consider the twelfth-grade teacher presenting theoretical probability to students who have had no experience with empirical probability. That teacher should plan a laboratory designed to have groups toss coins, roll dice, and participate in other activities that cause them to calculate multiple probabilities—all as background for the work on theoretical probability. Intellectual maturity alone is not sufficient reason for a presentation divorced from the intuitive plane. It is clear that the teacher must analyze the content and its concrete prerequisites, must diagnose the students' intuitive background, and must plan instruction to close the gap between the intuitive and the abstract.

Another mistake made by too many teachers is based on the apparent belief that students who are formal operational will approach all school lessons in a sophisticated way. Of course they won't if they don't need to. For example, if the problems given are simple, if students are never asked to predict, if the teacher always demonstrated the solution, then students will not use formal operational thinking. There is some evidence that suggests that formal thinking may, as a result, become a less preferred mode of thinking. This possibility is perhaps the most disastrous result of some teaching—the unintentional discouragement of formal operational thinking in those who are capable of it. Teachers who intentionally attack this problem use some or all of the following: "What if" questions; tree diagrams to encourage a systematic approach to many combinations; a "Predict, explain your prediction, now verify, modify predictions" sequence; or classwork on a problem for which the teacher has no answer.

Look again at the systems analysis model on the opening page of this chapter. Notice that the flow of the arrows from the top boxes suggests that you will have to synthesize what you know about the nature of the content as well as what you know about the intellectual development of your students before specifying objectives and, in the long run, before selecting appropriate instructional strategies. Piaget's data and an understanding of his theory will help you to establish reasonable broad guidelines about eleventh- or eighth-graders even before you step into the classroom. The results of recent research (Renner and Stafford, 1972) that attempted to identify the number of youngsters in the formal operational stage, those in the concrete operational stage, and those in an intermediate stage show that almost one-third of tenth-grade college-bound students may *not* yet be clearly formal operational thinkers. Although there is no guarantee that your tenth-graders will fall into a similar pattern, there is sufficient evidence to suggest that you should plan to include strategies that introduce concrete instances of a concept or concrete applications of a rule. Ms. Sosnick, a geometry student teacher, tried to do just that when introducing congruence proofs in the case of overlapping triangles. She constructed a large poster paper model of a triangle and used paper clips to attach smaller varicolored triangular cutouts to the master figure. After employing the demonstration model to illustrate coincident sides, corresponding angles, and the like, Ms. Sosnick gave each student a smaller version of the model to use if needed while working on several proofs. After a few minutes of pseudosophisticated remarks about the cutouts, *all* students were observed making use of the individual models. Some used a model to verify a conjecture; some, to locate corresponding parts; and some, to suggest a potential problem-solving strategy. The concrete models not only got and kept attention, but served as a stimulus to novel problem solving.

The above illustration is just one of the many ways in which instructional modes may be used to advantage in assisting interaction and in support of Piaget's data on stages. In ensuing chapters you will be reading many more Piagetian-based applications to teaching and you will be asked to construct similar strategies for the students you will teach.

3.6 CROSS-REFERENCES TO A SAMPLE CBTE EVALUATION INSTRUMENT

Without a doubt, competency 2.2 is the one most directly connected with the component of "The Intellectual Development of Students." But now as you read through category 2.3, you should also be able to identify modes that are generally appropriate if you're teaching seventh-graders and similarly, those you would expect to observe being used less. In particular, if 2.32 were used in large doses and if 2.37 and 2.38 were avoided with general mathematics students in the ninth grade, we would begin to wonder about the lack of match between the teacher's choice of modes and Piagetian data. In our experience, the teacher would soon realize all was not well, since student feedback to this pattern generally ranges from outright rebellion to dull passivity.

Perhaps 2.38 is a competency you won't be expected to demonstrate if you teach honors students in grade twelve? Think back to that pendulum activity with college students, and you have the answer to that conjecture. You may not use the laboratory mode as often with honors students of that age group and you surely won't utilize a laboratory to illustrate a relationship that the students have already learned. However, you should expect to design, adapt, and utilize the laboratory mode in a meaningful manner in selected areas of the content over the long term. In general, to consistently and effectively perform competency 2.2, you must select combinations of modes that are promising from the point of view of Piaget's work. Refer back to Chapter 1 for the operational definitions of these modes and to Chapter 2 for the relationship of modes to feedback.

3.7 SUMMARY AND SELF-CHECK

This chapter was entitled "Adolescent versus Adult Thinking" in an attempt to emphasize the large intellectual gap between you and your students. You are the adult in the title, and even if your students are just three years younger than you, they are the adolescents. You, presumably, have learned the subject matter you will be teaching. However, you have transformed it so completely from your initial experience with it that your resulting knowledge is in a form inappropriate for the initial learning of your students. Somehow you have to retrieve that initial experience and help your students to transform it successfully. To do this, you must first learn all you can about the adolescent intellect. How do your students learn? What has been their history of knowing? What do their questions and errors tell you? These and related questions have been the subject matter of this chapter.

According to Jean Piaget, intellectual development depends on the interaction of several factors. Mental structures develop in an orderly pattern with time being the major variable so that it is possible to describe maximum intellectual development of humans in terms of four distinct major stages—each characterized by specific intellectual changes. In this chapter an introduction to Piaget's work and aspects of his stage-dependent and stage-independent theories were outlined. In addition, some implications for teaching based on Piagetian research were described.

Now you should be able to:

1. Operationally define interaction, adaptation, accommodation, and assimilation.

2. List some differentiating characteristics of the four major Piagetian stages.

3. Justify student behavior in terms of Piagetian theory after viewing a "live" or "canned" lesson in which students were involved in a laboratory problem-solving activity.

4. Rank teaching modes from promising to unsuitable in terms of Piagetian research given a specific set of modes and description of the students.

5. Suggest *at least* three different methods that could be used to obtain some initial clues as to your students' Piagetian stage levels.

The exercises that follow include some additional ways to deepen your understanding of Piagetian theory as well as activities designed to help you identify your own grasp of the material in this chapter and your ability to translate that grasp into classroom applications.

3.8 SIMULATION/PRACTICE ACTIVITIES

A. Ms. Pringle has her eighth-grade "lower-level" class cut out the three squares constructed on the sides of a right triangle (see Fig. 3.4). Then they are directed to cut the smaller squares into five sections using the dashed lines and to try to "cover" the largest square with these five pieces.

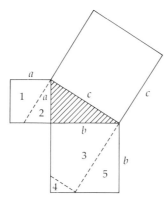

Figure 3.4

Eventually all the students are able to complete the task. Next she asks the class to imagine cutting one or more of the five sections so as to obtain six (or eight or ten) total sections.

"Would we still be able to 'cover' the large square if we used the new set of six (or eight or ten) sections? Why or why not?"

Most agree that the square will still be covered with six sections, but as Ms. Pringle increases the number of sections they are to think about, some aren't sure what will happen. Two students argue loudly that if the pieces were small enough, there would be so many of them that they would overlap the largest square.

Explain the reaction of these two students and the confusion of others in the light of Piaget's theory.

B. View the film "Piaget's Developmental Theory: Conservation," a 28-minute production from Davidson Films. Then try one of the experiments from the film on several children in the age range of 7 to 14. What results did you get? Compare these with those obtained by your classmates.

C. From your subject field choose a topic that is usually taught to bright eleventh-graders. Describe a demonstration, an AVA, or a lab you might use to teach this topic. List questions that would be asked in conjunction with the chosen mode in order to assist those students who were just on the verge of formal operational thinking and to identify those who were still concrete operational thinkers. (Choose a topic that involves abstractions, multiple variables, and is not perceptually obvious.)

SUGGESTIONS FOR FURTHER STUDY

Adler, I. Mental growth and the art of teaching. *The Mathematics Teacher* 59 (December 1966): 706-715.

Adler's article has become a classic introduction to Piaget's theory. It has the advantage of focusing on highlights, identifying and disposing of major misconceptions about the theory, and specifically pinpointing implications for all teachers in both the choice of strategy and of content. Since all the examples are mathematics/science examples, the article can be read with most profit by teachers of those disciplines. The reader will do well to obtain a basic understanding of Piagetian theory from *other* sources and to read Adler for his interpretaion of Piagetian contributions to the "Art of Teaching." They are excellent!

Furth, H. *Piaget and knowledge.* Englewood Cliffs, N.J.: Prentice-Hall, 1969.

Furth's text is recommended to students who have read Gorman or a similar source and one or more Piaget-related articles. Furth is a philosopher and does much to dissect Piaget's concept of knowledge. His treatment of the constructs of interaction and adaptation is similarly both scholarly and illuminating.

Gorman, R. M. *Discovering Piaget: a guide for teachers.* Columbus, Ohio: Charles E. Merrill, 1972.

This small paperback is probably the best introduction to Piaget thus far on the market. The author intends that the book will promote a guided discovery approach and so encourage that interaction at the core of Piaget's theory. For this reason much of the material is read with ease. The exception is the section on the INRC group (pages 44-57), which is highly technical and could well be omitted in favor of concentration on other sections.

Higgins, J. Piaget's analysis of intelligence. In *Mathematics teaching and learning.* Worthington, Ohio: Charles A. Jones, 1973.

This unit, like all others in the Higgins text, consists of several "modules" ranging from an investigation with pupils, to an elaboration of one of Piaget's most famous articles, to an application module, and finally to a study module. Even if the reader is unable to perform the investigation, a careful reading of the task is an excellent specific instance of Piagetian theory and is further clarified by the reading module. Don't miss the reading module and the section on discovery teaching in the application module! Piagetian theory is explained beautifully and then made more concrete via the do's and don'ts of discovery teaching. Of great practical value!

Rosskopf, M. The main states of cognitive development (Piagetian). *New York State Mathematics Teachers Journal* 21 (October 1971): 138-147.

This article carefully details the stage-dependent theory of Piaget by means of descriptions of children's reactions to Piagetian tasks. A novel part of the article is a section on experiments that emphasizes ratio and proportion and a lucid explanation of the INRC group. All readers would find the first part of the article either a good advance organizer to a more thorough exposition or an excellent summary. Mathematics teachers who have studied groups and/or symbolic logic will appreciate Rosskopf's handling of the INRC group and his explanation of the construct of operation in the final pages of the article. Although the article is brief, the author carefully keeps intact the complex ideas in the theory, but also communicates those ideas to the nontheorist reader.

CHAPTER FOUR

PRODUCTS AND PROCESSES
THE STRUCTURE OF MATHEMATICS

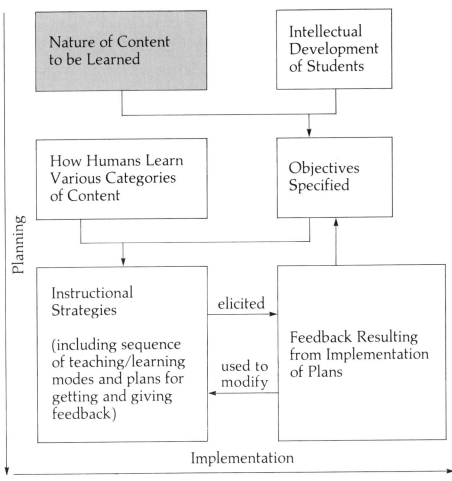

ADVANCE ORGANIZER

It is truism that we learn more about a subject when we teach it. Every tutor, swimming coach, and scout leader knows the experience. So be prepared for some rude shocks to your mathematical belief system. Such shocks are part of teaching and can be learned from and capitalized on. In this chapter we'll pose a range of questions designed to help you learn (and relearn) mathematics from views other than that of a student.

Why is a thorough analysis of subject matter content and structure so important? Why pursue this matter at this particular point in our study of teaching? Let's see if you can begin to formulate answers to these questions by looking back at the opening page of this chapter. Note that the shaded box, "Nature of Content to be Learned," couples with the previously treated "Intellectual Development of Students" portion of the systems analysis model to form the bases for specifying objectives. In any systematic approach to instruction it is vital to formulate clear objectives that closely match both the students they are presumed to affect and the subject matter content they are claimed to represent. Why vital? As the rest of the model depicts, these objectives will be used to design instructional strategies, guide the collection of feedback, and then be reassessed in the light of resultant learning. Thus the teacher's comprehension of the subject matter of mathematics will have profound effects on all aspects of instruction. Now is the time to face the issues. College students typically have spent nearly all of their time in three to four years of studying mathematics courses committing to memory the mechanics of the subject, dutifully completing exercises in how to obtain the correct answers and learning how to prepare for final examinations. A few college students have experienced courses in the history, philosophy, or structure of the discipline of mathematics where the focus has been on the analysis of various subcategories of knowledge within the subject matter and the interdependence of the processes used to generate knowledge within this specialized field. If you are one of the fortunate few with this type of background, a quick reading of the early sections of this chapter should suffice to check the completeness of your own comprehension. If you are one of the great majority who have not yet concerned yourself with the structure of mathematical knowledge and the processes used to generate it, you will need to give immediate and concentrated attention to this chapter since major contemporary seventh- to twelfth-grade mathematics curricula projects claim to reflect a valid picture of the discipline. It is with you in mind that this chapter is written.

We begin by asking you to take a stand on an an issue that intrigued scholars for centuries. The issue is phrased as an either/or, but not both statement.

Mathematics was invented by man

or

Mathematics was discovered by man.

You must choose a stance now. So stop and consider the implications of each statement. Let's be clear about the meaning assigned to the key words. *Invented* is to be thought of as it is used in a sentence such as "Alexander Graham Bell invented the telephone," while *discovered* is to be used in the sense of "Mary discovered her diamond ring under the counter." Notice that in both cases the person may have had an insightful idea, but in Mary's search the product—a ring—was not altered or brought into being by her activity. Take a few minutes of thinking time to decide on your position before reading any further.

Did you choose "invented"? Then how do you explain the novel work in the calculus by both Newton and Leibniz, who independently came to similar conclusions? There was much bitterness in seventeenth-century England when Leibniz published his findings first and thus received acclaim throughout Europe. Did the basic concepts of the calculus always exist and lie in wait for the insightful searching of Newton and Leibniz? On the other hand, if the concepts existed in some "discoverable" sense, like Mary's ring, they could be seen, touched, heard, or perhaps tasted. However, Newton's formulation of the concept of an "infinitesmal" was startling to him and to mathematicians for the next two centuries. After all, all basic concepts were, they thought, supposed to be self-evident truths and there just was no way in which the infinitesmal corresponded to the physical world. Thus Newton "invented" the concept of the infinitesmal.

If the cornerstones of mathematics were invented to satisfy a practical or theoretical need, then the teacher must point out the reasons behind a definition, a label, or a rule, explore with students alternative beginnings, and encourage mathematizing. Why is 5/0 undefined? Mr. Harris might say, "That's an agreement made by mathematicians. We must just accept it as fact." Ms. Marple might say, "Let's see what division means. What do you think the answer should be? If 0 is the answer, it should check. Let's try it. Let's try Jack's idea, too. Does 5 check? Now will any number times 0 equal 5? You see, that's the reason: because our definition of division is such that none of our numbers will work."

What stance do you think Ms. Marple or Mr. Harris would have taken on the invented/discovered question? What is your stance now? The question is not closed. There is much more to learn. In the next sections you'll do some mathematizing yourself. That should help you defend or assess your stance.

4.1 INTRODUCTORY ACTIVITY

Mathematics is a verb, as well as a noun.

What do we do when we mathematize? What do we obtain? In this activity you are asked to identify as many processes and products of mathematics as you can. Think junior/senior-high mathematics, and complete a chart like the one portrayed by Fig. 4.1. Write nouns under the "products" column and the "ing" form of verbs under the "processes" column.

Products of mathematics	Processes of mathematics
Corollary Formula Number	Generalizing Computing Assuming

Figure 4.1

A. Think geometry, algebra, arithmetic, and so on and try to formulate as complete a list as possible.

B. Compare your chart with those of at least two other classmates and try to resolve any differences you have found.

C. Which of the listed processes are employed primarily by the user of established mathematics? Which are used primarily by the person who needs to develop novel mathematics?

D. Try to rank, or to classify, both the products and the processes in the order in which they might occur in the development of mathematics. (Did you remember to include deducing and inducing?)

E. Which of the processes are likely to be emphasized in junior-high mathematics? Which in senior-high mathematics?

As you read through this chapter, you may want to revise your original chart. In this and later chapters we'll consider ways of teaching these aspects of the subject matter.

4.2 PRODUCT ASPECTS

If you included postulate and theorem in your list of products, you were probably thinking tenth-grade geometry. However, some of you may have learned that there are postulates and theorems in arithmetic and algebra. Most contemporary high-school texts are designed to introduce students to theorems in the real number system, such as $(-a)(-b) = ab$. This emphasis on structure in which many general results are deduced from a small number of accepted statements is a key aspect of the nature of mathematics. Since you will be teaching about structure in algebra, geometry, and even in junior-high mathematics classes, it is important that the nature of the products directly related to structure be crystal clear. That's "old hat" to a mathematics major, isn't it? Maybe it should be, but if you're not quite ready to distinguish postulate from theorem, don't be discouraged. Even Euclid, the genius of geometry, found that these distinctions were not easily made. We'll review just a few of his decisions and their consequences.

Don't get the wrong idea! Euclid was, indeed, a genius who is credited with the initial, monumental task of systematizing much of the mathematics used in his time. Thus, he might be considered the father of the present-day emphasis on structure. Strange though it seems now, most of the relationships in Euclidean geometry were known and used *before* Euclid. A cuneiform tablet dating from 1900 to 1600 B.C. provides evidence that the Babylonians were familiar with at least 15 sets of what are now called "Pythagorean Triples" (NCTM, 1969). However, Euclid accomplished the creative work of arranging these results in a hierarchical fashion, in which *d* followed *c* if *d* could be derived from *c* alone or else from *c* and any of *a* and *b*.

However, one of Euclid's first mistakes was that he tried to define everything—an impossibility that yielded some ridiculous results. Here is his "definition" of a point. *A point is that which has no part.* It was not until the work of nineteenth-century mathematicians, such as Gauss, Bolyai, Lobachevsky, and Riemann, who initiated a reexamination of the foundations of mathematics, that the necessity for beginning with *undefined concepts* was universally accepted. If you wrote point, straight line, or plane in your list of products, you were identifying *undefined concepts* from plane Euclidean geometry. We can give meaning to such concepts by asserting relationships that we wish to exist among them. For example, a familiar relationship called a *postulate* is: Two points determine a straight line. That postulate asserts something about the nature of a straight line. Other postulates add different shades of meaning to the notion we wish everyone to share. However, we no longer try to define *straight line*. (Unfortunately, there are still some junior/senior-high texts that include so-called definitions of *each* of these terms. One of your responsibilities as a mathematics teacher will be to correct errors such as this in instructional materials.)

The second "error" made by Euclid would seem completely reasonable if you were one of his contemporaries. Postulates were considered self-evident truths and, thus, were accepted without proof. As history makes clear, Euclid's view of the nature of postulates (and the nature of mathematics) was upheld by most mathematicians until the nineteenth century. The famous story of the "parallel postulate" reads like science fiction. We can only highlight a small portion of it here, but we urge you to read the entire tale.

Euclid's fifth postulate is wordy and sounds complicated. Surely a postulate ought to be a *simple* statement of the self-evident! Something so complicated should be proved, so throughout the centuries, famous mathematicians tried to prove the fifth postulate (or an equivalent version of it). Nothing worked! The attempt of Saccheri, who used the indirect method of proof, is an object lesson. It is said that Saccheri, who was not able to deduce a contradiction from both of his negations of the postulate, ignored the possibility that perhaps there was no contradiction and altered his work so that it would seem to "vindicate Euclid."

What is Euclid's fifth postulate? Most of you learned it in the form proposed by Playfair: On a plane, through a point external to a given line, there is one and only one line parallel to the given line. The statement seems obvious and is a good match to experience. It is not surprising that mathematicians hesitated to consider alternatives—none of them had your vantage point. You have vicariously traveled in spaceships and have seen (via television) that the earth is not a good physical model of the Euclidean plane. It is all the more remarkable that three nineteenth-century mathematicians turned their attention *from* trying to prove the fifth postulate and instead proposed substitute postulates.

When Bolyai proposed his substitute postulate—that through a point not on a line infinitely many lines may be drawn in the plane, each parallel to the given line—the geometry he derived from this postulate and other postulates was ridiculed by many as the work of a madman. Such was also the case with Riemann's non-Euclidean geometry, which was based on another substitute postulate: Through a point not on a line, no line may be drawn parallel to the given line. Yet Riemann's geometry later was utilized in the theory of relativity developed by Einstein.

What is the nature of postulates? *Postulates* are, indeed, *statements accepted without proof, which together with defined and undefined concepts are used to prove theorems by means of deductive logic.* Some may seem obvious, but others may appear to flaunt the evidence of our senses. That possibility makes mathematics seem very much like mental gymnastics. Is Bertrand Russell's comment that mathematics is a game, "where we never know what we are talking about, nor whether what we are saying is true" (Russell, 1929, p. 75) a correct description of mathematics? It's not quite that capricious a subject. The criteria for a mathematical system, criteria built into Euclid's formulation, are simplicity, parsimony, sufficiency, and consistency. The choice of postulates should not violate these criteria. Above all, a set of postulates must be consistent.

When we defined *postulate*, we, at the same time, defined *theorem*. Why can we make that claim? Take a second look at the definition of *postulate* and convert it into one which begins: A theorem is . . . . You should have obtained a statement equivalent to the following: A *theorem* is *a statement which is proved by deductive logic on the basis of postulates, defined concepts, undefined concepts, and other theorems.* Notice that we did not use the verb form *can be proved.* Sometimes mathematicians and textbook writers assert statements which *can* be proved as postulates. Proof may not be included for reasons such as simplicity or efficiency (or even because the mathematician has not yet been able to contrive a proof). Theorems are remarkable products of mathematics. Given the same set of postulates and concepts, they are statements that are true for all time.

The area of a triangle is equal to one-half the product of its base and the height to that base.

This theorem holds for all triangles and will never be rejected on the basis of future research in mathematics. To put it another way, the probability that some future researcher will find a triangle (remember, we must agree to accept the postulates and concepts on which this proof is based) whose area isn't equal to (1/2)*bh* is zero. Why are we so certain? If you're struggling with an answer like "You can't *find* a triangle!" or "A triangle isn't something you can touch or see," you've correctly identified the source of our certainty—the nature of mathematical concepts. Without saying much about it so far, we've been gradually introducing you to this other major product of mathematics.

What is a concept, whether defined or undefined? If you included terms such as *triangle, integer, pi, locus, congruence, set,* and *inequality* in your products list, you were referring to a few of the defined concepts of mathematics. When you learned the concept of triangle, you may have been shown triangular shapes—cardboard cutouts, three pipe cleaners tied together, or pictures of triangular grids on bridges. Eventually, you learned that all these objects and drawings were representations or physical models of a triangle, not the triangle itself. In fact, you probably learned the concept of triangle before you were taught to recite a definition, and you may have even learned quite a bit about the concept before anyone told you its name. So a concept is not its label, nor is it any physical model or single example.

What is a concept? Some people use as synonyms words such as *idea, thought model,* and *classification.* We prefer *classification,* since the first two terms have meanings broader than that of *concept.* We have synthesized the following definition as one that seems consistent with psychology of learning usage and, at the same time, applicable to mathematics content. A *concept* is *a classification of objects, object-properties, or events into a set by the process of abstraction.*

The concept itself, then, exists in the mind as an abstraction. Mathematical concepts, as we've seen, can be only imperfectly represented by physical models. Therein lies one of the challenging instructional problems for you as teachers. Bella insists that two different lines can intersect in more than one point and demonstrates that concept by placing two pencils together, as shown in Fig. 4.2. Think about Piaget's description of intellectual development as you reflect on Bella's conclusions. Bella has a confused concept of line, and perhaps of intersection. It will be difficult to make her realize that distinct lines, whether parallel or not, occupy distinct locations in the plane if Bella is totally reliant on concrete objects for learning. A variety of physical represen-

Fig. 4.2 Bella's model of two intersecting lines.

tations, analogies, and questions will help to disturb her complacence and are the best hope for helping to correct her view of these mathematical concepts.

Have you learned the concept of a concept? Let's test you on a novel instance. Which of the following would you classify as concepts?

1. tangent lines

2. isosceles trapezoid

3. $C = \pi d$

4. greater than

5. infinite

6. area

You should have selected all but item 3 as concepts. Notice that items 1 and 2 are formed by adding another essential attribute to an existing concept. Item 4 describes a relationship between members of another concept (for example, 7 is greater than 2). Sometimes this kind of concept is called a *relational concept*. Why is item 3 rejected? Although several concepts (diameter, π, equals, circumference) are referred to, this statement is more than the sum of these concepts. When concepts are chained together to result in a standard procedure, the result is known as a rule or a principle. All theorems and postulates fall into this category. Some are abbreviated into statements called formulas (e.g., $C = \pi d$). Some are described by a set of sequential steps and are called algorithms (the division algorithm).

Why bother with different labels for the same mathematical product? The major reason is that we are reminded to emphasize something different in our instruction. When we treat $C = \pi d$ as a theorem, we are focusing on proof. When we treat it as a formula (one kind of rule), we are emphasizing application in type or novel problem solving. Both aspects are important parts of mathematics instruction.

The final product of mathematics subsumes all those dealt with so far. A *mathematical model* refers to *a set of mathematical terms and statements which appears to be an idealized, but faithful, reflection of data and/or events in the physical world.* Often, a discrete product of mathematics (such as a formula, a graph, a number system, or even a collection of points and lines) is called a mathematical model of a particular phenomenon.

Much of the mathematics taught in junior/senior-high classes would be more meaningful to students if the concept of a mathematical model were introduced. Consider the two situations that follow and ways in which each offers the teacher an opportunity to help students sense the relevance and power of mathematics.

In an eighth-grade class, Jack Wiseguy raises his hand and offers the following contribution: "My father says 1 + 1 isn't always equal to 2. Put one drop of water together with another drop of water. You get one drop of water." Jack may be showing off, but he's also posing a legitimate question. How would you respond?

Jack Wiseguy might be nonplussed if you showed him the following demonstration. Add 1 cup of water to 1 cup of alcohol. (You get *less* than 2 cups of liquid.) Here is an example of a mathematical model, counting (natural) numbers, which behave as we'd like them to when we apply them to some physical objects but not to others. It would be difficult to explain all that to Jack, but this would be an excellent opportunity to have the class try to find other examples in the physical world where only certain kinds of numbers are used. A good "homework" assignment is the direction to write down the kind and size of numbers on cans in the cupboard, signs on the street, appliances in the kitchen, and so on. The possibilities for the next day's post-mortem are limitless.

In the next situation, a graph (Fig. 4.3) has been flashed on the overhead projector screen in a ninth-grade class.

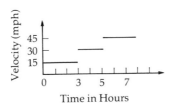

Fig. 4.3   Graph of velocity of a car.

The teacher has described this graph as the picture of the velocity of a car traveling for eight hours and now asks a series of questions.

"What do the steps represent? What happens at the end of the third hour? There's something wrong with this graph. It doesn't quite match what happens in a moving car. Where are the points on the graph where these discrepancies show up?"

This teacher's set of questions suggests an entree to the topic of mathematical models, the idealized situation which here distorts reality badly if we're concerned with acceleration. Yet if average velocity over a time period is our only concern, then the graph isn't a bad picture of the data. Whenever we abstract from a physical situation, distortion must occur. If the distortions are too great, then the chosen mathematical model will not be a faithful predictor of real-world events and must be rejected. The teacher has an opportunity to give students an illustration of the choices which must be made and, in this case, a practical rationale for doing so.

The consideration of mathematical models leads us to the second aspect of mathematics you were asked about in the Introductory Activity: What do mathematicians do?

4.3 PROCESS ASPECTS

Remember, "mathematics is also a verb." Deductive processes associated with actions such as assuming, computing, hypothesizing, and proving were referred to in the previous section. But so were inductive processes implied by actions like testing, conjecturing, and generalizing. Were these listed in the chart you formulated in Section 4.1? Which did you omit? If you included all the above, you have a head start on the work of Chapter 5, where we will consider the objectives of mathematics instruction.

However, we also portrayed mathematicians as those who sometimes idealized physical situations by *abstracting* and *symbolizing* both objects and relationships among objects—two more processes of mathematics. The language of mathematics has been a powerful, historic asset to developments in knowledge. Consider the classical Greek version of

$$a^2 + 2ab + b^2 = (a + b)(a + b).$$

Since the symbols in the above statement had not yet been devised, the Greeks could not easily handle problems involving the square of a binomial. Yet they used this relationship by operating with the dissection and rearrangement of square and rectangular shapes. Figure 4.4 illustrates one way they may have depicted the dissection of the square region with sides of length $(a + b)$ into the sum of four regions, a square with edge a, a square with edge b, and two rectangles each with edges a and b. Imagine attempting to solve complicated equations by such methods. Yet they did. But, it is also understandable why some apparently simple areas of mathematics were delayed by the lack of the symbolic tools that many ninth-grade algebra students now manipulate with relative ease.

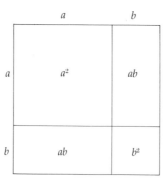

Fig. 4.4   The dissection of $(a + b)^2$.

The Greek use of regions is not just a historical curiosity, but also an excellent way to introduce some of the algebraic formulas we teach. Have students use cutout square and rectangular shapes and attempt to generate different ways of picturing the same algebraic phrase. You will have to define $(a + b)$ as the placing of two strips of differing lengths next to each other, while

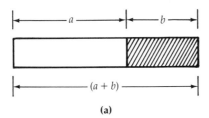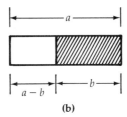

(a)   **(b)**

Figure 4.5

$(a - b)$, where a is longer than b is shown by overlapping the strips (see Figs. 4.5a and 4.5b).

Next, demonstrate to them the product of two quantities as a rectangular region whose dimensions are the two factors, as in Fig. 4.4. Students will easily obtain the usual equivalent for $a(b + c)$, but will find that transforming $a^2 - b^2$ requires more ingenuity (see Fig. 4.6). Remind them that while you (and perhaps some of them) used algebraic tools to obtain the equivalent results, the Greeks obtained these by testing and merely utilizing their agreed-upon assumptions. They used a result if it appeared to be helpful or simpler. Your students may obtain results that the Greeks chose to ignore.

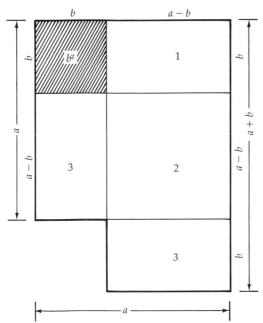

Fig. 4.6   The dissection of $a^2 - b^2$.

In the third century B.C. a Greek mathematician named Eratosthenes worked on the problem of measuring the circumference of the earth. His fame resides in his creative use of data and the very simplicity of the processes he applied to solve this problem.

Eratosthenes was the director of the university at Alexandria and, as such, collected data he later used to obtain his final measure. He knew that at noon on the day we know as the summer solstice, the rays of the sun were reflected from the water in a deep well near Syene (now Aswan). At this time when the sun casts no shadow, it is at its zenith. Eratosthenes had learned the following relationships, all of which he used:

1. When a heavenly body, such as the sun, is at its zenith, a line joining that body to an observer passes through the center of the earth.

2. The rays of light coming from such a distant source appear to be parallel.

3. At noon, the sun is located directly over the observer's meridian of longitude.

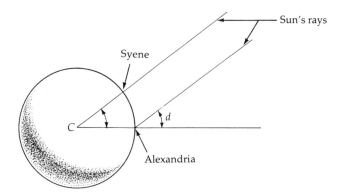

Fig. 4.7   Diagram for estimating circumference.

Eratosthenes reasoned that if he knew the measure of angle C (see Fig. 4.7) and the distance from Syene to Alexandria, then he could use a simple proportion to find the circumference of the earth. (By this time, the use of 360 degrees as the measure of a complete revolution was standard among astronomers.) Eratosthenes used an estimate of 5,000 *stadia*, or about 500 miles, as the distance from Syene to Alexandria. Now all he needed was the measure of angle C, an angle at the center of the earth. Yes, he knew, as tenth-grade geometry students learn, that the measure of angle d would be equal to that of angle C. So in Alexandria, at noon on the summer solstice, he measured that angle by measuring the length of the shadow of a pillar of known height—the shadow method of triangulation often taught to junior-high students. His estimate for angle d was about $7°12'$. The rest was easy. He solved the problem by employing the reasoning illustrated by the proportion:

$$\frac{7°12'}{360°} = \frac{5000}{x}$$

Eratosthenes thus obtained a circumference of 250,000 *stadia*. This is equivalent to a diameter of 7,850 miles, only 50 miles less than the polar diameter used today.

Did you notice the use of observing, collecting data, and reducing data to a minimal set in the work of Eratosthenes? Just as surely, he used agreed-upon assumptions and deduced by calculation his various estimates. Why were we careful to say he employed the "reasoning illustrated by the proportion . . ."? Remember, symbolism so natural to us was not available at the time of Eratosthenes. The history of the contribution of Eratosthenes is a valuable way to introduce tenth-graders to the study of circles, the central angle, the measure of arcs, and more.

Apparently, Eratosthenes used both inductive and deductive processes in his measurement of the earth's circumference. Induction or deduction—which kind of thinking did Sherlock Holmes use? From the clues he found, he arrived at conclusions as to the killer, motive, and method. When Watson asked Holmes how he reached his conclusions, Holmes would reply, "Deduction, Watson, pure deduction." Was it pure *deduction*, or did Holmes actually use *induction*? Stop here and formulate an operational definition of inductive thinking and of deductive thinking. We'll warn you that the terms are frequently misused in the literature and by some scholars. If you said *inductive reasoning proceeds from the particular to the general*, you are correct. Inductive reasoning must be in operation when we generalize (arrive at conclusions) from specific instances. So Sherlock Holmes used inductive reasoning after all. He behaved very much like the researcher who collects data, orders and classifies it, and then mentally "jumps from" the data to a conclusion. *Deductive reasoning*, on the other hand, *occurs when we move from an accepted generalization to specific instances*. You'll be helping your students learn through illustrations of both kinds of reasoning.

Inductive reasoning is an excellent way to help students see for themselves some of the rules and concepts of mathematics. Sometimes we call it *looking for patterns*. However, inductive reasoning must be treated with care. In the transcript of Ms. Pauli's lesson, several excellent tactics are employed, but she makes one serious error. Identify both the positive approaches she used and the error.

(Ms. Pauli hopes to have her eighth-grade students "see" a pattern, the rule for finding the base when the rate and percentage are known: b = p/r. They have already worked with b × r = p.)

"Class, as I write each example on the board, raise your hand if you know the answer."

    1. 12 is 50% of ?

2. 14 is 25% of ?

3. 10 is 20% of ?

(Ms. Pauli gradually obtained answers of 24, 56, and 50.)

"Now look carefully at the examples and try to find a pattern, a rule using b, p, and r."

"Jack? Do you have an idea?"

"I think you divide the percent, r, into the percentage, p."

"That's a good idea. Let's test it and see. Does it work for example 1?"

(Many hands wave agreement.)

"Good. Let's test the rule on example 2."

(More slowly, the class agrees.)

"Now, class, all of you work out Jack's rule on example 3. If it works there, we've proved that Jack's rule is the correct one."

Did you notice the sequence in which Ms. Pauli presented the three examples? She wanted her students to observe data, classify that into a pattern based on previous knowledge, and be able to express that pattern in symbolic form. Look at the examples again and try to identify the characteristics which seem to be positive aspects of pattern constructing.

Where is the error? Ms. Pauli's class did obtain a generalization from three instances but they didn't *prove* it. Nor will they prove it by further tests on other instances. Further positive tests will simply improve the *probability* that Jack's rule is the desired one. *Inductive reasoning never results in proof.* Could Ms. Pauli have *proved* Jack's rule? Yes, quite easily, if she had merely let Jack defend his rule. He might have argued that it followed from the earlier assumption, $b \times r = p$, and from the definition of division as the inverse of multiplication. Jack may have used different words, but if his message was a paraphrase of that explanation then he would be using deductive methods of proof. Should eighth-graders bother with proof at all? Yes, whenever possible, for the essence of mathematics is lost without it, but teachers should encourage informal arguments and avoid formalistic rigor. In this case most of the students could defend this rule *if* the earlier rules were clearly understood. That's a necessary *if*. Without such spiraling between topics, the students are simply playing a guessing game and learning that mathematics is a collection of meaningless, disconnected statements.

But what can we do when the students just haven't enough mathematical background to even informally defend a generalization? Notice how Mr. Cain handles the finale in his seventh-grade laboratory lesson on π.

The students have measured the circumference of various cans with string and the diameter with metric rulers. All groups have calculated C/d.

"The group leaders should come to the board and fill in the data table."

After all have completed recording their results, Mr. Cain and the class begin to talk about the pattern they see in the resulting computation.

Mr. Cain: *"We noticed that all groups obtained a result of about 3.14. Why do you think there were some differences in the hundredths place?"*

Mark: *"It was hard to use the string and sometimes we were careless."*

Rosie: *"Some of the cans had rough spots on the top. Maybe those bumps caused trouble?"*

Mr. Cain: *"That's very good! Measurement inaccuracies are always with us. In fact, even high-powered scientific tools have measurement inaccuracies. Measurement is never exact. So we don't prove anything by using measurement, but we do use measurement to test ideas. As a matter of fact, your data seems to show that C/d is always a little more than 3. Would measuring more cans, bigger and smaller ones, prove that?"*

Johnny: *"No, but I bet we'd get the same thing."*

Mr. Cain: *"Aah, you're saying it seems very convincing! You're right. More data is what the scientist would need to test that pattern. But the mathematician always wants to prove rules. Do you know that centuries ago mathematicians did prove that, for every circle, C/d is always exactly the same number? They used an argument you'll study in tenth grade. They also found out that the number they obtained was different from the ones we get when we measure objects. So later mathematicians gave that number a new name. They called it π (pronounced "pi"), a letter of the Greek alphabet. But they did want to measure circular objects, just as we do. Think of the circular shapes found in sports alone—the markings on the basketball court, the cross-sections of baseballs and golf balls, and the outline of an archery target. So we use an approximation of a number we can measure with, a number you obtained in your lab, 3.14. Sometimes we'll use 22/7 as an approximation of π. That's sometimes easier to work with, but it's still only an estimate. Now let's see how to use this famous number."*

Mr. Cain will have to return to these ideas often, because the concept of π is difficult to grasp. Why didn't he introduce the label *irrational*? Did you notice that he never stressed the difference between the physical models of the circle in the can shapes and the mathematical model that they represent? He'll have to be alert in follow-up lessons for the appearance of a common misconception, that 22/7 actually equals π. Since 22/7 is approximately equal to 3.142, some mathematics students believe that a computer program in which 22 is divided by 7 will result in the decimal expansion of π. "But π is irrational!" we hear you cry. Mr. Cain and his senior-high colleagues have a lot of spiraling to do in helping students understand this concept.

Now look back at the systems analysis model that appears at the beginning of this chapter. The shaded "Nature of Content to be Learned" box is connected to that called "Intellectual Development of Students." To what extent might Mr. Cain's choices be explained in terms of this link in the systems analysis model? What match is there between Ms. Pauli's choice of mathematics processes and the probable intellectual development of her students? As you interact with these and later illustrations in the text, look for goodness of fit of these two aspects of the model to what you have learned.

Unlike inductive reasoning, deductive reasoning *is* the method of proof, but it has its own constraints, which are important facets of this mathematical process. Mr. Greenberg's explanation of a^0 to his senior-high class contains a subtle error which clouds a major process used by mathematicians. See if you can find it.

(The class had used the rules for $a^m \times a^n = a^{m+n}$ and $a^m/a^n = a^{m-n}$ where $m > n$ and where m and n are natural numbers.)

Mr. Greenberg: *"$a^6/a^6 = 1$ since any nonzero quantity divided by itself is 1. But using our rule for division of like bases, $a^6/a^6 = a^{6-6} = a^0$. So, since quantities equal to the same quantity are equal to each other, $a^0 = 1$."*

Have you found Mr. Greenberg's error? That's right—he ignored one part of the assumption for utilizing the division rule. He missed a great opportunity to illustrate the way mathematicians alter assumptions and construct definitions so that rules may be extended. Here m is equal to n and a zero exponent will result from application of the division rule. Hence the mathematician makes a *tentative* assumption that m and n may be whole numbers with $m \geq n$ in the case of the division law and checks to see if that assumption will lead to contradictions. In this case, no contradiction was found so the next step, formally defining $a^0 = 1$ if $a \neq 0$, was taken. Finally, it is proved that the extended laws hold under the new definition.

In all three of the previous illustrations both products and processes of mathematics were inextricably related. You must have found this when you developed your chart of processes and products. In the next section we synthesize these two aspects of mathematics in a single schematic, a diagrammatic representation of our thought model of the nature of mathematics.

4.4 A MODEL OF THE NATURE OF MATHEMATICS

By this point in the text, you've read about mathematical models and physical models of mathematical entities. You've also been repeatedly directed to study the systems analysis model that appears at the beginning of every chapter. Of course, the actual systems analysis model doesn't appear on paper; but its representation does. The thought model, which we call a systems analysis model, exists in the mind and thus cannot be accurately captured by any representation but the selected representation, we've found, stimulates an understanding of the complex processes of instruction. Every representation, or physical model of a thought model, has a similar reason for its construction—to clarify, to simplify, and to suggest novel relationships. The schematic thought model shown in Fig. 4.8 is a synthesis of the process/product views of mathematics in relation to the physical world and the idea world. Study it carefully.

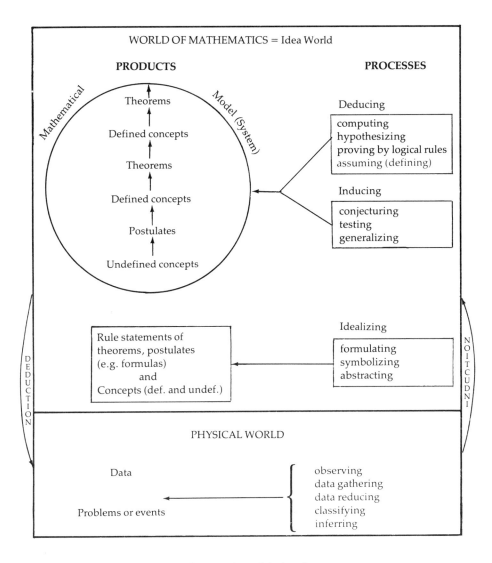

Fig. 4.8 A model of mathematics.

Each of the lines and shapes in the schema is there to convey an aspect of the thought model of mathematics. For example, we drew a solid line rather than a dashed line between the space representing the idea world and that representing the physical world. We wanted to emphasize the nature of mathematical concepts as inventions, constructs, "creations" of the idea world. At the same time the motivation for mathematical development often begins in the physical world and the models of mathematics are utilized constantly to solve real-world problems—hence the physical world appears in the schema. But note that the space assigned to it is much smaller than that assigned to the idea world.

There are two places in this model where inductive and deductive reasoning are in operation. Note the curved arrows between the physical world and the idea world. The arrows are curved toward one another to suggest a cyclic movement in and out of the two worlds.

There's also a branch connecting the inducing/deducing clusters in the idea world. What does this signify? Think of the times you've grappled with a tough mathematical proof. What were your working and thinking procedures? Did you test (perhaps with numbers?) some special cases, make a conjecture, test again, formulate a hypothesis, and try deducing the required result? If you came up against a stone wall and hadn't yet thrown in the towel, you probably stopped, did some more testing and conjecturing, and so on. Implicit in that problem-solving effort was the seesaw use of induction and deduction; but in this instance, the specifics on which you employed inductive reasoning were mathematical concepts or statements—inventions of the idea world.

Don't be concerned if you find some places in the schema which just don't convey an accurate picture of the corresponding thought model. You are experiencing first-hand the impossibility of a one-to-one correspondence between thought model and physical model. One problem which we struggled with was a way to convey the dynamic nature of a mathematical system, in which new definitions are added to the existing system in order to deduce new theorems. The chain of arrows within the circular shape was our final result. There are other places which we leave for you to identify as you continue to interact with our thought model of mathematics.

One way of studying both the thought model and the related schema is by contrast with the sister discipline of mathematics, science. Recall that the inventions of mathematics have been used to communicate scientific knowledge, to represent data, and to pave the way to scientific development by simplifying, clarifying, and, in particular, inspiring new hypotheses. Read the fascinating tale of Watson and Crick's decision to use the double helix as a thought model of DNA. Is it any wonder that some mathematicians call mathematics the Queen of the Sciences? Perhaps E. T. Bell's double appellation would be more illuminating—Mathematics, Queen and Servant of Science.

However, note that there are some aspects of this model of mathematics which differ sharply from any model of science. Two, in particular, are worth critical consideration for you as a mathematics teacher—theorem and concept. The label *theorem* appears in the model of mathematics. A similar-sounding, but vastly different, label, *theory,* appears in any model of science. Whereas theorems are proved in the idea world by deductive logic, theories are tested by returning to the physical world and checking their correspondence with real-world data. Theories are *not* proved; they are tentatively verified. What processes would have to be eliminated in any model of science? What happens to the idea-world and physical-world notions in a model of science? If these questions disturb your confidence, then join the crowd. There are college graduates who think that science has "laws" which have been proved. Next, consider the second aspect of the model of mathematics which was identified as being worthy of further study—mathematical concepts.

Are mathematical concepts different from scientific concepts? Well, unlike other kinds of concepts such as cow, dog, glass, ant, and the like, you cannot see or subject to the other senses exemplars of triangles, points, pi, or congruence. But we write numbers, don't we? No, we write symbols which some prefer to call *numerals*, names for numbers. Now you have the key to another difference between mathematics and science. For scientific concepts include *both* those whose exemplars can be perceived by the senses, such as insect and flower, and those whose exemplars cannot be perceived by the senses, such as atom and gravity. As in the case of mathematics, these latter concepts are taught by using physical models or representations of the concepts. Now you should be a little closer to an understanding of some of the differences and some of the similarities between mathematics and science. You should also be starting to sense the enormous possibilities for helping students learn mathematics through the vehicle of science.

4.5 FURTHER IMPLICATIONS FOR MATHEMATICS INSTRUCTION

An examination of the classroom illustrations in this chapter makes it evident that a mathematics teacher's view of the nature of mathematics must affect that teacher's instruction. In the long run that instruction has a lasting effect on the mathematical understandings of future citizens—for better or for worse. In this section we will consider, in detail, some further implications of the model of mathematics for the junior/senior-high mathematics classroom.

From the Products Aspect Although the concepts and assumptions which form the basis for every mathematical system are inventions of the human mind, they are not the thoughtless gibberish which might be assembled by a robot. There is a rationale behind each such product of mathematics. Sometimes that rationale takes the form of a motivating force, a need in the physical world, or a desire for a simpler way of handling a chore. Sometimes the rationale has its roots in history, in common usage, or in the etymology of a word. At other times, the rationale for teaching a particular concept at this time and in this way may be explained in terms of its sensible match to the previous learnings of your students. In the illustrations which follow, one or more of these views of rationale are detailed.

The teaching of the trigonometric functions usually occurs at two distinct levels of complexity. In elementary algebra, the student may be introduced to the trigonometry of the right triangle in a unit on indirect measurement. In a later algebra or trigonometry course, the trigonometric functions, their graphs, and the trigonometric identities are all considered in depth. At each point in this instructional spiral, the problem of the student being introduced to the topic differs.

A promising approach to the introduction of right triangle trigonometry begins with a measurement lab. Provide each student with a metric ruler and a grid on which has been drawn a known angle, say 40°. Have the students measure the legs of right triangle BP_1A_1 and record the data in the table below the grid (see Fig. 4.9).

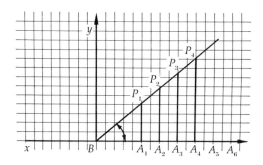

	Measure of vertical leg	Measure of horizontal leg	Measure of vertical leg / Meas. of horizontal leg =
1			——— =
2			——— =
3			——— =
4			——— =

Fig. 4.9   Worksheet for tangent lab.

Then they should compute to the nearest hundredth the ratio shown in the last column. Allow the students to check each other's work for accuracy and encourage them to look for a pattern. Here is a case where your preparation can help or hinder the students' progress. Carelessly chosen placement of P_1, P_2, P_3, and P_4 may result in avoidable measurement obstacles. A word to the wise is sufficient! Plan ahead and always dry-run the lab material before you use it in class.

Once the class has agreed on the apparent pattern, they can be helped to justify its presence from a consideration of similar triangles. Now the problem posed by A_5 and A_6 makes sense. Have the students measure BA_5 and BA_6 and find the length of the corresponding vertical leg without drawing or measurements. With more questioning, the rationale for the usefulness of this ratio is apparent. It's also clear that a ratio this useful needs a name, and now the label *tangent* can be associated with this defined ratio. The students will agree that an acute angle other than 40° would result in a different numerical ratio. All the teacher needs is to superimpose a 50° angle on the grid and use the same length BA_1 for the first right triangle. The rationale for always using the name of the angle as in "tangent θ" has been provided. The provision of a table of values to circumvent repeated measurement and the substitution of the usual "opposite leg" and "adjacent leg" for "vertical" and "horizontal" to allow for triangles in varied positions is a natural development.

Notice how the label *tangent* appeared relatively late in this introduction, and *sine* and *cosine* were deliberately ignored. It is best to continue to ignore them until the students have become masters at the varied uses of tangent. Then sine and cosine can be introduced together and without many of the above preliminaries.

If you are not careful, a presentation like the previous one can be convincing evidence that the trigonometric ratios only make sense with acute angles. This problem can be avoided if you challenge the students to measure the lengths of the vertical and horizontal components from points on the terminal side of an obtuse angle (as in Fig. 4.10). The rewording of the definition to tan $\theta = y/x$ seems reasonable now, and the extension of that definition to any angle, θ, is an easy next step.

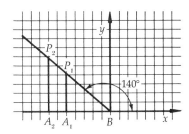

Fig. 4.10   Obtuse angle sheet for lab.

You are also a stone's throw from providing a rationale for the labels assigned to the first three trigonometric functions. If we place a unit circle on the grid so that its center is at the origin, we need measure only one length for each ratio (Fig. 4.11). Then tan $\theta = P_2A_2$, but $\overline{P_2A_2}$ is a tangent segment.

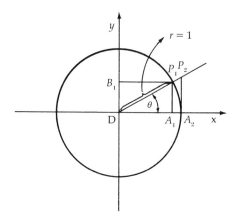

Fig. 4.11   Unit circle diagram.

Mathematicians didn't really run out of vocabulary—P_1A_1 is labeled *sine* θ. Why? In this case, a meaningful label got garbled in its translation from one language to another. The Hindus called the function *ardhajya,* meaning half-chord. (Check Fig. 4.11 again for the reasonableness of this label.) An abbreviated version of *ardhajya* became *jya,* which later was transliterated into three Arabic characters, *jyb*. However, European translators read *jyb* as *jayb,* Arabic for "pocket" or "gulf," and thus gave us the Latin term "sinus." (If you ply yourself with decongestants in humid weather, you're familiar with those pockets called "sinuses.") The etymology of the term *cosine* seems dull by comparison. Cosine is just a shortened version of *complementi sinus* or complement of sine—an obvious choice since $\cos \theta = DA_1 = P_1B_1 = \sin (90 - \theta)$.

The history of mathematics as well as the hierarchical nature of its products is a gold mine of these connecting links, which illuminate the mathematics you are teaching. In some cases, careful analysis of concepts and principles will help provide you with ties which must be embedded in your instruction. In many instances, a search into the multiple sources available to mathematics teachers will be necessary. If you are one of the lucky ones who has studied the history of mathematics, you have a head start on utilizing this type of instructional resource.

From the Process Aspect   You've surely noticed that even though the previous section was labeled "From the Products Aspect," there were processes of mathematics both implicitly and explicitly specified in most of the illustrations. As you found from the very beginning of this chapter, the two are considered separately only for the purpose of focusing more thoroughly on one aspect at a time.

Generalizing was required in the introduction to tangent described earlier and in several of the classroom illustrations of prior sections. You will find that this process is indispensable in all problem solving, whether in mathematics or in everyday life. However, there are dangers inherent in generalizing—dangers which you must help students recognize and compensate for. The most obvious danger is *generalizing from insufficient data.*

Consider an exercise based on a generalization obtained from a powerful mathematical product, Pascal's triangle.

$$
\begin{array}{c}
1 \\
1\ 1 \\
1\ 2\ 1 \\
1\ 3\ 3\ 1 \\
1\ 4\ 6\ 4\ 1 \\
1\ 5\ 10\ 10\ 5\ 1
\end{array}
$$

Ask the students to add the numbers in each row of Pascal's triangle as far as they like.

$$
\begin{array}{ll}
1 & = 1 \\
1+1 & = 2 \\
1+2+1 & = 4 \\
1+3+3+1 & = 8 \\
1+4+6+4+1 & = 16 \text{ and so on.}
\end{array}
$$

Then ask them to find, without adding, the sum of the numbers in the 15th row, the 24th row, the nth row. Next challenge them to continue the following number pattern and compare results with that obtained in summing the numbers in the rows of Pascal's triangle.

$$
\begin{array}{ll}
11^1 = 11 & \text{and}\quad 1+1 \quad = 2 \\
11^2 = 121 & \text{and}\quad 1+2+1 \quad = 4 \\
11^3 = 1331 & \text{and}\ 1+3+3+1 = 8 \\
11^4 =
\end{array}
$$

At which term does the apparent generalization fail? It is important that the students are not told ahead of time that, in this case, the pattern will break down. A few examples like this mixed in with others where generalizing does yield a valid result will suffice to warn the students that care must be taken in the use of data.

The same illustration pinpoints the second danger in the use of generalization. When students fail to verify a presumed generalization, they fall into the trap of abusing this inductive process. Results must be checked against more data and, in mathematics, proved by deductive processes. Stop right here and test yourself. In the first example, what generalization did you obtain for the nth row? (See exercise B in Section 4.8, where a method of proof is considered. Work out a proof and check your argument against that of a classmate.) Then return to the second example and compare the data from this exercise with that from the first. Try to pinpoint the arithmetic behind the lack of a continuous, common pattern.

Just as the search for generalizations is an aid to problem solving, so the processes labeled *idealizing* in the model of mathematics are ways to simplify a sometimes unwieldy problem. The solution of a well-known problem, called the Königsberg bridge problem, is a good example of idealizing in action.

In the eighteenth century strollers in the German university town of Königsberg walked along the shores of the Preger River and over the seven bridges which connected two islands to each other and to the mainland (see Fig. 4.12).

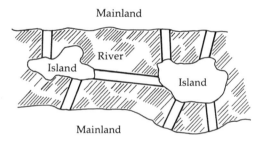

Fig. 4.12 Königsberg bridge diagram.

The problem which one of these strollers is said to have posed is in the form of a question: "How can you take a walk so that you cross each of our seven bridges exactly once?" The problem was eventually solved by Leonhard Euler, who idealized the situation in the following manner: The bridges will be drawn as line segments; the islands and shore, as points of intersection (see Fig. 4.13). Try to solve the problem yourself before reading further.

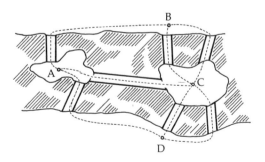

Fig. 4.13 Modeling the Königsberg problem.

In the process, Euler invented networks, an aspect of that branch of mathematics called topology. Euler's analysis of networks led him to characterize points of intersection as either "odd" or "even vertices." *A* is an odd vertex since 3 arcs intersect at *A*. There are no even vertices in the Königsberg bridge network. He found that the number of odd vertices is limited to two or less if the network is to be traveled without retracing any arcs. So Euler's answer was, "It can't be done!" Euler's use of idealizing processes eliminated extraneous data which had only confused others who attempted to solve the problem. Idealizing sometimes results in graphs, equations, and geometric shapes. In your teaching of mathematics, you have many opportunities to point out the wisdom of some choices and the blind alleys resulting from others.

Generalizing is a valuable tool of the scientist, as is idealizing, but the processes most characteristic of the mathematician are those which result in a mathematical system. Somehow a study of Euclid's geometry only reveals part of the picture. Until students get their own hands dirty in the task of constructing an axiomatic chain, they have only partially learned these basic deductive processes. An exercise within the grasp of geometry students who have studied the family of parallelograms and learned how to deduce the properties of each member of the family centers around a quadrilateral called a "kite." A kite is defined as a quadrilateral with two pairs of congruent adjacent sides. *ABCD* is one example. The students are asked to deduce the properties of a kite and state these as theorems and corollaries. They may define special kites. For example, a "right" kite could be defined as a kite with a right angle included by a pair of congruent sides, as in *MNOP*. Of course, it could also be defined in other ways. Here is an opportunity for student discussion of alternative choices for definitions and the usefulness of each. The kite exercise can do much to dispel the mystique of axiomatic thinking.

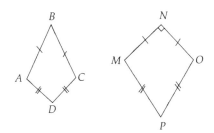

There are multiple examples in professional journals which can be utilized by the teacher to help students grasp more fully the nature of mathematics. How do you decide which ideas to use? That depends on the objectives you have set for your lesson. It now becomes clear that setting these objectives must depend both on the intellectual development of your students and on the nature of mathematics. As you interact with the material in the next chapter, keep checking your ideas with both Piagetian theory and the model of mathematics presented in this chapter.

4.6 CROSS-REFERENCES TO A CBTE EVALUATION INSTRUMENT

Even without the heading next to category 3, a quick reading of competencies 3.1 through 3.12 identifies this cluster as matching the component of "Nature of Content to be Learned." Notice that 3.1 and 3.2 are directed at helping students experience some of the processes of mathematics, while 3.3 involves the teacher's need to keep straight, in students' work, the difference between induction and deduction, two of the major processes. Several of the classroom scenes in this chapter embody examples of teacher behavior consistent with, and sometimes inconsistent with, competencies 3.4, 3.5, and 3.6. In Mr. Cain's lesson on π he behaved in ways consistent with these three competencies. Reread that illustration and identify the data supporting this assertion. As always, check out your ideas with a classmate. Did Mr. Cain also demonstrate competency 3.7? Yes, when he chose to utilize physical models of circles and measurement tasks to introduce π, he chose a valid way of translating the concept of π into a solid (spatial) form. His choice of modes and materials is also data supporting criterion performance on some other competencies. Which ones? Think back to earlier chapters. If your selections included 2.2, 2.37, and 2.38, you have the idea.

We did not devote particular attention to illustrations of 3.8 and 3.9, but these are perhaps the minimum ingredients for a successful mathematics teacher. While 3.9 concerns the teacher as a spokesperson for mathematics, 3.8 is directed toward the teacher as a listener and a viewer. Don't assume that a high-priced film is mathematically correct or that a much-touted curriculum series has been edited and made error-free. We can assure you that the buyer must indeed beware! However, this competency does not demand a verbal attack on a student who has carelessly slipped in a response. Your job is one of guiding the student into correct mathematical language.

Where have you heard the word *spiral* before? In this text, we used it back in the Preface and, in this chapter, we referred to spiraling after the presentation of Mr. Cain's lesson. Can you locate other examples? If you suspect that competence in 3.10 requires an understanding of the model of mathematics and continual learning on the teacher's part, you are right! We never said the task was an easy one. As for competencies 3.11 and 3.12, we've only touched the surface of these two areas. We will provide many more opportunities to grasp their power in Chapter 10.

Notice that in order to act in accord with these competencies you must choose modes, use feedback, and frame all in the context of the intellectual development of your students. Such interconnections are a must in a systems analysis approach to instruction. You may have also observed that "choice" is a necessary ingredient in the approach to planning. What shall you stress and what can you expect of your students? This is the role of the next chapter—to help you relate objectives to all the previous components.

4.7 SUMMARY AND SELF-CHECK

In this chapter we opened with a question: Mathematics—Invented or Discovered? and asked you to make an initial response and to test that response against the data in later sections. The processes and products of mathematics were explored by considering the known history surrounding several notable mathematical developments. The final synthesis of process and product into a model of mathematics crystallized the invented nature of mathematics by drawing attention to the nature of the theorems and the concepts of mathematics.

Throughout this chapter, you have been asked to interact with classroom illustrations of teachers attending to this model or distorting it. The power and weaknesses of generalization and the various ways to provide a rationale in mathematics were captured in their application to specific junior/senior-high mathematics content.

Now you should be able to:

1. Operationally define each of the processes and products in the model of mathematics.

2. Explain physical world/idea world and induction/deduction loops in the model of mathematics.

3. Identify errors, of omission or commission, in communicating the nature of mathematics during either a "live" or a "canned" lesson and give reasons for your judgments.

4. Provide rationale of differing kinds for selected content in junior/senior-high mathematics.

The exercises in the next section are intended to provide you with further opportunities to test your understanding of the model of mathematics. As you read, analyze, and perhaps research topics, continue to extend these activities into the total body of junior/senior-high mathematics. Remember, you are more than a plugger-in of formulas.

4.8 SIMULATION/PRACTICE ACTIVITIES

A. Each of the objects listed in column 1 is sometimes used as a physical model of the mathematical concepts in column 2. In what way(s) do each of these physical models communicate and fail to communicate the mathematical concept?

Column 1	Column 2
a) baseball	sphere
b) point of pin	point

c) jet-stream of an airplane	curve
d) flashlight reflector	paraboloid
e) edge of a table	straight line segment
f) wall of a room	plane

B. In this chapter, we have emphasized that inductive processes do not result in proof. Yet the process called *mathematical induction* is used to prove some generalizations. Explain this apparent anomaly. Be sure to defend the use of the term *induction* in this connection.

C. Although contemporary mathematics texts use *axiom* and *postulate* as equally good synonyms for the mathematical assumptions which form the basis of deductive chains, Euclid characterized certain statements as axioms and others as postulates. Identify some examples of each, according to Euclid. Try to explain his rationale for two different terms and contemporary mathematicians' rationale for disregarding this distinction. (Hint: An excellent source is Heath's *Euclid's Elements*, Vol. 1. Dover Publications, New York, 1956.)

D. Read the article "What We Can Do With $a \times b = c$," by Ruth Erckmann in the March 1976 issue of *The Arithmetic Teacher*. Then select any *other* model from arithmetic and devise and/or locate as many examples of the practical applications of this model as possible in mathematics, in science, and in everyday life. (See the reference by M. S. Bell listed in the Suggestions for Further Study section of this chapter for one source of ideas.)

E. Refer to the Resource File modular assignment in the Appendices. This is an excellent time to begin collecting ideas in one area of that assignment. As you read or hear of ways to include in your instruction illustrations communicating the process/products of mathematics, excerpt the key features (sketches, historical notes, instructional ties) and begin filing these on individual sheets or cards in the appropriate file folder. Ask your instructor to check your classification scheme and your utilization of it. As you approach student teaching, extend your file to each subject you will teach. Continue collecting ideas from experienced teachers, as well as from professional journals and texts (see Chapters 9, 10, and 11 for recommended sources).

F. Each of the following illustrations was drawn from actual classroom observations of student teachers. Each of the students is expressing honest confusion. If the teacher is unable or unwilling to unveil the apparent mystery, these students will have evidence that mathematics is unreasonable, illogical, or at best, incomprehensible. Assume the role of teacher in each of the five illustrations and formulate a response both mathematically accurate and consistent with the view that mathematics is a meaningful product of the human mind.

1. Mrs. Wampole has explained the rule for multiplying radicals to her ninth-grade algebra students. The sample example she presented is:
$$2\sqrt{2} \times 3\sqrt{5} = 6\sqrt{10}$$
Mike reacts in confusion: "But those are unlike radicands, and before you said we couldn't add or subtract the numbers in front of unlike radicands. How come we can multiply them?"

2. Mr. Conners is reading the answers to a set of exercises dealing with the multiplication of algebraic fractions. He reads: "The answer to example 21 is $2(x - 2)/3(x + 5)$."

Joan: "I have a question: I wrote $(2x - 4)/(3x + 15)$ because you're always saying: 'Leave the answer in simplest form.' But you didn't."

3. Ms. Sutliffe's intermediate algebra class has been studying systems of equations. As a result of classwork on the system

$$3x + 4y = 18$$
$$2x - y = 1,$$

Herbie says that that system is equivalent to the system

$$x = 2$$
$$y = 3$$

Sue disagrees: "How can that be? $x = 2$ is not equivalent to either $3x + 4y = 18$ or $2x - y = 1$."

Who is right? How should Ms. Sutliffe respond?

4. Paul, a tenth-grader, asks if all circles are similar. What is your answer? Are all parabolas similar? Ellipses? Why?

5. Wally, irritated with teacher insistence on symbolism, argues that $f(x) = x^2 + 3$ means exactly the same thing as $y = x^2 + 3$ so why bother with the symbol $f(x)$? Is he right? Why or why not?

SUGGESTIONS FOR FURTHER STUDY

Bell, M. S. *Mathematical uses and models in our everyday world* ("Studies in Mathematics," Vol. XX). Stanford, Calif.: School Mathematics Study Group, 1972.

This valuable paperback is aimed at providing the teacher with everyday examples of applications of mathematics accessible to those with little mathematical background. Bell has achieved his purpose in a noteworthy manner. This is not a book of readings nor a dissertation on mechanics, rocketry, or computer science, but carefully selected sets of problems, questions, and tasks. The range of situations is vast—from questions on area codes with the phone book as the data source to questions dealing with automobile safety, drug abuse, the dieting fads and so on. The approach to all problems is in the context of the concept of a mathematical model. Chapter 1 is highly recommended reading for further examples of simple, everyday uses of mathematical models. The entire volume is recommended as a source to be used repeatedly by the teacher of junior/senior-high mathematics.

Polya, G. *Induction and analogy in mathematics.* Princeton, N.J.: Princeton University Press, 1954.

The title of Polya's book might be a subhead from the section on the processes of mathematics in this chapter. An extension of an earlier book, *How to Solve It* and the first volume of *Mathematics and Plausible Reasoning,* this entire book is highly recommended. Illustrations of generalization, specialization, and analogy will delight the teacher who seeks ways to spark student attention. Polya includes sections on the use of induction and examines the process of mathematical induction. Although the problems posed are more directly related to the content of senior-high mathematics than to junior-high, all 7-12 mathematics teachers will find Polya's work on the processes of mathematics an invaluable reference. Readers should appreciate the section titled "Solution to Problems."

Polya, G. *Mathematical methods in science* ("Studies in Mathematics," Vol. XI). Stanford, Calif.: School Mathematics Study Group, 1963.

A well-known lecturer and a prophet of mathematical intuition, Polya plays historian and teacher in this text. He shares with the reader the struggles of Newton, Galileo, Kepler, and others, who utilized simple concepts of mathematics to develop monumental results. The style of writing is such that the lectures are transformed into attention-keeping tales. Polya tells the teacher how to use the past to teach present students about triangulation and the method of successive approximation. Here the reader will find illustrations of the model of mathematics in action, and ways to translate the model into instructional strategies.

National Council of Teachers of Mathematics. *Historical topics for the mathematics classroom.* (31st Yearbook). Washington, D.C.: NCTM, 1969.

This yearbook should be the personal property of every mathematics teacher. If tomorrow's lesson includes an introduction to radian measure, check the index for background material. In a three-page article entitled "Angular Measure," the origin of the term, the development of its use, and a contrast with degree measure are outlined. These brief articles, called "Capsules," contain a wealth of historical information as well as some applications of mathematics.

Sawyer, W. W. *Prelude to mathematics.* Baltimore: Penguin Books, 1955.

Sawyer's opening line provides an apt characterization of both the content and the style of this little paperback: "This is a book about how to grow mathematicians." He then proceeds to outline the qualities of a mathematician and describe aspects of some branches of mathematics. Throughout, Sawyer introduces the reader to the surprises inherent in mathematics, the recurrent patterns, the power of generalization. To assist the reader, he includes a chart illustrating optimum chapter reading sequences.

Wilder, R. *Evolution of mathematics concepts: an elementary study.* New York: John Wiley and Sons, 1968.

This little book will be of particular interest to the teacher of mathematics, for it emphasizes the cultural forces that affected the timing of major mathematical developments. Moreover, Wilder's emphasis throughout on the evolution of number helps to avoid mathematical complexities and eases the reader into a fascinating exposition.

The first chapter is required reading for those who need to delve further into the invented/discovered question. The entire volume may be read with profit for illustrations of the rationale for certain mathematical conventions, the creative decisions made by mathematicians and the ways in which real world needs affected the production of mathematics.

CHAPTER FIVE

TARGETS OF SYSTEMATIC INSTRUCTION
INSTRUCTIONAL OBJECTIVES

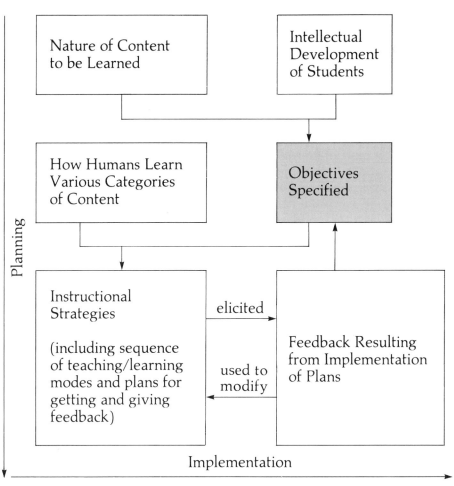

ADVANCE ORGANIZER

The comic leaves the stage as the guffaws and cheers finally die down. Will he be asked back? Was he effective? The surgeon slowly and grimly leaves the operating theater. She must find a way to tell Mr. Mack that his wife died on the table. Will the surgeon be sued for malpractice? Was she incompetent? Professor Seedwell lectures to over two hundred students on the field properties. His voice is clear and his articulation good. Printed notes that accompany the lecture are well organized and easily adapted to incorporating additional material from the lecture. As the class ends, some students can be heard griping about his cold, impersonal style. "He never even looks at us." Does this necessarily mean that Professor Seedwell is incompetent?

By now you should be wary, for all these situations are characterized by a hidden agenda. Effectiveness seems to be tied to some result or outcome. But what outcome? If Professor Seedwell had lectured in such a fashion that many students left with smiles on their faces, would his lecture have necessarily been effective? If Mrs. Mack had lived after the operation, would that be firm evidence that the surgeon had been skillful? If the audience had been quiet after the last joke, should the comedian have been fired? You're right if you tended to avoid straightforward yes and no answers and found yourself responding in terms of, "It would depend. . . ." If the lecturer's intention was to get students to smile, then he and the comedian have both been effective to the extent that their audiences were left with happy grins on their faces. But we all expect more of Professor Seedwell. His students are supposed to learn something about field properties. Let's look, in some detail, at Professor Seedwell's approach to teaching this topic.

Professor Seedwell had decided to spend two weeks on the field properties. He directed his graduate assistants to continue lecturing in the small-group sections so that more theorems could be proved for the students. No questions were to be directed at the students, and no time was to be provided for questioning by them. Is this choice of modes appropriate? If your response is no, formulate some conditions under which such a sequence of modes might be appropriate.

Next Professor Seedwell asked each of his graduate assistants to propose test questions based on the field property material of this unit. Joe Martin formulated questions in which the students would have to write proofs of theorems not considered in any of the classes but Art Cary thought that was unfair and proposed instead questions that required the reproduction of proofs presented by one of the instructors. Sally Beers's questions were of an entirely different kind. She proposed that the students be asked to test for field properties given a set and two binary operations new to them. Which test would provide better evidence of the effectiveness of the instruction? If you lean toward Sally's test but you wouldn't want to take it under the described circumstances, you'll have lots of company.

There surely ought to be a connection between instruction and testing. That connection might be characterized as "knowing where you're going." If Professor Seedwell had decided what he expected of his students with relation to the field properties, he might have proposed different uses of small-group time and he should have then been able to be quite clear about the kinds of evidence needed from a test.

Where are we going? How will we know when and if we get there? What are the objectives? Good for what? For whom? All these questions have been posed in the first four chapters of this text. The position of the shaded "Objectives Specified" box as an output based on analysis of the content of mathematics and analysis of the intellectual development of your students is a reminder that these two aspects of planning are precursors to defining "where you are going." In the activity that follows, be alert to the effects of variation in the mathematics content being studied and in the intellectual development of students on choice of destination (objective) by the teacher.

5.1 INTRODUCTORY ACTIVITY

*Where are we going, and how will we
know when we get there?*

In order to perform this activity, you need to observe a junior-high or senior-high mathematics class.

A. Before the class, ask the instructor *only* the topic(s) to be developed in the class. While observing the class, take notes on specifics in the following areas:

1. What do you infer that the students are expected to be able to *do* as a result of instruction? Cite evidence, verbal and nonverbal observables, on which you base your inferences.

2. At what point in the lesson did you become aware of these possible objectives?

3. What did the students say and/or do that seemed to illustrate that they knew what was expected of them?

4. When and how did the teacher get feedback as to the extent to which the inferred objectives were met?

5. Cite any verbal and/or nonverbal cues that might indicate student confusion about the objectives.

B. After the observation, ask the teacher to tell you the intended objectives. Share your observations and attempt to clarify areas of confusion.

C. Compare the results of your observation with results of at least two other classmates who observed different classes in your subject field. In classes where objectives seemed to be clear to both you and the students, which of the following characteristics were present?

1. The teacher verbally related past learning to today's lesson.

2. The teacher specifically told and/or showed students what they were expected to recall, how they were to apply a rule/principle, for what specific problem types they would be responsible.

3. The teacher introduced each new concept and successive skill by verbally tying these to earlier concepts and skills.

4. At times in the lesson the teacher attempted to get specific feedback on student achievement thus far, and the teacher gave oral feedback to the class relating to progress.

5. Instead of telling students only that they must "understand," the teacher followed up with expressions like "be able to give new examples of this concept" or "be able to use a formula to solve type problems."

If you were able to find examples of observables supporting each of the preceding, you have identified a class where teacher and students know where they are going and when they have arrived. If, on the other hand, the students began asking questions such as "What are we doing?" "What good is this anyway?" or "Why do I have to write all this?" then these are cues that the students were not tuned in on the objectives and, as a result, may have missed important developments in the lesson.

Your observations in this activity should be convincing evidence of the importance of having clear objectives and communicating these objectives to the students. Before you work through the next section of this chapter, try writing what seem to be suitable objectives for the class you observed. Be sure they are expressed clearly enough to communicate to students exactly what is expected of them after the lesson. As you interact with the reading in Section 5.2, modify your objectives as needed to make them consistent with the criteria you will learn to apply to such statements.

5.2 INSTRUCTIONAL OBJECTIVES VERSUS OTHER KINDS OF GOAL STATEMENTS

What are instructional objectives? In Chapter 2, you learned about diverse feedback strategies—ways to get and give feedback on progress toward objectives. In all those illustrations, whom was the teacher observing? The students. Then instructional objectives must be phrased in terms of *student* performance, not in terms of *teacher* performance. Remember Professor Seedwell? If he intended to lecture in an organized fashion, he apparently accomplished that goal. But he could have accomplished that goal in an empty lecture hall! There is no way to use that kind of teacher performance as a measure of *student* learning. But Professor Seedwell's graduate assistants were concerned about student learning and were still at odds over the nature of the test they should give. Suppose they had been told that the students were expected to "understand the field properties." Would that have helped? You're on the right track if you're still dissatisfied with this statement of an "objective." The verb *understands* is too broad in its scope to be helpful to either the students in class or the instructors as a clue to what the test should demand. On the other hand, if the students were expected to be able to "test any set under two binary operations for the presence of the field properties and to justify their results," then the character of the test questions becomes much clearer. Notice that student performance in this case is described in terms of *observable behavior*.

Stop and check your understanding by studying the four statements of possible "objectives" below. Which of these represent clear statements of instructional objectives that could be used as a guide toward planning of instructional and feedback strategies? Which are defective, and why?

1. The teacher will illustrate the use of the slide rule by a demonstration on a model slide rule.

2. The students will observe a teacher demonstration of the use of the slide rule.

3. The students will understand how to use a slide rule in multiplication/division work.

4. The students will write the product or quotient of any pair of three-digit numbers, given the use of a slide rule.

If you characterized the first statement as defective because it describes *teacher* behavior rather than *student* behavior, you have correctly focused on the first critical feature of instructional objectives. The last three statements are all written in terms of the student, so these seem to pass that test. Did you quickly reject statement 3 for its use of the word *understand* as a description of performance? Just as in the illustrative example, *understand* has too broad a range of meanings to be useful as an indicator of performance. Statements 2 and 4 should look good by comparison. Now put each of them to the dual test of feedback and content. How could a teacher get feedback on whether the students were able to perform the described task, and is performance of each task evidence that the students have learned slide rule-related skills? Now statement 2 should seem less promising. On the one hand, there is some obvious difficulty in assessing whether students are "observing." If all students are looking at the teacher during the demonstration, are they "observing"? On the other hand, even if we could agree on specific feedback cues that would assess observing, there is a lack of match between observing, in this case, and appropriate outcomes of a slide rule lesson. The teacher may want the class to observe the demonstration as a first step in a sequence of teaching modes designed to help students attain the objective described in the last statement, but the ability to observe a slide rule demonstration is not one of the important terminal outcomes of a slide rule lesson. Can you think of content areas where specific observing skills *would* be important terminal outcomes? If your ideas tend to cluster in the areas of biology, chemistry, earth science and the like, you're probably reflecting your own experience as a junior/senior-high student. If you can't think of any in mathematics, it's time to refer back to the "Process" section of Chapter 4. But for now let's examine the fourth statement more closely. It is *specific*, is in terms of *student performance*, and does include *necessary conditions* under which the student must perform. Does the word "any" disturb you? It needn't. The teacher can sample the student's ability by selecting a variety of pairs of three-digit numbers. If the sample is small or ill-conceived (such as 3.20×55.0), the teacher may infer erroneously that the student has met this objective. Effective sampling procedures are an important aspect of feedback collection and evaluation. You've already been alerted to sampling problems with respect to feedback collection and use in Chapter 2. You'll be reading a systematic approach to sampling for evaluation purposes in Chapter 8.

By now you should be able to write your own operational definition of an instructional objective. Take a minute and try it. Check your definition against the nonexamples represented by the first three statements of "objectives" as well as the example illustrated by the final slide rule statement. Now match your definition against the one we have devised. An *instructional objective is a statement that describes a desired student outcome of instruction in terms of observable performance under given conditions.* If your definition contained most of the above features, then you're off to a fine start. Why did we include the modifier *observable* before *performance?* The answer resides in the need to get feedback. You've got it. Feedback can only be obtained by observing what students do or say.

As a first check on your understanding of the definition, react to each of the following statements written by Mr. Jackson. He was planning to teach a unit on areas of various special quadrilaterals to his seventh-grade class. The following statements were some of those that he wrote as objectives for various lessons within the unit.

1. The student will be able to state the formulas for the area of a rectangle, a general parallelogram, a square, and a trapezoid, given either the label or a diagram of each quadrilateral.

2. The student will discover that the area of a parallelogram can be obtained from the area of a rectangle by a paper-cutting laboratory.

3. Given diagrams of various quadrilaterals and appropriate dimensions, the student will be able to state the area of the quadrilateral.

4. Given word problems involving physical regions in the shape of a quadrilateral, the student will write the matching area formula which could be used to solve the problem.

5. The student will appreciate the varied parallelogram shapes used in design and construction.

6. The student will understand that *area* means a "covering" of an enclosed shape.

7. The student will construct a square or a rectangle, given a compass and straightedge.

Which of Mr. Jackson's objectives satisfy all the criteria specified in the operational definition? You should have listed statements 1, 3, 4, and 7. If you rejected statements 5 and 6 on the basis of the nonobservable verbs (*appreciate* and *understand*), you were absolutely correct. Statement 2 may have seemed less of a problem since the students will be performing in a paper-cutting laboratory. But how will Mr. Jackson know they have "discovered" anything? Statement 2 fails to meet the criterion of observable performance, but it does describe a desirable teaching mode aimed at meaningful learning of an area formula. If you are somewhat uncomfortable about eliminating statements 2, 5, and 6, you agree with Mr. Jackson and the authors that attention to underlying concepts (as in 2 and 6) and to attitudes toward mathematics (as in 5) are vitally important overall goals of instruction. Try rewriting those three statements so that they do satisfy the definition. Then check your efforts against ours as we consider various classes of objectives in the next section.

5.3 DOMAINS OF INSTRUCTIONAL OBJECTIVES

A cursory examination of Mr. Jackson's "objectives" makes it clear that he was concerned about learning in distinct areas of emphasis and at various levels of significance within some of those areas. The name given to the three major areas of emphasis into which all instructional objectives may be classified is *domain*. The *cognitive domain deals with recall or recognition of knowledge and the development of intellectual abilities and skills.* Most of Mr. Jackson's statements dealt with this domain, as will most of your objectives. However, remember that Mr. Jackson *did* try to write an objective aimed at attitude. *Changes in interest, attitudes, values, and the development of appreciation belong to the affective domain.* The third domain, the *psychomotor domain, refers to the manipulative or motor-skill area.* Did Mr. Jackson's objectives include any in this domain? Yes, statement 7 was an objective from the psychomotor domain. All of these domains should be considered sources of potential objectives.

Within each of these domains, there have been attempts to devise categorization systems called taxonomies. The purposes of such categorization are to make more explicit the varied levels of instructional objectives, to improve testing and research efforts, and to clarify communication among professionals in education. Why was the term *taxonomy* selected as the label for these systems of classification? You probably remember the biological taxonomy used in junior- and senior-high science classes wherein all living things are classified into categories such as kingdom, phylum, class, order,

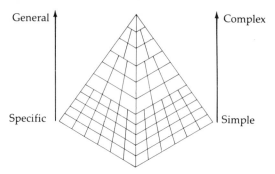

Fig. 5.1 Model of a hierarchy.

family, genus, and species. So a taxonomy is a set of standard classifications, and these categories are related in a hierarchical fashion. Figure 5.1 depicts a thought model of a hierarchy. At the base of the pyramid are the simple bits of knowledge. As one proceeds upward, ideas become increasingly complex. Similarly, one might think of the base as representative of the specifics of a field while the apex represents the generalizations. Since generalizations are built from specifics and since complex knowledge depends on simple units related in various ways, each layer of the hierarchy subsumes lower layers. Does the concept of hierarchy remind you of the nested stages model of Piagetian theory? You should be alert for correspondences between Piagetian theory and kinds of objectives as we build our taxonomies of instructional objectives.

Although all three domains are sources of potential objectives, taxonomies in two of the domains, the cognitive and the affective, are particularly useful tools for the secondary-school mathematics teacher. Since most of your work will deal with cognition, we will start with a classification of objectives in this domain.

The cognitive domain The most widely used taxonomy of cognitive objectives is that devised by Benjamin Bloom and colleagues (Bloom *et al.,* 1956). Bloom's taxonomy contains six major levels, or classes. We will use a version of Bloom's taxonomy as illustrated in Table 5.1.

It is important to keep in mind that a taxonomy is a hierarchy. Therefore, a level II objective implies that command of related level I behaviors is assumed. For example, if a student is expected to be able to use the formula for the area of a square in computing areas given the length of a side of a square, that student is implicitly expected to be able to recall the formula for the area of a square. However, the reverse does not hold. If a teacher plans lessons only on level I, then luck alone will be responsible for student learning of the level II *use* of recalled material.

Table 5.1
Cognitive taxonomy

Level	Descriptive label	Description of level
Above III	Evaluation	Judgments about the *value* of material and methods for given purposes.
	Synthesis	Putting together parts so as to form a whole pattern or structure of ideas not clearly there before.
	Analysis	Breaking down of material into its parts so that the relationships among ideas are made explicit.
III	Novel application	The selection and use of a learned rule, concept, method in a situation *novel* to the student.
II	Comprehension	The use of a specific rule, concept, method in a situation *typical* to those used in class.
I	Knowledge	The recall of material with little or no alteration required.

Before we get too involved with all this background, let's look at some specific examples. If Mr. Jackson had written at least one objective at each level of the taxonomy, his list might have included some of the following:
The student will be able to:

Level I   State the formulas for the area of a square, a rectangle, a trapezoid, and a general parallelogram, given the name of the shape.

Level II   Compute the area of a square, a rectangle, a trapezoid, or a general parallelogram, given appropriate lengths.

Level III   Solve novel word problems which involve physical regions shaped like any of the special quadrilaterals, and applications of the areas of these.

Levels
above III

(Analysis)   List the facts, the relevant relationships, and the irrelevant information contained in a novel world problem.

(Synthesis)   Write a novel word problem which includes the use of at least three of the four area rules in the unit.

(Evaluation)   Write an assessment of the advantages and disadvantages of two different solutions to a word problem, where accuracy and economy are desired criteria.

Study each of the preceding objectives and the descriptions of the corresponding taxonomic levels. Notice the modifier *novel* in the third, fourth and fifth objectives. If the word problems in either case are "type" problems, then the student is being asked to perform at level II. When you are attempting to identify the level of a particular objective, this element of *novelty to the learner* is a crucial characteristic that separates levels I and II from all other levels. The first question to ask of yourself, then, is "Is the student expected to perform a task that has some element of novelty?" A negative response leads you to the next question: "Can the objective be satisfied by recall alone?" Then, an affirmative response leads to the identification of level I as the correct one for the objective (see Fig. 5.2). For most classroom purposes, it is sufficient for beginning teachers to be able to distinguish among levels I, II, III, and above III. Table 5.1 gives descriptions of the subdivisions above III for clarification only.

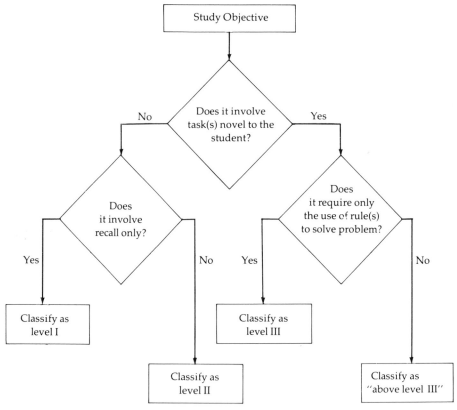

Fig. 5.2   Classification flowchart for cognitive domain.

Sometimes the idea of novelty is mistakenly equated with the notion of complexity. The ability to use 15 rules in a long calculus problem is still level II behavior *if* the problem is a typical one, whereas the ability to select the correct rule, however simple, in a situation new to the student is level III behavior. An illustrative example of such a situation is given below.

Find the area of the circumscribed square in the diagram.

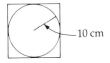

Notice that the student must realize that the radius is one-half the length of a side of the square—a step that requires a mental restructuring of the diagram! This is a level III task. A second flow chart (Fig. 5.3) schematizes this aspect of problem solving that distinguishes level III and above tasks from the lower levels of the taxonomy.

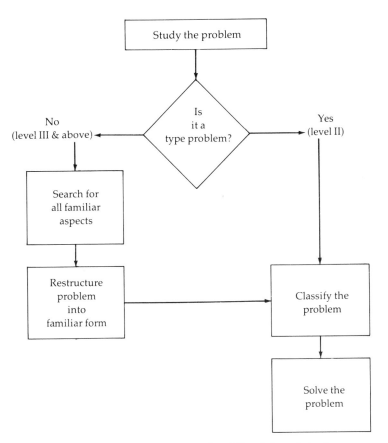

Fig. 5.3 Flowchart of level II versus level III and above behavior.

The "search" component of the flowchart may be a matter of seconds, minutes, or days. In any case, the objective represented by the problem is at or above level III. On the other hand, even though students cannot solve a type problem because they have forgotten needed formulas or are applying rules in an incorrect manner, the objective that is represented by the type problem is still a level II objective.

Examples and explanations are important ingredients in developing understanding, but there's no substitute for experience. Now it's time for you to check your own ability to classify objectives. Study each of the following objectives written by Miss Bonner for her eleventh-grade unit on trigonometry and classify each at one of the *four major* taxonomic levels described in Table 5.1.

The student will be able to:

1. Calculate missing side lengths or angle measures in right triangle problems, by means of the six trigonometric functions, given a table of trigonometric values.

2. State the formulas for the sin, cos, and tan of $(A + B)$, $(A - B)$, $2A$, and $\frac{1}{2}A$.

3. Verify that two unfamiliar trigonometric expressions are or are not identical by substituting appropriate formulas in the sequence taught in class.

4. Solve a trigonometric equation of first or second degree for a set of angle measures, within a given range, given a table of trigonometric values.

5. Explain the effect on the trigonometric functions if they are redefined for (x', y') on some nonrectangular grid.

6. Calculate a missing bearing in a navigation problem by application of oblique triangle and right triangle trigonometric formulas, given a table of trigonometric values.

Notice that Miss Bonner took the same shortcut Mr. Jackson used in the last set of objectives. Since all objectives are in terms of student performance, both teachers wrote "The student will be able to:" only once before the entire set of objectives. As a matter of fact, some teachers omit that phrase entirely since it is understood to be present. Did you perhaps want to reject some of Miss Bonner's statements as instructional objectives because verbs such as *calculate, verify, solve,* or *derive* suggest mental activity rather than observable performance? If so, you've been interacting carefully with the reading thus far. But we also slipped some others like these in earlier. Go back to Mr. Jackson's second set and note that some of his statements *seem* to be characterized by the same defect. If you're thinking that these objectives are perfectly clear and that to say otherwise is to make mountains out of molehills, we agree with you. Some instructors and curriculum writers would prefer to insert "in writing" or "orally" in each of the above statements. Whenever there is any ambiguity, we recommend this. However, as a general rule of thumb, highly specific, content-related verbs that are typically understood to imply observable performance may be used without any additional modifying phase. By the way, Miss Bonner's third statement would have been less acceptable if it had ended after the word *identical*. If you're not sure why, talk this over with a classmate or your instructor.

Be sure that you have written down your classification of each of Miss Bonner's objectives before you read further. We classified objective 2 as the only objective at level I. It is the only one of the six statements where recall *alone* is sufficient to meet the objective. Granted, some students may forget the double-angle formulas and, recalling the derivation techniques, may reconstruct the derivation in order to obtain the formulas. Although these students might be performing at level II, Miss Bonner, as most eleventh-grade teachers, desired the eventual memorization of this formula and intended this as a level I objective. If you classified most of her other objectives as level II, you are catching the idea. We classified objectives 1, 3, 4, and 6 as level II objectives and objective 5 as above level III. As written, each of the four level II objectives refers to the solution of *typical* problems for this unit. Even in the case of verification of identities, Miss Bonner will have taught a series of strategies which the student is expected to use. If you classified objective 5 as level III, rather than above III, don't be overly concerned. You correctly identified the element of novelty. However, we see that objective as requiring *more than* the application of rules to a novel task. It also entails contrasting and comparing of the old with the new and thus we would more specifically classify this objective as at the analysis level. For further practice on this kind of task, see the Simulation/Practice Activities in Section 5.6.

Now you're ready to start on the first written component of each lesson plan—the stated instructional objectives. We asked you to test yourself on this earlier by rewriting Mr. Jackson's defective statements from Section 5.2. Two of these, statements 2 and 6, were cognitive, so it's appropriate to consider those now. Compare your versions with those of the authors.

Substitute 2.   *The student will write a generalization for computing the area of a parallelogram in terms of its base and its altitude after completing a paper-cutting laboratory and tabulating the data obtained by all students in the lab group.*

Substitute 6.   *The student will compute the area of unfamiliar shapes using the results of "covering" the regions with smaller cutout shapes of given area.*

If your versions differ substantially from the above and you are not clear as to their suitability, check with other classmates and/or your instructor.

The steps in writing objectives for either a daily plan or for a unit are essentially the same.

1. Choose the unit or topic to be taught.

2. Study curriculum materials, syllabi, and student texts with relevance to the unit.

3. List the major content items (such as concepts, formulas, principles, and the like).

4. Write instructional objectives directed toward each of the designated *major* content items.

5. Classify the written objectives by level and check these results against your assessment of the nature of the content and the intellectual development of the students in the designated class.

It should help to refer back to the above guidelines as you work out Simulation/Practice exercise C at the end of this chapter. That exercise is designed to provide some needed practice in writing objectives that you will actually use to guide instruction. Like most beginners, you may encounter initial difficulty in selecting performance verbs that accurately convey your intentions. Thus we have prepared a sample list of those verbs that have proved particularly useful in your subject matter field (see Table 5.2). Use it to the extent it proves helpful, but don't assume that including one verb from this list *guarantees* a well-structured objective.

Table 5.2
A sample list of performance verbs

add	describe	hypothesize	list	solve
calculate	design	identify	multiply	state
cite	diagram	induce	plan	subtract
contrast	divide	infer	plot	translate
criticize	evaluate	interpret	predict	verify
deduce	explain	interpolate	propose	write
defend	extrapolate	justify	prove	
derive	graph	label	select	

The affective domain Remember Mr. Jackson's defective objective that was rejected because it included the nonobservable verb *appreciate*? Mr. Jackson attempted to include an objective in the affective domain (feelings, attitudes, and values) that would help students to relate mathematical concepts to physical world applications. He was struggling to communicate in a domain studied and classified by David Krathwohl and associates (1964) who constructed a taxonomy of five levels—commonly called "Krathwohl's taxonomy." We use a version of this taxonomy that includes a modification of the second and third levels of Krathwohl's taxonomy. See Table 5.3 for a description of these three levels.

Table 5.3
Affective taxonomy

Level	Descriptive label	Description of level
III	Valuing	Commitment to a value shown by consistent and stable response to objects, people, phenomena, etc.
II	Responding	Voluntary participation in activities, or selection of one activity out of several.
I	Complying	Passive acceptance of role assigned by teacher. No overt avoidance of activity.

As in the taxonomy of cognitive objectives, this taxonomy is also a hierarchy. At the lowest level, the student may, for example, merely sit quietly in a condition of apparent listening during a lecture, rather than chatting with neighbors, doing homework, or reading *Popular Mechanics*. When question/ answer begins, those students who volunteer are exhibiting behavior beyond level I. But the teacher must refrain from the immediate assumption that these students are interested in the subject. They may only be interested in being "on stage" momentarily. As you may have guessed, it is relatively easy to write objectives at levels I and II and far more difficult both to write a level III objective and to have it attained by many students. Yet there is little point in being a mathematics teacher, in the opinion of the authors, if you do not work toward motivation based on the acceptance of some of the values inherent in or related to the subject matter. Let's see what Mr. Jackson might have written if he had wanted at least one objective at each of the three levels of this taxonomy. Recall he was teaching seventh-graders a unit on area.

Level I   Participates in the lab-group role assigned by the teacher.

Level II   Attempts an optional challenge problem on areas of regular polygons.

Level III   Volunteers illustrative examples from nonassigned outside sources during a class discussion on the application of geometry to the real world.

As you compare each of the above objectives with the description of the matching taxonomic level, you will probably realize that levels II and III cannot be attained with any certainty if the teacher "motivates" tasks by either the threat or promise of a grade. In these cases, students may merely be complying with instructions—thus behaving at level I. The students may, in fact, be interested in the lab activity, but obedient participation gives the teacher no feedback as to this interest. The word "voluntarily" or a synonymous word or phrase will be found in a level II objective. However, a major distinction between level II and level III is that in a level II objective the teacher has defined the choice(s) to which the student can *respond*. Notice that in Mr. Jackson's level II objective, he will have to devise a challenge problem and state the free choice aspect of completing it. However, in his level III objective the reference to "nonassigned outside sources" suggests that students would have to voluntarily do some investigating and have formed some judgment as to the value of applications, rather than responding to a specific task posed by Mr. Jackson. Over the course of the entire two-week unit, Mr. Jackson would be looking for repeated instances of similar commitment before he could feel any confidence in the attainment of level III objectives by any one student.

This simplified affective taxonomy should be represented by some objectives in the total set written by each teacher as a new unit is planned. Miss Bonner included these in the set she wrote for the trigonometry unit. Study each statement carefully and classify each at one of the three taxonomic levels of the affective domain.

1. Voluntarily attempt one or more of the problems in set C each time these are identified as free-choice challenge examples.

2. Volunteer to put homework derivations on the board.

3. Do all assigned homework neatly.

4. Work quietly at task when assigned group work on identities.

5. On his or her own initiative, investigate library resource materials on surveying equipment after a classroom demonstration of the equipment.

6. Raise questions in physics class about the mathematical contributions to physics of men such as Kepler, Galileo, and Newton.

We classified objectives 3 and 4 as level I, objectives 1 and 2 as level II, and objectives 5 and 6 as level III. The major difference between the level II and level III objectives seems to be evidence of student-initiated interest over time. It is surely true that objectives at level III have little chance of success unless the teacher has consistently worked toward this level of performance. It is also true that objectives in this domain necessitate matching assessment forms. We will be developing this area in Chapter 8.

You're almost ready to begin writing your own affective objectives. Check your first attempt, the revision of Mr. Jackson's defective objective 5, with our version.

Substitute 5. *The student will elect to do an outside project which illustrates the role of parallelograms in design.*

Have you observed the intrinsic connections between the cognitive and the affective domain? For instance, the student who is able to meet the above objective will, at the same time, be exhibiting behavior at level II in the cognitive domain. You will observe similar overlap in other objectives in both of these domains. However, where the intended *emphasis* is on the development of feelings, attitudes, or values rather than on recalling, comprehending, or applying intellectual skills, the objective is classified as affective.

In the not-so-distant past, affective objectives were rarely included in curriculum materials. It was hoped that the intrinsic overlap of cognitive with affective domains would somehow become actualized. Yet we have generations of mathematics students who have become skilled in writing deductive proofs but vocal about their dislike of geometry—solid evidence that there's something wrong with that hope.

"All right," we hear you say, "attitudes are important; and I do want my students to like mathematics, or at least not to dislike it. But why does a mathematics teacher need to include psychomotor objectives?" Read on, and see.

The psychomotor domain Check out the label given to this third major category of human learning capabilities. Note that the *psycho* conveys the notion that motor (muscle) learning is intertwined with mind, soul, and spirit. Earlier we called attention to the interconnections among the cognitive and affective aspects of learning and stressed that it was largely a matter of the major *emphasis* of a given objective that allowed classification into one domain versus another. So it is with targeting instruction on manipulative skills important to mathematics. These kinds of objectives *do* require specific and intentional efforts on the part of teachers in selecting those to be included as important goals of the course and in designing special strategies to promote their attainment if students are to learn them effectively and efficiently. Novices are prone to assume that this category of objectives is inappropriate for the junior- and senior-high mathematics classroom. One might be tempted to agree with them after observing groups of eighth-graders wielding little more complicated than paper and pencil for days on end, or tenth-graders passively watching a teacher write out a proof on the chalkboard. However, by this time you should be disabused of the notion that teaching modes that systematically exclude laboratory activities are a desirable match to either the intellectual development of your students or the nature of the content. Measuring activities, paper-folding laboratories, and model building have as much to contribute to the learning of mathematics as the more often seen protractor and compass. But surely the *use* of such devices needs no special attention. Observe an eighth-grade class where students are supposed to be applying compass-use skills to produce copies of elaborate designs. Walk around and tabulate the number of students holding the compass in the center of the palm, placing the pencil at an awkward height, unable to adjust the compass to match a desired radius, unable to draw a complete circle, and other evidences of the *absence* of psychomotor skills. Then believe the evidence instead of the pious hopes of teachers and curriculum guides, and take the first step toward heading off such disasters in groups you will teach. That first step is specifying objectives in the psychomotor domain. The time to begin is now.

We have constructed a few sample objectives to help you get started. Read each carefully. Then decide whether or not each (1) contains a specific performance verb, (2) communicates clearly, (3) places primary emphasis on coordinated muscle movements, and (4) makes obvious the materials and equipment that would be made available to the learner. (It is our experience that one could write "given access to all required equipment and materials" after almost every objective in this domain. This seems senseless to us so we *assume* that phrase is understood to be part of each objective *unless* there is some good reason to state more specific limitations on these conditions for performance.)

1. Measure rectilinear shapes with a metric ruler to within ± .5 mm.

2. Describe how to operate a hand-held calculator to obtain sums, differences, products, and quotients of integers.

3. Adjust a georule to illustrate geometrically shaped letters of the alphabet.

4. Fold wax paper according to instructions to obtain models of the parabola, the hyperbola, and the ellipse.

5. Prepare a working model of Watt's linkage, given a set of precut and prepunched links and all other required materials.

6. Manipulate a set of cutout rectangular and square shapes to illustrate equivalent algebraic expressions.

Now let's see how closely your analysis of these sample objectives agrees with ours. We consider measuring shapes, describing procedures, adjusting a georule, folding wax paper, preparing a model, and manipulating shapes to be specific performances, and we think that each of the six statements communicates clearly. However, did you pick out the fact that *describe* in objective 2 emphasizes the cognitive aspects of learning and does *not* call for a demonstration of the indicated motor skills? We certainly hope so, for there is a world of difference between talking about manipulative skills and actually performing them, though comprehension is often a helpful prerequisite to motor learning. Perhaps this very thought occurred to you when you considered objective 6. It did to us when we wrote it. Why did we state conditions for objective 5 and not the others? We simply wanted to delimit the expectations by ruling out the measuring of suitable length links and the positioning of joints in appropriate locations as being part of this particular objective. Does objective 4 seem to represent a motor skill that definitely does not require specific instruction? We have observed college mathematics majors bungling these folding instructions. But don't take our word for it. Just observe and believe.

You should now be ready to try constructing a few objectives on your own. Think in terms of the manipulative skills associated with the use of devices such as tangram shapes, geo-d-stix, compass, string models, geo-board, transit, slide rule, graduated cylinders, stop watch, and pan balance. Then write at least three objectives relevant to a subject you expect to teach and submit them to a critique by your instructor or a classmate.

In the cases of the cognitive and affective domains we concerned ourselves with classifying objectives at a variety of levels within each domain. However, we will *not* deal with a taxonomy within the psychomotor domain. Why? Isn't there any such thing? Yes, there is a taxonomy for this domain, but it appears much more relevant to the elementary-school age group and physical education content than to secondary-school mathematics instruction.

This is *not* to say, however, that we should or can ignore varying complexities of objectives aimed at motor skill development. Consider some practicalities of planning instruction. When an algebra teacher decides to include a unit on the application of the slide rule, the expectation is that the students will learn to manipulate the rule in order to solve problems involving a series of operations. This requires facility in the use of all scales and several combinations of scales. This complex aim incorporates a number of more specific psychomotor objectives, some of which depend on the prior achievement of others. Thus, the probability of the major goal being attained will rest in large part on the teacher's ability to identify and correctly sequence the learning of component skills. A number of other similar examples ought to become obvious as you gain experience as an observer, planner, and director of learning activities.

5.4 CROSS-REFERENCES TO A SAMPLE CBTE EVALUATION INSTRUMENT

The fact that selecting objectives from the three domains is the content of the first four competencies on the instrument underscores the central position of objectives in any systematic approach to instruction. Notice that 1.2 and 1.3 deal with different levels of the cognitive domain. Which levels? You are correct in matching 1.2 with levels I and II and 1.3 with levels III and above. The teacher who is evaluated on the basis of these competencies is expected to select objectives in all three domains and at least above level II in the cognitive domain.

What other competencies are directly related to specifying objectives? You should have identified 2.1, 5.1, and 5.2. Why include 5.2? Think taxonomy and it will become clear that the "emphasis of instruction" is described by the selected taxonomic levels of the instructional objectives. If you included other competencies such as 2.34, 2.310, 2.63, and most of category 3, you have the right idea. The level and kind of objective selected affects the quality of questioning, assignments, and the extent to which the instructor attempts to communicate the nature of the subject. The same can be said for many other competencies, but in particular for 2.2, since strategies matching the intellectual development of your students are necessarily shaped by the objectives you assume those students are capable of meeting. As you continue working through the next two chapters, be alert for additional guidelines which you will use to select objectives.

5.5 SUMMARY AND SELF-CHECK

Instructional objectives can be considered analogous to route directions written on a road map. Without these the teacher can easily fall into the trap of jousting at every windmill sighted along the way. The route map must be well designed so that an equally trained professional, another teacher of the same subject, would agree on the nature of specific behaviors represented by the objectives. We emphasized the need for clear communication of ideas without excessive verbiage. Recall how their later objectives did not contain the commonly assumed phrase, "The student will be able to:" nor conditions that ought to be obvious to any other trained teacher. You may find it difficult to write both clearly and succinctly. If it comes to a choice, there is no argument. Clarity should win out.

Throughout this chapter, you were provided illustrations of objectives in each of three domains—the cognitive, the affective, and the psychomotor. Three criteria to be employed in writing instructional objectives of any kind were stressed and categorizations within the cognitive and affective domains were described and illustrated.

At this time, you should be able to:

1. Operationally define instructional objective, cognitive domain, affective domain, and psychomotor domain.

2. Operationally define each of the levels in the cognitive and affective domains.

3. Identify given statements as either meeting or failing to meet all criteria for an instructional objective and give reasons for your decisions.

4. Classify given instructional objectives according to domain and according to level for those in the cognitive and affective categories.

5. Write instructional objectives for each level of the cognitive and affective domain.

6. Write instructional objectives in the psychomotor domain.

The exercises in the next section are designed to provide you with some further experience in identifying, classifying, and writing instructional objectives. Skills developed in these areas will give you a head start in designing lesson plans and in constructing appropriate assessment measures—both everyday tasks of the classroom teacher.

5.6 SIMULATION/PRACTICE ACTIVITIES

A. Below are ten statements purported to be instructional objectives. Assume that each is of educational value to its subject matter area. React to each *only* in terms of whether or not it satisfies all three criteria for instructional objectives. Use an X to indicate your judgment.

	No	Uncertain	Yes
1. Identify the primes smaller than 100.	———	———	———
2. Know the Pythagorean theorem.	———	———	———
3. Derive the formula for the sine $(A + B)$ when A, B, and $A + B$ are each acute.	———	———	———
4. Appreciate the symmetry forms in physical objects.	———	———	———
5. Explain the "delta process" used in obtaining the first derivative.	———	———	———
6. Factor the difference of two squares, given that each square is a monomial.	———	———	———
7. Graph any function of the form $y = ax + b$, a not equal to zero.	———	———	———
8. Draw a diagram and write the "Given" and "To Prove" for any "If, then" statement from plane geometry.	———	———	———
9. Understand the nature of a circular argument.	———	———	———
10. List four ways to prove triangles congruent.	———	———	———

If you checked any of the above "no," revise those statements so you would check each "yes." Then compare all your responses with those of two other students. Discuss differences in judgment, attempt to arrive at a consensus, and then ask your instructor for feedback on your efforts.

B. The statements below are instructional objectives based on junior/senior-high-school mathematics content. Classify *each* according to domain (cognitive, affective, or psychomotor) and then by level for those within the cognitive and affective domains.

1. Measure an angle to the nearest degree, given a protractor (Math 7).

2. Generalize a nth term relationship given the first six terms of a numerical sequence (Algebra).

3. Define parallelogram, rhombus, and square (Geometry).

4. Raise questions in social studies class about the reasoning heard in television political debates and the instances of circular reasoning, reasoning from a converse, and so on studied in geometry class (Geometry).

5. Solve an equation of the form $ax^2 + bx + c = 0$, $a \neq 0$ by means of the quadratic formula for roots in radical form (Algebra).

6. Select mathematic-related reading material when given "free" reading time in English class (any level).

7. Write at least two inferences based on the line graph of a set of data (Math 8).

8. Move "spiro" gears to trace a polar curve (Algebra).

9. State the domain and range of a function given a graph of the function (Algebra).

10. Write a one-paragraph "story" in which mathematical concepts are used, given a topic sentence such as "I dreamt I was a point in a non-Euclidean plane" (Geometry).

C. Choose a topic from the subject matter of junior/senior-high mathematics. Check textbooks and syllabi to ensure that the selected topic would be one developed over at least a two-week period. Study student exercises and resource materials related to the topic. Then write 15 instructional objectives in the cognitive domain which would be appropriate for this topic and the anticipated students. At least 8 of the objectives should be at level II. Include some at level III and at least 1 above level III. Arrange these in the order in which you would sequence them for instruction.

D. Use the topic selected by you in exercise C to write at least *five* instructional objectives in the affective domain. Include some at level II and at least one at level III.

E. Use the topic selected by you in exercise C or choose another topic and write at least *three* instructional objectives in the psychomotor domain

F. Return to the "Summary and Self-Check" sections in Chapters 1-5. Classify each of the objectives listed there by domain and by level within domain where appropriate.

SUGGESTIONS FOR FURTHER STUDY

Avital, S. M., and S. J. Shettleworth, *Objectives for mathematics learning: some ideas for the teacher.* Toronto: The Ontario Institute for Studies in Education, 1968.

This brief pamphlet includes a modification of Bloom's taxonomy slightly different from that illustrated in Chapter 5 of this text. However, the authors' emphasis on higher-level cognitive objectives and their explanation of analysis and synthesis reinforce the illustrative examples provided by Farrell and Farmer. Although Avital and Shettleworth include more examples of mathematics test items than of objectives, the explanation of the taxonomic level of each item and of its match to a given objective should help to further clarify the distinctions among levels.

Bloom, B. S., ed. *Taxonomy of educational objectives, handbook I: cognitive domain.* New York: David McKay, 1956.

This landmark publication presents a hierarchical categorization of testable objectives which has achieved almost universal acceptance throughout educational circles. The utility of the schema for analyzing instructional objectives and test items alike has been well demonstrated. The definitions of each category (or level) are quite clear and numerous examples are provided throughout.

Krathwohl, D., B. Bloom, and B. Masia. *Taxonomy of educational objectives, handbook II: affective domain.* New York: David McKay, 1964.

This sequel to *Handbook I: Cognitive Domain* deals with the more difficult but equally important domain of feelings and attitudes. The format follows that of its predecessor and a quick overview can be obtained by reading Chapters 1, 3, 5, and Appendix A. Numerous examples are given, the relationship to cognitive domain is developed, and applications to both the classification of objectives and matching test items are described. Current interest in affective education underscores the importance of this useful book.

Mager, R. F. *Developing attitude toward learning.* Belmont, Calif.: Fearon Publishers, 1968.

Chapters 1-5 emphasize the importance of the affective domain for learning any school subject. Mager uses an easy-to-read interactive style to illustrate what he calls "approach behaviors" indicative of positive attitudes. Attention is also given to interrelationships with the cognitive domain, and readers are directed to reflect upon the development of their own system of values.

Mager, R. F. *Preparing instructional objectives.* Belmont, Calif.: Fearon Publishers, 1962.

This brief paperback is a simple introduction to writing behavioral objectives. Written in programmed format, most novices can complete the book in a little over an hour's time. The explanations are brief; the examples, clear. Feedback is immediate and those who catch on quickly are directed to skip repetitive materials.

Sund, R., and A. Picard. *Behavioral objectives and evaluational measures: science and mathematics.* Columbus, Ohio: Charles E. Merrill, 1972.

Behavioral objectives and their use in designing evaluation measures are viewed within the framework of a systems analysis approach to instruction. The authors include a rationale for the use of behavioral objectives, provide clear instructions for writing such objectives (including numerous examples from both mathematics and science), and relate these to the taxonomies of both the cognitive and affective domains. The use of objectives in curriculum design comes in for some attention but much more emphasis is placed on their use in developing evaluation measures.

CHAPTER SIX

STEPPING STONES TO PLANNING
THE LEARNING OF MATHEMATICS

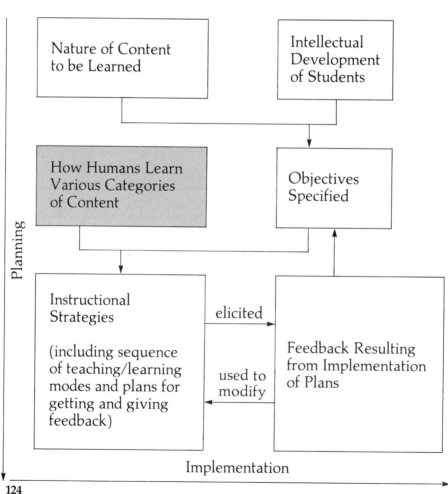

ADVANCE ORGANIZER

We have come a long way in our study of systematic instruction. Together we explored the modes used by teachers and the feedback strategies associated with those modes. We sought an answer to the question: "What are the intellectual capabilities of the typical junior/senior-high student?" and to the equally important query: "What is mathematics?" Armed with this background, we considered the nature of instructional objectives. Now we have mapped out where we are going, with whom, and for what purpose. But how do we get there? Instructional modes surely are part of that answer, but modes need to be chosen, sequenced, and implemented on a dual basis—the choice of specified objectives and the evidence illuminating the ways learning takes place.

How do humans learn various categories of content? That question should look familiar. Of course, you've been in a position to see it each time you begin a new chapter of this text. Notice the place of this component in the systems analysis model and be alert to the interconnections to lesson planning that will be foreshadowed in this chapter. Wait a minute. Haven't we already studied learning? Didn't Piaget provide us with all the evidence we need as to how learning takes place? You're making a common mistake if these are your thoughts, but you're not completely out in left field. Piaget studied the ways in which intellectual structures develop—the spontaneous development of knowledge as opposed to learning attributed to instruction. We have already stressed that his data and his theoretical constructs are invaluable inputs in our choice of objectives and in the eventual design of instructional strategies. But we must look elsewhere for data on learning that results from intentional attempts to change behavior.

Let's be sure we're on the same wavelength when we use the term *learning.* We concur with the definition used by most contemporary psychologists that *learning is a change in human attitudes, cognition, or psychomotor skills which is not due simply to the process of growth.* The definition excludes a change in human capabilities, such as increase in the size of the muscles, but includes specific eye-hand coordination abilities. And if learning is a *change*, then it is necessary for teachers to get feedback on the student's capability *before* instruction as well as *after* instruction. How many of you have had the experience of sitting through a college calculus course where at times you were bored by the repetition of concepts you had learned in high school? Were there perhaps other times in the same course when you were lost by an all-too-brief treatment of a content idea to which you had been exposed in high school but which you had never really learned? What accounts for such selective learning and remembering? Why is it that we seem to learn some ideas so thoroughly that they are easily recalled while there are others that we never could understand, let alone remember? Even stranger, why are there ideas we thought we had indelibly learned but can no longer recall? Remembering, like learning, becomes a mental process of more than passing interest.

Memorizing is the bane of most students. Yet probably every teacher of yours emphasized memory to some degree. (Too much, perhaps?) "SOHCAHTOA," chorused the class. Was this one of your memory crutches?

$$\text{sine } \theta \ = \frac{\text{opposite}}{\text{hypotenuse}} \quad \text{SOH}$$

$$\cos \theta \ = \frac{\text{adjacent}}{\text{hypotenuse}} \quad \text{CAH}$$

$$\tan \theta \ = \frac{\text{opposite}}{\text{adjacent}} \quad \text{TOA}$$

"SOHCAHTOA" is an example of a mnemonic, a mental trigger to assist the memorizer. The student who successfully applies the mnemonic is able to state the three definitions above—an essential prerequisite to use of these definitions. However, any human capable of imitating the verbalisms in the above statements and able to recall statements of this length can be taught the mnemonic and the related sentences. Animal learning is limited in this respect. The use of symbols of all kinds to represent ideas makes human learning far more complex than some psychologists once believed.

Experiments with pigeons, laboratory rats, and even higher animals have intrigued psychologists for years. Early attempts to equate human learning with animal learning led Thorndike, a psychologist, to write a methodology text (1921) which promoted the teaching of 100 addition facts as separate bits of unrelated material. Teachers were warned that $3 + 2 = 5$ must be given as much time as $2 + 3 = 5$. Little wonder that educational psychology left a bad taste in the mouths of teachers. Where was the meaning? Were there not relationships among parts? Which terms needed to be memorized? Which might be deduced, or even reconstructed when and as needed? Names and labels must surely be learned differently from concepts, skills, problem solving, proof constructing and the like. What are those differences? These are some of the questions we address in this chapter. We begin by engaging you in an activity that further explores the intricacies of memory and its relationship to understanding.

6.1 INTRODUCTORY ACTIVITY

You will need to enlist the aid of two or three acquaintances to serve as subjects for this activity. Seek out mathematics majors who are *not* presently in this course and have not taken it previously.

A. Provide each volunteer with a quiet place to study a typed copy of the list of words below. At the end of one minute ask the subject to look up from the list and recite the words *in order*. Stop the person at the first error and provide another minute of study time. Allow five such study periods of one minute duration each and record the number of correct words (in sequence) achieved by the student at the end of each study period.

List of Words

math	instructions	but
her	do	without
seeking	locus	she
could	young	a
defined	incurred	teacher's
always	all	constructions
bright	as	path
proofs	student	of

B. Using the same volunteer(s), substitute for the list of words a typed copy of the limerick below. Repeat five similar trials and keep comparable records of the results.

Limerick

A bright young student of math
always defined locus as path.
She could do all constructions
without seeking instructions,
but her proofs incurred teacher's wrath.

C. Compare the data you collected with that of at least five to seven of your classmates. Does a clear pattern emerge? If so, how do you account for the fact that one ordering (sequencing) of the same words was easier to learn than the other? Was there any factor other than sequencing that could also have made a difference in the ease of initial learning? Discuss your thoughts on these matters with classmates and your instructor and remain alert to further inputs in the sections that follow.

D. Design a simple way to check up on comparative retention of the list versus the limerick. We suggest allowing at least seven days to elapse between the initial learning and the remembering (retention) trials.

6.2 MEANINGFUL LEARNING

What made the limerick easier to learn than the same words in scrambled order? Doubtless your discussions generated ideas such as meaning, rhyme, and rhythm as likely causes of the observed results. If you further proposed that *meaning* was the most important factor, then your thinking is attuned to what has been clearly demonstrated both by research studies and the experience of effective teachers. Further, meaning of the material has been shown to be the prime factor associated with retention (remembering) of subject matter over time. Even the most cursory reflection on the nature and extent of the subject matter you are preparing to teach underscores the need to give careful attention to the central task of ensuring meaningful learning.

Unlike computers, the human intellect is *not* capable of storing or retrieving large numbers of discrete bits of information.

The term *learning* has already been defined in the Advance Organizer section and we urge you to review that definition at this point. Now let's see if we can nail down the key components of that all-important idea of *meaningful*. Meaningful to whom? The learner, of course. Recall that the directions for the Introductory Activity specified the use of mathematics majors. The words *locus, construction,* and *proof* all have special meanings to one trained in this subject field and these in turn facilitate comprehension of the whole idea of the poem. Persons who do not attach subject-specific meanings to these key words might learn to recite the limerick perfectly given sufficient time, but would respond differently to in-depth questioning about the student's plight as described in this poem. Thus new content becomes *meaningful* to the extent that it is *substantively (nonarbitrarily) related to ideas already existing in the cognitive structure of the learner* (Ausubel, 1968). This definition stands in clear contrast to that of *rote* learning wherein the *new content is arbitrarily (nonsubstantively) related to the existing cognitive structure of the learner.* If we imagine a continuum ranging from pure rote to highly meaningful learning, the task of memorizing the scrambled word list would be placed near, but not at, the rote (no-sense or nonsense) pole (see Fig. 6.1).

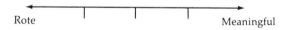

Rote   Meaningful

Fig. 6.1   Rote ⟷ meaningful learning continuum.

As an independent information bit, each *familiar* word did relate to existing cognitive structure. However, the list did not represent a meaningful whole. Had the scrambled list consisted of nonsense syllables (XTU, KLMA, DTTIZ), then we would clearly place the task *at* the rote pole. On the other hand, a competent mathematics teacher learning the limerick would be performing a task near the highly meaningful end of the continuum.

Consider a very real learning problem which occurs in all too many tenth-grade geometry classes each fall. Students are confronted with the task of learning a host of postulates and definitions, the names of numerous shapes, and relationships among them. Do these constitute *potentially meaningful content*? Of course they do. No one would argue against the importance of the understanding of the axiomatic structure of geometry. Recall, however, that the crucial point is to make new content meaningful *to the learner*. What already exists in the cognitive structure of these tenth-grade students to which the new material can be substantively related? Are concepts such as point, ray, segment, betweenness, and union already there in a stable form? If not, and time isn't taken to establish such prerequisites, students face a horrendous task of rote learning in this unit. Yes, much depends on what was meaningfully learned and retained as a result of junior-high mathematics courses.

By this time you have undoubtedly surmised that we believe that meaningful learning and retention depend primarily upon the *way* the learning occurs. Our constant reminders to you to interact with the material, to compare answers with those of other classmates, and to hypothesize and then check out your hypotheses are all specific learning processes that are supportive of meaningful learning. Some teachers and some curriculum writers have behaved as if all learning processes can be categorized as one of two types: reception learning or discovery learning. Indeed, some have gone one step further and have equated reception learning with rote learning and discovery learning with meaningful learning. We'd better take a closer look at the discovery versus reception, rote versus meaningful relationships before we make unwarranted conclusions.

We've already explored the difference between meaningful and rote learning. But to what do reception and discovery learning refer? The term *reception* might remind you of a telephone receiver, a receptionist, a receipt. *Reception learning* thus occurs when *the entire content of the intended learning is presented to the learner in its final form and the learner incorporates the content into his or her cognitive structure*. If the new content becomes substantively related to ideas already present in the individual's cognitive structure, then meaningful reception learning has occurred. If this condition is not met, then *rote reception learning* would be the appropriate descriptor.

Which modes would you identify as potentially promising ways of promoting reception learning? Which of those modes might not be employed to promote discovery learning? If you chose lecture as a potential way to promote reception learning, your judgment is in accord with ours. However, if you also decided lecture could not be employed to promote discovery learning, you doubtless have an erroneous concept of discovery learning. In contrast to reception learning, in *discovery learning the learner must generate the desired content end-product or construct a missing interrelationship*. As in the case of reception learning, discovery learning may be *either* meaningful or rote, depending upon the kind of relationship (arbitrary or substantive) established between the new and previously learned content.

It is true that a lecture aimed at student discovery would be carefully structured to pose a problem and highlight conjecture and would not fill in all the gaps for the student. If you've already decided that a mixture of lecture with question/answer, laboratory, or some other mode would be even more successful in promoting meaningful learning of either type than lecture alone, you're absolutely correct. And while lecture might be used to promote either reception or discovery learning, the use of the laboratory mode is not a guarantee of the coexistence of discovery learning, although it should be. A frequent abuse of the laboratory mode in junior/senior-high mathematics classes is to tell the students what results they will obtain *before* they have begun to perform any of the manipulations. Imagine how that kind of teacher behavior affects meaningful learning. How are those students likely to view the process aspects of the nature of mathematics? Does that mean that the teacher stands back and lets the students fumble away at their own pace? Not likely. *Pure* discovery learning is almost never a viable approach. There simply isn't enough time to have students rediscover all they need to learn.

Furthermore, junior- and senior-high mathematics students do not yet possess the background knowledge, intellectual skills, and sometimes even the laboratory techniques required to learn without guidance from teacher and text. In fact, the discovery learning recommended by teachers and scholars is more correctly called *guided* discovery learning. The teacher sets the scene, cues judiciously, and carefully structures the sequence of events so that the students need not reinvent the wheel. This kind of guided discovery has received much attention from Jerome S. Bruner (1962, 1966, 1971), a Harvard psychologist, whose research in school learning has resulted in both scholarly treatises on the nature of learning and essays to teachers on the nature of "going beyond the given."

However, ensuring that learning will be meaningful is not as easy as it sounds. Remember all those *potentially* meaningful postulates we mentioned earlier? Making content meaningful to the learner is a two-edged sword. The teacher must consider characteristics of the content as well as the match between the new content and the student's existing mental structure. We already learned that mathematical content could be classified in diverse ways—by process and product; by concept, postulate, and theorem. To what extent do these differences in content affect learning? What ways exist for categorizing various types of learning tasks and what conditions facilitate the learning of each type of task? These are the considerations to which we next direct our attention.

6.3 CATEGORIES OF HUMAN LEARNING

Unlike the early twentieth-century learning theorists, Robert Gagné rejected the notion of classifying all learning into a single category. Instead he added to and modified the theories of Thorndike and Skinner and selected aspects of the theory proposed by the Gestalt school of psychology. And he has done all this in the context of school learning. Not only has he continuously conducted experiments with human subjects, but much of his data have been collected in existing classrooms as opposed to the carefully controlled and contrived environment of a pseudo-class in a college laboratory. As a result, his inferences make practical instructional sense and are translatable into instructional strategies. In the following paragraphs, we summarize two of Gagné's major contributions on which we will base instructional strategy design to be considered in Chapter 7.

Learning Types

Gagné (1970) classified all human learning into eight major types, which are related in hierarchical fashion. His hierarchy of learning types appears in Fig. 6.2. The hierarchy depicted in Fig. 6.2, like the others you have studied, contains categories ranging from simple to complex, with each successive layer depending upon and subsuming those directly under it. However, unlike either the nested stages model or the taxonomies of instructional objectives, Gagné's hierarchy of learning types contains a branch midway up. Apparently, type 5 learning may depend just on related type 2 and type 1 learning. Branching of diverse kinds is prevalent throughout the learning of any subject. For example, learning to divide natural numbers may depend on first learning to subtract them *or* on first learning to multiply them. This concept of prerequisites *necessary* to future learning is the second of Gagné's contributions which we will study and relate to mathematics learning.

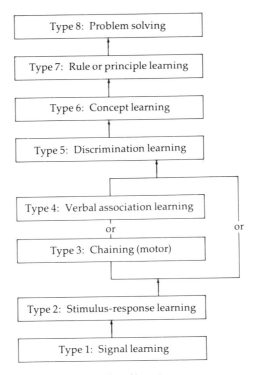

Fig. 6.2 Hierarchy of learning types.

But for the time being, we need to pay closer attention to the levels in Gagné's hierarchy. What does he mean by "concept"? Does his meaning agree with what we learned in Chapter 4? Would rule include the idea represented by such statements as the commutative principle for addition or the square root algorithm? The answer to these last two questions is a resounding yes. And although you will be primarily concerned with the learning of discriminations and the types dependent on these, the hierarchical nature of learning makes it imperative to have at least an acquaintance with all eight types of learning.

Signal learning Gagné was not the first to distinguish between signal learning and stimulus-response learning. Probably the most commonly known experiments in signal learning are those of Pavlov, whose dog learned to salivate in response to the signal of a buzzer. The shell-shock symptoms demonstrated by war-weary soldiers on hearing a car backfire or firecrackers explode is another example of signal learning. The buzzer or the noise of the backfire serve as signals (conditioned stimuli) in these illustrations and the response of the learner—salivating in the case of the dog and fear symptoms in the case of the soldier—are conditioned responses of a reflexive, emotional nature.

Stimulus-response learning From signal learning to stimulus-response learning we begin to move toward more differentiated responses. Now the signal, or stimulus, must be connected with the response and the response itself is a *specific* terminal behavior that provides satisfaction to the learner. The training of animals and, indeed, the shaping of human behavior to respond in specific ways to verbal or nonverbal cues are good examples of these $S \rightarrow R$ (stimulus-response) connections. Some aspects of classroom management (for example, lab groups being taught to cease talking when the lights momentarily are switched off-then-on) are considered matters of stimulus-response learning. There have been entire texts written on the subject of behavior modification that deal with this kind of learning. Suffice it to say that (1) the correct response must result in some reinforcement—perhaps praise, a reward, or a success experience; (2) the reinforcement must occur fairly close in time to the response; and (3) the stimulus must be repeated often with partially correct responses being identified and positively treated. As you interact with Chapter 12, look for cues such as "using Grandma's Rule" and "be consistent." These are specific applications of $S \rightarrow R$ learning to classroom control. "You mean learning theory is practical?" You bet it is.

Chaining Learning to swim requires a whole series of coordinated responses—each subunit being an $S \rightarrow R$ connection. The movement of arms and legs and the breathing behavior must all be appropriately linked, or *chained* together. So chaining is the name given to the type of learning represented by these nonverbal sequences.

Think back to the behavior described by the sample instructional objectives in Chapter 5. In which cases did the objectives clearly refer to the learning type Gagné has called "chaining"? If you fumbled over the answer to that question, you'd better go back and reread Section 5.3 with special attention to the psychomotor domain. Yes, psychomotor objectives all require that the student chain together individual $S{\rightarrow}R$ units. (1) These units must be well learned and (2) then integrated in the correct sequence. (3) Again, immediate reinforcement of the final response and repetition of the entire chain and pieces of it in close time succession are conditions required in order that chaining may occur. In the example given—learning to swim—the reinforcement for most people is success—an example of an intrinsic or internal reward. Some novices mistakenly believe that only extrinsic rewards (such as prizes, verbal praise, or grades) are satisfying to the learner. Initially, it is true that the struggling student may need the satisfaction of an external reward. However, if some internal reward does not gradually replace the external one, the learning has less chance of permanence.

Verbal association learning Gagné's type 4 learning encompasses much the same bonding of stimulus and response as that described by chaining. However, in this case, the response is a verbalism. Naming geometric shapes, labeling constants such as π, even the memorizing of verbal information such as the square roots of 3 and 2 are instances of the learning called verbal association. All the conditions required to assure the learning of motor chains are also required here. Of course, the stimulus situation now includes a drawing, an object, or a printed expression and an initial repetition of the desired response by the teacher. The teacher may also include cues, such as the mnemonic "SOHCAHTOA," which must be learned and then recalled to assist the learner in obtaining the desired longer verbal response. As you remember from the Introductory Activity, long verbal chains are easily forgotten unless they are learned in a meaningful context. Verbal association learning in mathematics includes fewer long verbal chains than social studies or even science, but it is complicated by a written and spoken language of symbols and groups of symbols.

Discrimination learning Closely related to this last illustration of verbal association learning is the learning of discriminations, Gagné's type 5. The ability to perceive the differences in shape between { } and ( ) may follow the learning of the respective verbalisms "set-braces" and "parantheses," but is a prerequisite if the student is to correctly distinguish one printed mark from the other. Similarly, the ability to perceive the *distinctive* features of objects observed, sounds heard, materials touched, foods tasted, and odors smelled are all examples of discrimination learning. Even intelligent adults are sometimes tricked by ignoring one perceptual change or overemphasizing another. How many smart chefs know this and slice all sandwiches on the diagonal? The sandwiches look bigger. The importance of discrimination learning to the learning of concepts cannot be underestimated. Some students never learn the *essential* features of a class of objects and are at a loss when novel exemplars are proposed. Geometry teachers have experienced this phenomena when diagrams involve overlapping triangles or shapes turned in an unusual orientation. Thus, the most important condition needed to promote discrimination learning is the presentation of diverse exemplars of the feature to be learned. Perceptual problems are often the source of difficulty at this level of learning—a difficulty which then blocks progress at the levels above.

Concept learning Concept learning is the basic coin of the realm in areas usually described by words such as *thinking, understanding*, and *problem solving*. In Chapter 4 several kinds of concepts were used as examples—triangles, a defined concept; point, an undefined concept; greater than, a relational concept—and an operational definition of *concept* was provided. Reread that definition and note the use of the term *classification*. Classification, as used by Piaget, by logicians, and by all scientists, is the key to the conditions under which a concept is learned.

Suppose Mr. Mapes had written a definition on the board, had placed next to it several triangular shaped cutouts, and had drilled on the oral recitation of the label and definition? Would his sixth-grade students be likely to have learned the concept of triangle? If not, what could they be credited with learning? If you said they probably learned the *name of the concept* and a verbal association describing that name, you're in agreement with what experience has taught us. Suppose the students had been tenth-graders? Would it have made a difference? Yes, *if* they already had concrete referents to attach to the presented verbalisms and diagrammatic representations. Otherwise, no. In general, the conditions for concept learning include (1) the availability of prior prerequisite discrimination and/or verbal association learning and (2) the presentation of gradually differentiated examples and nonexamples. In the case of the concepts of mathematics (which exist in the idea world with rare exceptions, such as numeral), the examples and nonexamples take the form of physical models or the names of prior concepts (for example, 2, 3, 5, 7, and 11 might be given as examples of primes with 4, 9, 15, and 16 labeled as non-examples). Sometimes an analogy is used to exemplify a mathematical concept. "2x and 5y are illustrations of *unlike* terms. It's as if you have two apples and five oranges."

Whatever method the teacher employs to establish the conditions for learning, the evidence that a concept has been learned is *not* the recitation of a definition, nor is it any other feedback that might signify nothing more than accurate recall—such as responding correctly to exemplars used by the teacher during instruction. But if the student responds correctly to exemplars not used in instruction and can produce novel exemplars when asked, then the teacher has obtained positive feedback on concept learning. *Concept learning*, then, is manifested by *the ability to generalize beyond the instances used in the learning situation and beyond the physical similarities present in some instances.* For example, a student who has learned the concept *triangle* will not be misled whether shown the intricacies of the geodesic dome at the United States Pavilion in Expo '67 or the lacy web constructed by a spider. Thus learning of concepts frees the student from total reliance on the physical world and makes it possible for rules and principles to be meaningfully learned.

Rule/principle learning Rules and principles, the heart of mathematics, depend on concept learning. Why? Let's look at an example of rules. The formula for finding the area of a triangle might be memorized without any prior learning but the formula itself is only a *statement of a rule* just as the label "triangle" is the *name of a concept*. What is a rule? Identify the learning you would desire with respect to the given area formula. If you'd expect the ability to *use* the formula in a variety of situations, you've captured the essence of rule learning. Now to use the formula, Mark (one of your eighth-grade students) would need to have learned the concepts of triangle, area, equals, base, height, and one-half. He would also have to know how to multiply (another rule!) and finally would have to use all those concepts and in this case, an earlier rule, in the proper sequence and manner. *Rule learning*, then, is *the capability to respond to a class of situations with a class of performances where the situations and performances are related by a chain of concepts.* The rule itself is a chain of concepts. In this instance, given diagrams of a variety of triangles and some way to obtain the lengths of the bases and the corresponding heights, Mark will have to be able to compute the area. It is impossible to tell what he might be thinking, but if this final performance results in the statement of the correct areas, then it is inferred that the rule has been learned.

Notice that his ability to state the formula is not even mentioned. However, you would probably want Mark to be able to state the formula to avoid constant searching for this repeatedly used tool. That's fine, but that's a verbal association which must be learned differently and tested separately. Just don't make the mistake of assuming that instant recall of the formula guarantees the ability to use that rule. Furthermore, don't assume that success, given the base and height lengths as the only lengths indicated on the diagram, will be repeated if the teacher labels the length of a median or of a noncorresponding base. Are you getting the idea? Your instructional objectives with regard to any rule may range from low-level and restricted use to complex and widely generalizable use. If, however, there is to be any subsequent problem solving (type 8), the rule that has been learned must be of the latter class.

What conditions must be present so that widely generalizable rule learning may take place? As in the example given: (1) The concepts which are to be chained must be separately mastered first. (2) The instructor must clearly let the students know what kind of terminal performance is expected. (Mark should be told that he'll be asked to find areas given triangular shapes and any necessary measuring instruments *if* that is what his teacher has decided.) (3)

Verbal cues, concrete examples, or a carefully structured exercise can be used to (a) help students recall essential concepts and to (b) encapsulate the structure or main idea of the rule. (4) The use of the rule should be demonstrated in the format desired by the teacher. (5) Finally, the students must be asked to demonstrate the rule in *diverse* situations. If they can demonstrate its use successfully, they have learned the rule. Retaining it is another matter. Now the teacher must work at frequent meaningful reinforcement, drill of a novel kind, relatively short practice sessions spaced out over several classes. Examples of novel practice sets are considered in Chapter 7, as are further characteristics of meaningful drill.

Problem solving Rules, no matter how complex the situation they encompass, are useful only if the problem can be characterized as belonging to a particular kind. However, throughout much of our in- and out-of-school life, the questions which intrigue us most are not typical problems. They belong instead to Gagné's type 8 learning, problem solving. Gagné is referring here to the moves toward the solution of *novel* problems—that is, problems *novel* to the student. (You should be reminded of the cognitive taxonomy.) Since the student cannot classify the problem as a typical one, a search behavior which includes defining the problem, formulating hypotheses, verifying these or altering hypotheses, and then verifying the modifications continues until the problem is solved or at least restructured. At this point we part company with Gagné, for he seems to imply that learning does not take place if the problem is not solved. However, we would assert that thoughtful strategies engaged in by a student faced with problems like the mind-boggler below are behaviors characteristic of type 8, whether or not the problem is solved.

> *Prove that if the lengths of two angle bisectors of the angles of a triangle are equal, that the triangle is isosceles.*

The modifier *thoughtful* eliminates erratic trial and error attempts or blind algorithmic tactics. Moreover, the unsuccessful student who has engaged in problem solving may have been successful in intermediate stages. Some approaches may have been rejected after testing them. Additional data may have been deduced. The problem is definitely restructured now, even if it has not yet been solved. What conditions enhance the occurrence of this kind of learning? (1) The problem must be a problem to the learner—that is, the learner must conceive of the problem as posing a difficulty, an obstacle. Many so-called problems posed by teachers are considered simplistic questions by the students. (2) The relevant prerequisite rules and concepts must be recalled by the learner. John Dewey (1910) pointed out that some of these can be readily learned when the stumped problem solver realizes that an impasse has been reached. Thus, instruction leading to the learning of a rule which is in immediate demand is seen as totally relevant by the motivated student. (3) Cues to help the learner recall appropriate rules and to suggest approaches to the hypothesizing, testing, and modifying processes are assets to problem solving. However, if the cues become a step-by-step exposition of a solution, then the student is off the mental hook and has lost out on this most important aspect of mathematics teaching. The student must be left time to fumble, to explore unpromising paths, to make mistakes. Indeed, the final condition to be considered by us emphasizes this aspect of student performance. (4) The instructor must stress the nature of the task and carefully distinguish expectations here from those in rule learning. It is *not* the case that a short-cut, algorithm, or formula is the important output. The objectives here are to seek sensible strategies, to develop systematic ways of checking out conjectures, to be open-minded about both possible solutions and potential routes to a solution. Subsequent to work on the problem, an examination of the profitable and unprofitable approaches is in order to clarify further the potential of this new strategy. Terms like *hypothesizing* and *systematic approaches* were used in this text in an earlier chapter. Remember the solutions problem and the approaches of the formal versus the concrete operational student? Piaget's data clearly illustrate the effect of intellectual development on the nature of problem-solving abilities. In particular, more guidance and cues would be required if the concrete operational student were facing a problem having multiple potential factors. What are some other possible differences? Compare your ideas with those of a classmate before turning to the Suggestions for Further Study section.

Task Analysis

One of the most striking characteristics of the subject matter of mathematics is its hierarchical nature. Small wonder that Gagné's emphasis on prerequisites has been found to be especially useful in mathematics instruction. An outgrowth of Gagné's analysis of learning types led to the development of a technique called *task analysis*. Very simply, Gagné recognized the fact that instructional objectives at the concept, rule, and problem-solving levels depend upon the attainment of subbehaviors and then the integration of these. He devised a method by which complex objectives can be analyzed for instructional purposes—a method which consists of asking the following question: "What *must* the student be able to do in order to meet the terminal objective?" Suppose the terminal, or major, objective is the following:

> *Solve a quadratic equation of the form $ax^2 + bx + c = 0$ where a, b, c are real numbers, $a \neq 0$ and $ax^2 + bx + c$ can be factored over the rationals.*

A task analysis might result in the partial learning hierarchy depicted in Fig. 6.3. Such hierarchies must be read from the top down. Each of the prerequisite objectives (2a, b, and c) must be mastered by the student if the terminal objective is to be met. Since all three are *necessary* prerequisites, they are joined by a three-way branch. These must be integrated in some fashion before objective 1 can be met so a single arrow connects the branch to the terminal objective. Why are these prerequisites on the same level? Surely a student might meet objective 2a before meeting objective 2c? That's true, but the task analysis is intended to illustrate *necessary dependence* and none of these three prerequisites depend on one another. Thus, task analysis helps the teacher identify the range of choices open for the sequencing of instruction. However, a task analysis does not depict the sequence in which a student must *use* these three skills in order to master the terminal objective. In this instance, the student must learn to use 2a, 2c, and 2b in that order—another cue that the whole is more than the sum of its parts in learning.

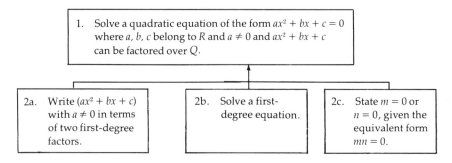

Fig. 6.3 A partial learning hierarchy.

Let's see if you have the idea. Below we've listed five additional objectives which are prerequisite to those in Fig. 6.3. Read each carefully and sketch out the extended learning hierarchy by asking of each upper-level objective the task analysis question: "What *must* the student be able to do in order to perform this complex task?"

1.   Transform first-degree equations into equivalent forms by the addition and/or multiplication property of equality.

2.   State that "or" is equivalent to union.

3.   State that $0 \cdot c = 0$ and $c \cdot 0 = 0$ where $c \in R$.

4.   Factor any second-degree polynomial into prime factors over Q.

5.   List the members of a solution set by using the relationship that "or" is equivalent to union.

We recommend the writing of the essence of all nine objectives on cards and arranging them on a flat surface, such as the floor. Then test each one by asking the task analysis question and checking carefully the lower objectives. Our response to this exercise can be found in the Simulation/Practice Activities, where we ask you to go one step further with this partial hierarchy.

6.4 ADVANCE ORGANIZERS

What is an *advance organizer?* Although we have not yet defined this term, you have encountered seven examples in reading through this text up to this point. Six were clearly labeled as such by side headings at the start of each chapter. Where was the seventh? The "To The Student" section which preceded the beginning of the text was also intended to incorporate the essential attributes of an advance organizer. It should be obvious from the word *advance* that an advance organizer comes prior to something else and from the word *organizer* that it is designed to facilitate putting things together in a meaningful way. Thus the term itself tells when an advance organizer occurs in an instructional sequence and sheds light on its overall purpose. David Ausubel (1963, 1968), the cognitive psychologist who originated this idea, has identified the attributes essential to this concept. First, *the advance organizer must present relevant content ideas which are of a higher order of abstraction, generality, and inclusiveness than the new material to follow.* Advance organizers typically take the form of broad concepts, rules/principles, thought models, theories, or conceptual schemes (themes) which subsume the more detailed knowledge to be learned next. Thus it becomes clear that teachers must have command of the structure of the subject matter, as presented in Chapter 4, if advance organizers are to be designed correctly. For example, the thought model of "group" could be taught as an advance organizer for related subsumed concepts/rules such as set, binary operation, closure, Abelian group, associative property, permutation group, ring, field.

Second, *the advance organizer must be presented in terms of what is already known by the learner.* In other words, the teacher must find out what relevant knowledge the learners possess and then use it to teach the generalization(s) which will serve as the advance organizer. For example, assume that the teacher finds out that the students already know (either from life experience or from previous instruction) that the order of operations makes a difference in some computations and activities (as in subtraction of natural numbers or in putting on one's socks and shoes). The teacher can then use these ideas as prerequisites to build comprehension of the desired generalization (group), which in turn will subsume the related facts, operational definitions, concepts, rules/principles to be learned in the new work. Is it beginning to seem as if Gagné's task analysis technique would find further application in structuring advance organizers? We have found it to be essential to our own efforts in this regard.

You should be starting to get the idea of what an advance organizer is. Now let's consider a few potentially confusing nonexamples to further clarify what an advance organizer *is not.* Many mathematics texts begin each chapter with a short introductory or overview section which consists of anywhere from one paragraph to several pages of writing. Often the one- or two-paragraph variety is really a summary of the main ideas to be treated in the chapter. These overviews almost always fail to meet *either* of the essential criteria since (1) the ideas presented in capsule form are at the same level of abstraction as the content to follow and (2) little or no provision has been made to use the existing knowledge of the reader to teach the ideas summarized in such an introductory section. Similarly, several pages of introductory material tracing the historical development of a major mathematical idea may stimulate the interest of some students but fail to qualify as an advance organizer on the basis of one or both essential criteria. On the other hand, introductions consisting of long excerpts from original papers written by famous mathematicians of years gone by not only fail to qualify as advance organizers but also serve to kill student interest. This is not to say that anything using a historical frame of reference is always ineffective. Both the history and the conceptual themes of mathematics can be used to structure advance organizers.

How did we tackle the problem of providing you with advance organizers? We elected to employ our systems analysis model of instruction as a unifying (conceptual) theme throughout this text, beginning with the "To The Student" section. Our plan was to use this overarching idea to subsume all the other ideas and information in this book. Thus, both the diagrammatic representation of this thought model, and relevant written material appear at the start of each chapter. Then common background of the reader (both from real-life experiences and from preceding chapters) was used to teach meaningfully a new aspect of the model which in turn would constitute ideational scaffolding for the contents of subsequent chapter sections. To the extent that we were successful, each advance organizer should have provided you with anchoring ideas and bridged the gap between what you already knew and what you needed to know in order to learn the tasks at hand. Did our attempts work in your case? Feedback from past students has been very encouraging and therefore we continue to use this technique.

If some reflection leads you to conclude that our advance organizers have not been very effective in promoting your learning, the fault may be either ours or yours. In any case, we have two suggestions to help you derive maximum benefit from the advance organizers you will read in succeeding chapters. First, be sure you *study* each thoroughly *before* going on to the other chapter sections. Think about the two essential characteristics which should be present and attempt to identify specific examples of these elements. Second, check your understanding of each advance organizer with that of a classmate and ask your instructor for help to resolve any discrepancies.

Why do we place such emphasis on advance organizers? One of their major functions is to facilitate the initial learning of new material so that it is of the meaningful variety (as opposed to rote). There is a growing body of research evidence which testifies that the degree of meaningfulness of newly learned material correlates positively with both remembering and the ability to use that material in applicable situations. Since both remembering (retention) and future applied use (transfer) are major goals of education, any device with potential for promoting their achievement merits our careful attention and best efforts.

Think back to previously learned material which will help you structure effective advance organizers. We have already called your attention to the contributions of Gagné as they apply to this task. Now consider the research of Piaget summarized in Chapter 3. What comes to mind first? Most people think of the nature of the concrete operational student and cite the need to include many concrete referents in an advance organizer. This is good thinking, as far as it goes, but it does not go far enough. Can you identify yet another application of Piaget's work to making effective use of advance organizers? If not, turn back to the nested stages model in Chapter 3 (Fig. 3.2) and focus your attention on the thought modes available to formal operational students when they are working with content for which they do not have a base of concrete experiences. Always remember that in the long run all is lost if the initial learning is meaningless (rote) to the learner.

6.5 CROSS-REFFERENCES TO A SAMPLE CBTE EVALUATION INSTRUMENT

A thorough examination of the major categories on the CBTE Instrument in the Appendices should reinforce the "classroom" character of these categories. There is no classroom payoff if you are unable to translate knowledge about learning theory/research into effective instructional strategies. Hence, the major competency which depends on the component "How Humans Learn" is item 2.1. You will find specific illustrations of the stage of translation from learning theory/research to planning strategies in Chapter 7.

Which other competencies would you select as related to the emphases in this chapter? Look at the systems analysis model for a clue. If you selected all of category 2.3, we'd agree. Strategies are made up of sequences of modes. You may even be able to conjecture some likely modes to be used with rule learning as a result of the description of the conditions for learning rules in this chapter. If you have an idea, verify its validity as you interact with the reading in Chapter 7. In addition, we'd select competency 2.32 for its frequent use in presenting advance organizers and competency 2.310 for its connection to discovery learning. In a similar fashion 2.43, 2.44, and 2.45 depend on the use of the concepts of meaningful learning and learning hierarchies. We could identify many other competencies which stand no chance of being performed at criterion level without the use of an adequate learning theory/research base. See, for example, 2.55, 2.63, 2.64, and 3.10. The interconnections diagrammed in the systems analysis model, as always, are apparent in the dependency of competencies on one another and their dependence on multiple aspects of the model.

As you read the objectives listed in the next section, note the ways in which your knowledge base of learning theories/research has been delineated. These are the stepping stones you must cross before tackling the task of lesson planning.

6.6 SUMMARY AND SELF-CHECK

Learning is an activity we engage in throughout our lives, but we seldom stop to reflect on how we learn. However, successful teaching depends on thoughtful reflection and study. So we chose to begin with the learning of a jumble of words and then a limerick as a way to clarify some characteristics of meaningful learning. Discovery learning, learning hierarchies and learning types, and the role of an advance organizer were described and applied to the learning of mathematics.

You should be able to:

1. Operationally define learning, meaningful learning, rote learning, discovery learning, reception learning, Gagné's learning types, task analysis, learning hierarchy, and advance organizer.

2. State the conditions for learning *each* of the eight types of learning.

3. Construct a learning hierarchy in your subject field.

4. Identify concepts (defined, undefined, relational) and rules/principles by an analysis of syllabi, course outlines, or textbooks in your teaching field.

5. Evaluate the extent to which textbook presentations employ advance organizers and justify your decision.

6. Critique the sequence found in secondary-school texts or syllabi for closeness of match to a Gagné-type task analysis.

7. Assess the extent to which feedback in a "live," "canned," or simulated classroom scene would justify inferences that a concept or a rule had been learned.

The exercises which follow are designed to test your ability to meet the preceding objectives and to extend your acquaintance with related writings in this area. We encourage thoughtful consideration of as many of these activities as possible.

6.7 SIMULATION/PRACTICE ACTIVITIES

A. Obtain a state or local syllabus or course outline for a subject you are likely to teach. Read through the first two major units and make separate lists of concepts (defined, undefined, and relational) and rules/principles which are referred to as learning outcomes. Be sure to include those concepts prerequisite to each rule you list whether or not these were specifically identified in the syllabus. Compare your results with those of another classmate.

B. Most mathematics laboratory efforts are based on the assumption that guided discovery techniques will promote the meaningful learning of mathematics, yet some of these labs fail due to a confused interpretation of discovery learning and the conditions which promote it. Hendrix compares *three* existing discovery approaches in the article "Learning by Discovery" (*The Mathematics Teacher* 54 [May 1961]: 290-99). Read the article and contrast the three approaches that Hendrix describes in terms of their potential for achieving meaningful learning.

C. Obtain a copy of two texts which are used in secondary mathematics classes. Read the first few pages of several chapters and decide on the extent to which the authors have written advance organizers. Justify your decisions.

D. In each of the articles below, the authors have focused on problems associated with the learning of mathematics. Choose any *two* of these and prepare a written review of two or three paragraphs. Include in your review a summary of the writer's chief ideas and a consideration of the match among those ideas and the positions taken by the authors of this text.

   1. Bruner, J. S. On learning mathematics. *The Mathematics Teacher* 53 (December 1960): 610-19.

   2. Dewey, J. The psychological and the logical in teaching geometry. *Educational Review* 25 (April 1903): 387-99.

   3. Dienes, Z. On the learning of mathematics. *The Arithmetic Teacher* 10 (March 1963): 115-26.

   4. Wilder, R. L. The role of intuition. *Science* 156 (May 1967): 605-10.

E.

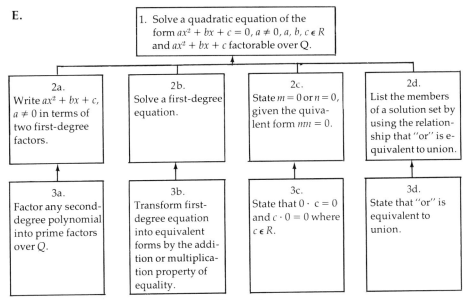

Figure 6.4

The solution to the learning hierarchy question posed in Section 6.3 is illustrated in Fig. 6.4 on page 143. Compre your version with the authors' solution and check out differences with your instructor. Then task analyze objective 3a and construct a learning hierarchy of prerequisites to that objective alone.

F. Refer to the Sequence of Content modular assignment in the Appendices. The careful completion of this assignment will give you a "leg-up" when it comes to the writing of well-sequenced lesson plans. Proceed as indicated in the directions, or as modified by your instructor.

G. Column 1 in Fig. 6.5 contains concepts or rules which are usually learned in junior- and senior-high mathematics classes. Column 2 contains feedback the teacher might obtain from the students. In each case, decide whether the feedback is *sufficient* to assure the teacher that the matching concept or rule has been learned. What else would you require in each instance of inadequate or inappropriate feedback?

Column 1 Concepts and rules	Column 2 Feedback obtained
1. Commutative property for addition.	1. States that "$a + b = b + a$."
2. Prime number	2. Labels 2, 3, 5, 7 as primes and 4, 6, 8 as nonprimes, when asked for examples and nonexamples of primes.
3. $A = \pi r^2$	3. Finds area of a circle given the radius.
4. Typical percent problems, involving use of $p = rb$	4. Finds the percent if given the rate and the base.
5. Parallelogram	5. Chooses only (d) when asked to select shapes which are parallelograms:

6. Graphic solution of systems of first-degree inequalities in two variables

6. Draws the graph of $y < 3x$ and $2y - x \geq 5$.

Figure 6.5

SUGGESTIONS FOR FURTHER STUDY

Gagné, R. M. *The conditions of learning.* 2nd ed. New York: Holt, Rinehart and Winston, 1970.

Gagné's overriding concern for the improvement of education shines through the pages of this volume. However, the verbiage of the psychologist is generally more suited to a reading audience with a psychology background than to the usual mathematics education student. For this reason, we recommend only Chapter 9 of the text, "Learning Hierarchies," as more accessible reading out of all the valuable material here. Pages 237-56 focus on the relationship of such hierarchies to instruction and illustrate hierarchies in the subject matter of mathematics. The examples given in this chapter should help you in the task of structuring your own learning hierarchies.

Higgins, J. L. *Mathematics teaching and learning.* Worthington, Ohio: Charles A. Jones, 1973.

All chapters in this book are directly related to issues raised in this chapter. Unit 1 begins with an investigation of mathematical memory from a slightly different view than that taken by the authors of this text. Stimulus-response learning and the use of hierarchies in lesson planning occupy a central position in Unit 3 while problem solving and concept learning are the major thrusts of the author in the last two units. We have already recommended the study of Unit 2 for its work in Piaget and now suggest a more critical analysis of the particular section entitled "Discovery Teaching and Mathematics Laboratories." The reading modules in each unit are almost a short course in learning from the masters themselves. And throughout, Dr. Higgins connects the theory and recent research based on it to classroom practice. This text was written for prospective teachers of mathematics and has been successfully used by them. We recommend it highly.

LeFrancois, G.R. *Psychological theories and human learning: Kongor's report.* Monterey, Calif.: Brooks/Cole, 1976.

This lively text is written from the point of view of Kongor, a being from another planet. We especially recommend Chapter 10, in which Kongor introduces the reader to aspects of the cognitive theories of Bruner and Ausubel.

Martorella, P. H., *et al. Concept learning: designs for instruction.* Scranton, Pa.: Intext Educational Publishers, 1972.

Although all chapters of this book are well worth time and study, Chapter 4, "Concept Learning in the Mathematics Curriculum," by Jensen, is highly recommended reading for the mathematics major. Pages 73-93 and 109-114 of Chapter 4 are devoted to succinct summaries of the contributions of Piaget, Gagné, Ausubel, Bruner, and Dienes. Jensen has illustrated aspects of the work of these men by reference to the learning of the mathematical concept of function. Concept-learning exercises based on several of these examples are outlined in Chapter 9, where Jensen and Martorella have collaborated on a semiprogrammed approach designed to deepen the reader's understanding of concept learning. If you are eager to avoid the pitfalls of inadequate concept learning by your students, then don't miss the opportunity to work through these exercises.

NCTM. *The learning of mathematics.* 21st Yearbook. Washington, D.C.: NCTM, 1953.

Although this yearbook is over one quarter of a century old, it is still a source of excellent illustrations of the special learning problems in mathematics. We recommend Chapter 1 and pages 228-47 of Chapter 8.

Wertheimer, M. *Productive thinking.* Enl. ed. New York: Harper, 1959.

In this classic written by the father of Gestalt psychology, the reader is taken into the classroom to observe students' responses to novel problems in geometry. The blind grinding-out behavior of one student is contrasted with the insightful tactics of another. Of particular interest is the chapter in which the problem-solving process is dissected from the moment of presentation to that of solution. This is a book worth reading in its entirety.

CHAPTER SEVEN

THE GAME PLAN
DESIGN OF INSTRUCTIONAL STRATEGIES

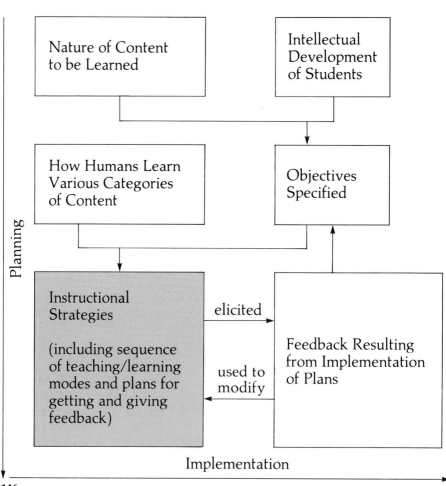

ADVANCE ORGANIZER

"Tom, I'm really glad you were assigned to do your student teaching with me this semester. You have asked many meaningful questions during your first few days of observing my classes. I have noticed you assisting individuals during class and working with students who come in after school for extra help. I know you are anxious to start teaching a class section on your own, and I have decided that the second-period geometry class would be best for your initial assignment. They are a typical sort of group, and geometry is your strongest area. Let's see. Today is Thursday and we'll be finishing the congruence topic tomorrow. Monday would be a good day for you to take over. Bring me a proposed lesson plan tomorrow, we'll look it over together, and I'll see what suggestions I can make. Then you will have the weekend to polish things up and be ready for a solid start on Monday."

You are Tom. Being an excellent student, you have thoroughly learned the material presented in the first six chapters of this text. You have learned to analyze the nature of mathematics and couple this with knowledge of the intellectual development of students in order to specify objectives that are clearly stated, accurately representative of the content, and attainable by the intended learners. You understand some of the major findings of cognitive psychologists with respect to human learning and can identify uses and abuses of nine main teaching/learning modes. And not only are you alert to the need to give and get feedback during lessons, but you also can design a variety of ways to provide for this all-important aspect of instruction. What else could you possibly need in order to whip that first lesson plan into good shape?

Note the shaded box in the systems analysis model on the opening page of this chapter and recall that this is the same box we directed attention to at the start of Chapter 1. At that time we concentrated on commonly employed teaching/learning modes and said very little about the design of instructional strategies—the overall theme of that part of the model. At that point in time we had developed neither the prerequisites nor the need for detailed attention to the task of strategy design. Now you have *both* background and need. Efforts to plan effective lessons will be stymied until you are able to synthesize instructional strategies based on all you have learned thus far.

But what is an instructional strategy? We define an *instructional strategy* as *a sequence of teaching/learning modes designed to promote the attainment of a particular type of objective.* Think in terms of the notion of *patterns.* It is neither necessary nor productive to think of teaching each bit of subject matter as an entirely novel task. Just as we can categorize many items of content as concepts, we can also construct an overall instructional strategy (pattern) to teach concepts. Likewise, we can design appropriate overall strategies for teaching other kinds of outcomes such as principles/rules, vocabulary, and psychomotor skills. The introductory activity that follows is intended to provide some initial experience in the critical matter of strategy design.

7.1 INTRODUCTORY ACTIVITY

Planning is a Bloom level III-VI activity.

This activity is based on the use of the ruler lab described in Section 2.2 of Chapter 2. If you've forgotten the details, take time to refresh your memory now. The activity refers to a quiz situation faced by mathematics education majors who, like you, had studied the ideas incorporated in Chapters 1-6. The students participated in the ruler lab and then wrote their responses to a follow-up quiz question. The question and one student's response are reproduced below.

> Assume you are teaching a ninth-grade mathematics class in which most of the students have been assessed as concrete operational. Write two objectives that would be appropriate for this grade level. Then list a sequence of modes which would be likely to achieve your stated objectives and which would incorporate the ruler lab.

Mary's answer

Objectives:

1. Choose a graphic way to display the data and justify the decision.
2. Extrapolate the coordinates of the next three points on the basis of the graph.

Modes used in sequence:

1. Divide Ss into groups and have them perform the ruler lab.
2. With total group watching, T has small groups record cumulative data on the board.
3. T lectures on feedback T observed.
4. T lectures on "best" way to graph this data and Q/A Ss as to reason it is "best" way.

After studying the above response, react to:

1. the aspects of Mary's answer that were consistent with what you have studied thus far,
2. the match between the strategy and the stated objectives,
3. the match between the type of Ss and the modes, and
4. the sequence of modes and choice of modes.
   a) Look for omissions.
   b) Identify places where students, rather than the teacher, should be at the focal point.

Check Simulation/Practice Activity A for a response we judged more adequate, *after* you have interacted with the preceding exercise.

7.2 DAILY PLANNING BASED ON SYSTEMATIC INSTRUCTION

How can plans be designed so as to attend to all aspects of the systems analysis model? How much should be written? What written format is helpful to the beginner? We have found that detailed written plans organized to correspond to key components of the model are invariably excellent bases for the beginning teacher. Such plans do not guarantee success, but their absence correlates highly with failure. In other words, they are necessary, but not sufficient, conditions for teaching effective lessons.

What format is amenable to the organized planning we have been emphasizing? We have found that the one depicted in Fig. 7.1 not only accounts for all aspects of the model, but also has proved workable in the hands of both novices and experienced teachers. Analyze the lesson plan format in Fig. 7.1 and note the categories explicitly represented in the outline. Some items—such as date, topic, and class period—are simply identifiers for the teacher or visitors. The routines box is included so that announcements, attendance notes, and other such items might be recorded. However, the remainder of the headings should make sense as a synthesis of modes, feedback, objectives, and characteristics of students and of content. Most novices find that the spacing in Fig. 7.1 is inadequate in certain areas—especially the instructional objectives box. You should adjust the spacing to fit your needs. Under content item are listed the labels of the concepts, the rule statements, the rule use ideas, and so on, in the sequence in which you expect to introduce these. Across from each content item in column 3 would be a description of the sequence of modes designed to achieve the stated objective(s) with respect to that content item. If materials other than standard chalk and chalkboard are to be used, these are listed opposite the related strategy in column 2. In addition, we added two columns labeled *feedback strategies* to give particular emphasis to

Topic _____   Date _____

Class/Period _____

Instructional Objectives:			Routines:		

1 CONTENT ITEM	2 SPECIAL MATERIAL, EQUIPMENT	3 INSTRUCTIONAL STRATEGIES	4 FEEDBACK STRATEGIES		5 TIME EST.
			GET	GIVE	

Fig. 7.1   Daily plan format.

the need to plan ways to *get* feedback as well as ways to *give* feedback. Be careful. These columns are not meant to be diaries of what happened, nor fond hopes as to what students might say. They should contain the strategies you have chosen in order to elicit feedback at critical times and to give certain kinds of feedback during the lesson. Finally, column 5 should contain your time estimates for a strategy aimed at a particular objective. After the lesson has been taught, these estimates should be checked against actual times to assure more realistic planning of time use in the future.

Skeletons may provide helpful clues to the trained archeologist, but beginners need to study the human body in the flesh before graduating to the inference-making of the science of archeology. In like manner, the teacher-to-be needs to analyze completed lesson plans before attempting to construct plans. In the subsections that follow, we have included some sample plans that deal with differing kinds of content. Study each of these plans carefully and look for patterns.

Planning for Concepts/Vocabulary

The plan that follows (Fig. 7.2) was designed for a seventh-grade class that had previously been taught an introductory lesson on circles and the various segments and arcs associated with them.

You should assume that this plan is, at least, a second draft attempt on the part of a student teacher. The cooperating teacher would have checked objectives, the match of strategies to objectives, the sequence, and the time estimates. Notice the time ranges indicated at several points in the plan. If the maximum times are utilized throughout, 44 minutes will be consumed. But the class is scheduled for 40 minutes. What implication does that possibility pose with respect to the need to give directions on Handout 2 near the end of the period? No, you can't keep students beyond the assigned class time. Suppose that the minimum times are utilized. You should have obtained a sum less than 40 and, if this were your plan, you should be worried about what might happen in that leftover time. If you don't plan an activity for all parts of the class period, you can be sure that students will innovatively utilize the time in ways you might prefer to consign to oblivion. This plan badly needs an *elastic clause*—an instructional strategy that can be inserted into the lesson at the teacher's discretion as a response to feedback. The nature of the elastic clause will determine its position in the plan, but we recommend that it be inserted somewhere in the plan rather than "tacked on" at the end. Do you see why? The old reward and punishment aura rears its ugly head if the elastic clause comes over like "You've done so well, you can do some more of these."

This sample plan deals with the concept π and the labels or vocabulary associated with it. Which objectives are aimed at the concept? Which, at the vocabulary or symbols? Objective 3 seems to be at least partially directed toward the learning of some mathematical processes. Which ones? Now analyze the instructional strategies column for strategies matching the conditions for concept learning outlined in Section 6.3 of Chapter 6. Why did the teacher plan to include a book, a square box, and an oval-shaped bottle along with different sizes and kinds of circular models? How did the teacher intend to channel student thinking in order to focus on essential features of the concept? If you've decided that the organization of the lab sheet is one critical factor and the Q/A on measurement errors, another, you are absolutely correct. When and how did the teacher plan to get feedback on the learning of the concept, π? The thought question on π versus 3.14 is surely one place where this should occur, but feedback sampling here is bound to be somewhat limited. The strategy which follows that Q/A is more promising in terms of feedback sampling, but will it match the objective? Only if the exercises in Part 1 of Handout 2 are appropriately constructed. The cooperating teacher, in this situation, suggested that the original draft of Handout 2 be modified in the following ways: (1) have a small number of exercises in Part 1 and (2) have many *Why? How? When?* and *Defend* questions. Two of the questions which were included in the final version were the following:

1. Astronomers measured the circumference and diameter of Mars. Geologists measured the circumference and diameter of Earth. How would you expect the quotient C/d to compare for Earth and Mars? Why?

2. The diameter of a hula hoop measures 80 cm. The circumference of the hula hoop is unknown. Mary claims she can still tell you exactly what C/d equals. Jim says, "If measures are taken, then C/d won't come out a constant." Who is right and why?

You should have been able to identify connections among the conditions for learning a concept and the instructional strategies outlined in this plan. But the specific instructional strategies in this plan can readily be generalized to an instructional strategy pattern common to all plans designed to teach concepts. Study the pattern outlined below; note its relationship to Gagné's conditions for learning a concept; and identify the specific strategies in the π plan which illustrate the generalizations in the pattern.

Instructional Strategy Pattern for Concepts

Show differentiated examples and nonexamples in sequence; get essential and nonessential characteristics identified; elicit operational definition; model by analogy; elicit generalization of the concept to a variety of specific instances not previously used.

If some of the strategies in the π plan seem to be "extra," you may have correctly excluded those which are aimed at the teaching of vocabulary and

Topic: Concept of π

Class/Period: Math 7, Period 3

Date: November 7 (Monday)

Instructional Objectives:

1. State that C/d is approximately equal to 3.14.
2. State that the symbol π, pronounced pi, stands for C/d.
3. Give reasons for measurement differences in the values of C or d obtained by different Ss.
4. Justify the use of π rather than 3.14 in specified applications of the concept.
5. Use the symbols π and $\approx$ correctly in written and oral work.

Routines:

1. Take attendance via seating chart as Ss enter.
2. Collect late work from Mary T. and Pete K. as they enter.

CONTENT ITEM	SPECIAL MATERIAL OR EQUIPMENT	INSTRUCTIONAL STRATEGIES	FEEDBACK STRATEGIES		TIME EST.
			GET	GIVE	
Recall of concepts of circle diameter circumference perimeter diagonal axis	bike tire, assorted cans, dinner plate, dixie cup (some of above with taped strips for d and r), book, square-lidded box, oval-shaped bottle	T demos with objects and conducts recall Q/A for concepts. T has Ss point out ex. and non-ex. "Where is the segment called diameter?" "How is radius related to diameter?" "If we measure around this can top, what do we call the result?" "Why do you think we give different names to the distance around these 2 objects?"	T spreads Qs. T takes some straw polls followed by Q/A for reasons.	T confirms some ans.; asks another S to confirm at times; praises thoughtful responses.	3-5 min.
Circle in history and informing Ss of outcome		T lectures on importance of circle to Greeks— both practically and in mythology; reminds Ss of everyday ex. they volunteered in last class and the measurement needs listed then. T states today's task: To engage in lab work in order to find for themselves a pattern which early mathematicians first guessed at after measurement.			2 min.

string, meter sticks, lab sheets, board copy of lab sheet	T has Ss pass out lab sheets and tells all to fill in heading. T gives lab instructions: a. Where groups and leaders will be assigned. b. Duties of each gp. member. c. Collection, use, and return of equipment. T demos measurement task with one can. Ss record data on sheets as T does at board. All Ss to perform division and all record results. T holds up one or two unlike objects and Q/A as to measurement problems. "Where to hold string?" "How to read ruler?" etc. T gives signal for preassigned groups to begin lab.	T tours. T spreads Q/A. T takes straw polls.	Designated S does $\div$ on calculator and gives result. T has Ss volunteers come front to demo.	5-7 min.
$C/d \approx 3.1?$	Ss work in lab groups. T gives time signal, reminds groups who are finished to seek a pattern and to note any exceptions.	T tours and asks Qs on procedures and result.	Ss can check $\div$ on calculator. T praises gp. progress.	10-15 min.
Measurement errors	As groups finish, T has leaders put final column results on board. T directs all others to compare results. T has leaders state group conclusions. T: Q/A differences in hundredths place; asks contingency Qs relating to measurement problems.	T spreads Q/A.	T confirms pattern. T circles hundredth place nos. on board.	2-3 min.

Figure 7.2

CONTENT ITEM	SPECIAL MATERIAL OR EQUIPMENT	INSTRUCTIONAL STRATEGIES	FEEDBACK STRATEGIES		TIME EST.
			GET	GIVE	
Inductive vs. deductive validity		T lectures on the inexact nature of measurement, the nature of proof, etc.			1 min.
Symbol π, symbol $\approx$, value of 3.14, $C/d = \pi$ vs. $C/d \approx 3.14$		T lectures on exact value of C/d for a mathematical circle, says "pi," and writes π on board. T tells Ss to copy π on lab sheets, states that 3.14 is a good estimate for π, writes symbol $\approx$ as in $\pi \approx 3.14$. Ss to copy above. T writes $C/d = \pi$ and $C/d \approx 3.14$ and asks for an explanation of difference.	Diverse Ss to say "pi." T checks some notes. T uses "wait time," calls on several volunteer Ss, asks others to defend responses.	T praises ans. T confirms and praises good ans.	3–5 min.
Generalizing to other cases	Handout #2	T has Ss complete Part 1 of handout in Supv. Practice.	T tours.	T cues and/or praises. Ss to help ea. other.	
		T reads all answers aloud.	T takes polls.		2 min.
Look ahead		T tells Ss to complete Part 2 by Tuesday. Results will be used in class. If time, start sheet in class.			0–4 min.

Figure 7.2 (Cont.)

symbols. Objective 5 was aimed at the learning of a verbal association, as well as the discrimination of a written mark. It is important to notice that the label was one of the last items attended to in the plan. There is an old adage, "Don't name the baby before it's born," which is worth applying to most concept/vocabulary lessons. An unfamiliar label for which the student has no concrete referents is meaningless if introduced early and can sometimes lead to a negative attitude on the part of the students. "Mantissa? What's that? Oh, who cares. More big-word math." Review the part of the π plan which deals with labels and look for specific strategies which match those in the pattern below.

Instructional Strategy Pattern for Vocabulary

Show object (or exemplify idea); then pronounce name and spell on board; have students do likewise; repeat (practice); give name and elicit statement of idea or description of object; have students use in context.

The two instructional strategy patterns outlined above are standards against which concept and/or vocabulary plans can be checked. They are also guidelines both novices and experienced teachers have found useful in constructing their own concept/vocabulary plans. However, important as the strategy patterns are, they are ineffective unless used in conjunction with the other components of the systems analysis model. Note that the choice of modes—laboratory and demonstration in particular—was heavily concrete, and even the exercises on the handout dealt with real phenomena. That kind of choice agrees with the Piagetian implications you studied earlier, just as the particular set of objectives reflects both the processes and products of mathematics at a level suitable for seventh-graders. If the students had been eleventh-graders, they might have been expected to consider the concept of "irrational," and perhaps distinguish between this irrational number and other kinds of irrationals, such as the square root of 3. What else is contained in the single plan? There is an attempt to produce an overall mix of guided discovery with reception learning, and *if* the teacher implements this plan effectively, the result promises to be meaningful learning. Have you identified an advance organizer? You should be able to find both conditions present in the two-minute lecture on the "Circle in History and Informing Ss of Outcomes." The designer of this plan put all these specifics together in a carefully sequenced package.

Have you wondered why dashed lines are used to separate strategies from one another? Lines are useful at the planning stage to help us check our own thinking and are very helpful during actual implementation as an assist to keeping track of where we are in the lesson. However, solid lines would tend to convey more separation among lesson segments than is intended. Dashed lines are meant to imply the need for the kind of connections that will enable the lesson to flow smoothly. Are such connections provided in the sample plan? Go back and read the end of each strategy and the beginning of the next and identify the connections. Are they explicit? What could the teacher say or do so that students will comprehend the connections? Be sure to satisfy yourself on this important concern before continuing. Perhaps you are thinking that sequencing decisions can either enhance the building of connections or make the task all but impossible. We agree! Try mentally rearranging the strategies on any lesson plan and then analyze the resulting effects. Sometimes you will find equally (or perhaps more) productive ways to structure the lesson. In other cases bizarre results will show up fast.

There certainly is a lot more to this business of planning lessons than is obvious to the casual observer. And just think—we have considered only concept and vocabulary instruction thus far. Does the task seem overwhelming when you realize that coping with plans aimed at rules, novel problem solving, proofs, and psychomotor skills lies ahead, and that many lessons will involve combinations of these? Don't let yourself be discouraged. The situation is not nearly as momentous as it may seem at this point. What you have learned about concept and vocabulary plans will serve as the base for all these other types of plans. Thus, the next task is to apply that which has just been learned and add some new ideas that have special relevance to various other types of target outcomes. Since mathematics is said to be a rule-governed subject, we consider planning for principle (rule) learning in the next section.

Planning for Rules/Principles

The π plan in the previous section ended with a reference to Handout 2, which would be utilized the following day. The section of Handout 2 which was to be completed contained a table that is partially reproduced in Fig. 7.3.

Part 2: Group 7 in Ms. Mason's class handed in the lab sheet below. The leader, Jerry Batt, complained that some joker had erased part of their work. Help Jerry by filling in the blanks.

Object	Circumference	Diameter	C/d
Soup can	24. _____ cm	8.2 cm	3.0
Jug	_____ . 0 cm	10.0 cm	3.1
Half dollar	_____ . _____ cm	4.0 cm	3.1
Disc (large)	620.0 cm	_____ cm	3.1

Figure 7.3

The plan which follows (Fig. 7.4) is that designed by the same student teacher for the second day of instruction with the seventh-grade class. Study it carefully for the earmarks of rule learning to be found in Section 6.3 of Chapter 6. This time the student teacher designed an elastic clause based on the two starred objectives in the plan. The elastic clause may be saved for another day if these seventh-graders are weak in basic calculation skills and become bogged down in computations, or if they are very bright and offer diverse explanations as to the nature of this mathematical model or if they ask about the existence of other patterns they observed with respect to the exercises of Handout 2. As you might guess, the cooperating teacher was shown a draft of the new handout, a list of the practice problems, and a complete blow-by-blow exposition of the material to go in the students' notes.

But the central issue here concerns the rule-learning objectives and the strategies that are related to the conditions for rule learning. You should have identified specific strategies which conform to the pattern below.

Instructional Strategy for Rules/Principles

Get students to recall/review prerequisites; indicate nature of expected terminal performance; cue (via questions, lab work, applications) students to find the pattern by chaining concepts; get the rule stated (by students, if possible); provide a model of correct performance; have students demonstrate instances of the rule in a variety of situations. *Fix and maintain skills by spaced and varied drill.

The key elements in the rule strategy pattern are prominent features in this or any other well-designed rule/principle plan. Notice, though, that one feature often merges with another. In this case, recall of prerequisites, indicating the outcome, and cuing to find the pattern are encompassed by the strategies from HWPM (homework post-mortem) up to the beginning of model problem 1. Yet, "Where we're headed" is explicitly treated by the teacher as well as implicitly handled in HWPM and the sequence of developmental questions. Although the explicit consideration of a goal should be present in every plan, early in a rule lesson the students must be alerted to the specific terminal expectations of that lesson. Then their outlook during the eliciting of the rule is not cluttered by musings such as where we're headed, what I'll need to remember, whether two questions in a sequence are related or not. They know what to listen for and concentrate on. When the teacher later "provides a model of correct performance" as this designer of a plan proposed to do with model problem 1, the details of the teacher's expectations for the student should be crystal clear. Finally, no rule instruction is complete if a lesson does not contain a well-structured strategy aimed at *student* demonstration of learning. Too many novices are trapped by lazy or confused students into doing "one more problem together" and thus demonstrating that the *teacher* has learned the rule. You were warned about this abuse of

Topic: Formula: $C = \pi d$

Class/Period: Math 7, Period 3

Date: November 8 (Tuesday)

Instructional Objectives:

1. State that $C = \pi d$ in symbols and words.
2. Compute the measure of either C or $d(r)$ given the measure of the other by applying $C = \pi d$ and using $\pi \approx 3.14$.
3. Express the measure of C in terms of $d(r)$ and π, or of $d(r)$ in terms of C and π.
*4. State that 22/7 is another estimate of π.
*5. Calculate as in #2 by using $\pi \approx 22/7$.

Routines:

1. Take attendance via seating chart as Ss enter.

CONTENT ITEM	SPECIAL MATERIAL OR EQUIPMENT	INSTRUCTIONAL STRATEGIES	FEEDBACK STRATEGIES		TIME EST.
			GET	GIVE	
C as a multiple of d d as a factor of C	Acetate copy of Part 2 on overhead projector (OH) with 1st two ans. written in	HWPM. As Ss enter, select diverse Ss to fill in acetate sheet. Tell others to ck. work and circle differences in pencil. Be ready to defend ans. or find error. Get attention of all on 1st and 2nd ans.	Tour and look for problems. Take poll. Q Ss for reason. Spread Q/A and poll selectively.	Praise all Ss for Mon. gp. work.	
		Q/A through other exs., helping Ss decide on correctness of ans.		T verifies errors and corrects ans.	5-7 min.
Where we're headed Rule: $C \approx 3.14 \times d$		T notes what Ss have been able to infer and states today's goal—to use pattern in systematic way. T uses dev. Q/A to elicit more general pattern. "Yesterday we agreed to use 3.14 as an estimate of C/d. So if d had been 1 cm, C would be about ? cm? Suppose d were 2 cm? 3 cm? 4 cm? etc." T gives more exs. until most Ss have idea. T elicits $C \approx 3.14 \times d$ and writes this on board.	Spread Q/A and seek hands. Straw polls.	Write correct ans. on board in orderly pattern.	2 min.

Rule: $C = \pi \times d$	T reminds Ss that 3.14 is only an estimate for C/d. T asks for exact value of C/d. T uses lecture and Q/A to have Ss treat π as they had been treating 3.14 above and gradually elicits $C = \pi \times d$.	T spreads Q/A until almost all Ss are involved.	Write correct ans. on board under previous ans.	2 min.
	T lectures on nature of formula, use to ans. Qs such as those on Handout #2.			
Model problem 1 (notation πd for $\pi \times d$ and rationale) — Acetate # 2	T reminds Ss to take notes as T and Ss work out model problem written on acetate. T introduces πd notation, asks Ss if they can guess why $\pi \times d$ might be confusing. T emphasizes the form desired at each step of the labeling, the change from $C =$ to $C \approx$; the use of ()() for "times."	T tours and scans several notebooks from time to time. / Asks one S to react to another's ans.	Praises esp. good rationale.	3 min.
Practice Set — Acetate # 3 / Find C, given d / Find C, given r	Ss to try some on their own in Supv. Pract. Selected Ss are sent to board to put good work on. Others are told to check work against ans. on board. If necessary, T stops classwork so all may Q/A procedure. If no particular problem, T encourages Ss to try "little stickler" ex. where r is given. Depending on progress, T assigns another with r given or immediately assigns "big problem," where C is given.	T circulates. / T circulates and warns all to *read* given.	T indicates how all are doing, points out common errors, etc. / T reads correct ans.	8-10 min.
Find d, given C				
Summary of d, C, r relationships	T emphasizes stating $C = \pi d$ in words, then has Ss repeat. Q/A for $d = 2r$, d in terms of C and π, C in terms of π and r. Ss to write major relationships in notes.	Ss to repeat aloud.		2 min.

Figure 7.4

CONTENT ITEM	SPECIAL MATERIAL OR EQUIPMENT	INSTRUCTIONAL STRATEGIES	FEEDBACK STRATEGIES		TIME EST.
			GET	GIVE	
Mathematical model $C = \pi d$		T poses thought Qs *re* need of $C = \pi d$ when $C \approx 3.14d$ seems to do.	T uses "wait" time, checks several resp.	Verbal praise for esp. good ans.	1-2 min.
Expressing C in terms of π and d		T notes that some exercises do not require use of 3.14 and Q/A, C in terms of π for $d = 4$ cm, 2 m, 3½ dm, etc., then d in terms of C and π.	Spreads Qs.	T writes ans. on board.	2-3 min.
Elastic clause: 22/7		T lectures on historical background as to estimates of π, gives Archimedes' values of π as between 3 10/71 and 3 1/7. Has Ss change to improper fraction and decimals. T states use of 22/7, rationale for this, and emphasizes 22/7 $\neq$ π. If time, T and Ss begin selected exs. using 22/7.		2 volunteer Ss give calculator result.	5-10 min.
Analysis of mixed ex.	Handout #3	Oral Q/A, e.g., "'Must 3.14 be used?" "Which exs. ask for the value of r?" etc.	Fast diverse Q/A. T asks for defense from other Ss, straw polls.	T confirms.	3-4 min.
Applying formulae with variety of problems	Puzzle Sheet Attachment to Handout #3	Ss to do as much of puzzle as possible. The HW for Wed. is to bring in 4 cutout discs based on pattern on bottom of puzzle sheet. Tell Ss they will be used in a lab exercise investigating surface.	T circulates.	T may ask ind. Ss to help one another.	5 min.

Figure 7.4 (Cont.)

feedback in Chapter 2. Sometimes, it's true, the students are honestly confused because the teacher didn't initially clarify the expected student outcomes and/or has in the past immediately given practice examples far more complex than the model problem. Once again, we emphasize the need for gradually differentiated practice examples and for in-class decisions as to the number and kind of problems to be assigned, based on feedback. We hope you noticed that this plan included more than one written practice sequence as well as oral practice. That final five minutes of written practice is a *must*. Unless the teacher obtains positive feedback on a substantial part of these mixed exercises, there is no solid evidence of student capability on such tasks.

Why did we star the last sentence in the instructional strategy pattern? The star is meant to identify the long-range nature of this aspect of the strategy pattern. Intellectual skills, however effectively introduced, are maintained only by means of selective practice in later lessons. Since learning that is not retained over time is of little interest, we'll come back to this topic in a later section of this chapter.

"Word" problems of algebra If you were to ask a typical ninth-grader how much problem solving occurred in algebra class, we'd bet that you'd hear: "Too much." But if you studied the usual algebra text, you'd find few illustrations of what Gagné calls problem solving. Instead there would be sections labeled *coin problems, mixture problems, digit problems, age problems, motion problems,* and so on. Even a cursory look at the sample problems would be enough to convince the reader that all of these are merely examples of rule/principle learning. Thus, instruction should correspond to that particular strategy pattern outlined earlier. Therein lies at least one explanation for our ninth-grader's negative attitude. Some teachers behave as if the students are engaging in type 8 learning, novel problem solving, until the end of the lesson when an algorithmic procedure, such as filling out a table for *d, r,* and *t* and *always* (well, almost always) setting the same two designated expressions equal, is presented as a *fait accompli.* The students who thought they were expected to be hypothesizing now find out that there's a single formula-like approach which they must be able to apply to similar problems. No wonder confusion and often antagonism results.

A second and even more significant cause of negative attitude and ineffective learning of this kind of "word" or verbal problem resides in the very nature of the problems. Just read the names of the types mentioned earlier. Who but a dedicated lover of mathematics would want to be proficient in the solution of age problems or digit problems? Mixture problems and motion problems at least have some potential relevance to the everyday world. *If* all these kinds of problems are to be taught (and we'd encourage you to review that question *after* you have had two or three years of *effective* teaching experience), the teacher must make a different case for each variety in the advance organizer. For example, age problems might best be treated as recreational puzzles which can be solved by means of equations, whereas motion problems should be presented in terms of an actual, interesting situation.

> *Pigeon racing, a development of the use of homing pigeons in the ancient Olympic Games (776 B.C.-393 A.D.), originated in Belgium.*

> *The longest recorded flight was estimated to be 7000 miles flown in 55 days by a pigeon owned by the Duke of Wellington. The flight ended when the exhausted pigeon dropped dead one mile from its home loft in London, England, on June 1, 1845. What average velocity in mph was attained by this record breaking pigeon? (Guinness, 1976)*

An attention getter like the pigeon problem above does much to create a positive mind-set toward the new topic. But the pigeon problem contains a characteristic even more important than that of simply provoking curiosity. It makes sense. It is a *bona fide* motion problem to which even we nonpigeons can relate. And that latter characteristic, the meaningfulness of the situation *to the student,* is the one most frequently ignored by textbook writers and thus by those teachers who rely solely on textbook problems.

However, even when a problem statement is meaningful to the students, understanding of the solution process is hindered if a teacher overlooks a second characteristic of type problems. Compare the two problems which follow. Consider especially the steps the student must take to solve each by means of equations.

1. The difference between two numbers is 24. Twice the smaller number is equal to the larger number minus 14. Find the two numbers.

2. A bottle containing 40 cc of tincture of iodine (iodine crystals dissolved in alcohol) is labeled as having 2% concentration of iodine. However, each time the cap is removed some alcohol evaporates, and thus the concentration of iodine is altered from the medically prescribed percentage. How much alcohol would have to be evaporated in order to double the recommended concentration? To triple it?

The difficulty becomes obvious, doesn't it? The straightforward left-to-right translation which works so well for problem 1 doesn't seem to get us anywhere with problem 2. We call problems like problem 2 *situation problems.* Since the solving of meaningful situation problems is one of the most important objectives of mathematics instruction, strategies directed toward this end deserve our special attention. But, first, let's take a longer look at the kind of instruction needed to learn translation skills.

How does a student become a skillful translator? There are clearly some prerequisite vocabulary/symbol learnings. Words such as *difference, equal to, twice, minus,* and the like—as well as their related symbols—should have been mastered in previous lessons. Then translation from English to mathematical phrases and sentences and back might be taught as a number game.

The following number game model is one way to make translation highly concrete. After challenging several students to think of a number, perform a set of operations on that number, and write the final result on paper, the teacher writes the same result on the chalkboard. Despite challenges with other students, the teacher continues to appear to read minds. Finally a set of plastic containers and a set of checkers is used to model one of the problems (see Fig. 7.5). The teacher might also contrive a demonstration with colorfully drawn pictures on an acetate or simulate the mental action associated with various steps by moving cutout shapes on the overhead, on a felt board, or even by means of masking tape on a chalkboard.

Direction	Teacher talk	Model
1. Think of a number	Since I don't know the number, let's assume that you put that many checkers in this box.	
2. Add 2	Place 2 checkers next to the box.	
3. Multiply by 4	That means 4 times your number and 4 times the 2 checkers or 4 boxes and 8 checkers.	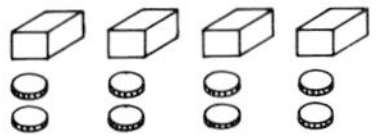
4. Subtract 4	Take away 4 checkers	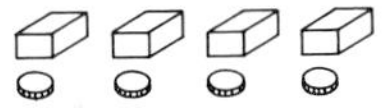
5. Divide by 4	That leaves one box and one checker.	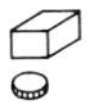
6. Subtract your original number	Take away the box. Your answer is 1.	

Fig. 7.5 Translation model.

Following lessons of this kind, students can be gradually introduced to translation exercises without the benefit of pictures or models. Now the teacher must be alert to evidence of reading deficiencies, special vocabulary problems (such as "exceeds") and the need to teach word-for-word reading of mathematics. See the Suggestions for Further Study section for a reference on the reading of mathematics texts.

However, translation skills are useless learnings if a student cannot apply them to solve situation problems. Keep in mind the two characteristics pointed out earlier.

1. The situation must be meaningful *to the student* and

2.   The student must be alerted to the need for an initial strategy other than translation.

The first characteristic is the key to what should occur in the construction of the advance organizer and the follow-up development of this kind of rule. The situation must be made meaningful in a concrete way.

The following paragraphs outline the idea concocted by Mrs. Armstrong, who was beginning to plan just such a lesson—an introduction to solution problems. The plan itself is not included, but you should be able to expand the development described below into an appropriate sequence of instructional and feedback strategies.

"I'll start with a demonstration that can be matched to a specific problem. I'd better start with a situation in which water is **added***; evaporating may befuddle their thinking," Mrs. Armstrong mused.*

After borrowing two 100 cc graduated beakers, two 10 cc graduated beakers, and a bottle of fluorescein dye from the chemistry teacher, she added drops of the dye to the 90 cc of water in one of the beakers until an intense color was present. She recorded the number of drops used and then observed the effect of adding 10 cc of water on the color. There was a perceptible change.

Mrs. Armstrong decided to do the final step of the demo in exactly the same way at the beginning of class and then question the students as to their observations. As a result of a last-minute brainstorm she decided to use all four beakers with the two 100 cc kind already containing identical amounts and intensity of the fluorescein solution, while the two 10 cc beakers would each contain 10 cc of water. When the students entered class, they would see the apparatus set up as in Fig. 7.6.

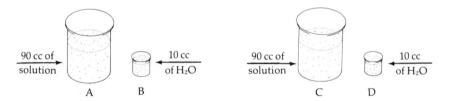

90 cc of solution → A   B   → 10 cc of H$_2$O   90 cc of solution → C   D   → 10 cc of H$_2$O

Fig. 7.6   Beakers set up for solution demo.

The students in the front rows would be able to read aloud the total number of cc of liquid in each of the four beakers, but it would be up to the instructor to reveal the nature of the ingredients and the concentration of the solution in beakers A and C.

Mrs. Armstrong intended to pour the water in beaker B into the solution in beaker A but leave the other two beakers as is, so that students might have a ready ref-

ference as to the "before" situation. She knew she must obtain answers to two major questions: (1) What changed? and (2) What remained the same? She would have to use contingency questions to elicit the hidden constant—that is, the constant volume of the fluorescein dye in beaker A both before and after the demonstration. Then she would help the students compose a list of varied practical situations where mixing of this kind might occur but where the desired final concentration would be known (in order to clarify where the class was headed).

If Mrs. Armstrong implements the above idea successfully, she should next provide students with a model problem. The model problem should be matched to the action illustrated by the demo (that is, one ingredient should be added to a mixture) although the unknown might differ. Then she will need to move gradually from the demo to the mathematical model in at least three steps (pictoral, English sentence, mathematical sentence). Mrs. Armstrong would probably cue students to help complete:

1. A labeled sketch of the situation in the case of the model problem with perhaps a "before" and "after" version.

2. An English sentence that describes what stays the same (for example, the cc of fluorescein dye "before" is the same as the cc of fluorescein dye "after").

3. The corresponding mathematical sentence with the appropriate designation of the unknown.

Although the new rule has been obtained at the equation-stating step, the students should still be asked to complete the problem and check the result in the statement of the problem as well as in the original equation. Moreover, as in all problem-solving instruction, it is vital that the teacher and the students review the process which led to this rule. Next comes student practice of the new rule, right? Yes, but it takes a relatively long time to completely work through even a small number of these problems. And leaving the bulk of them for homework is never a sound substitute for supervised practice. The answer is to assign several problems, direct the students to label the variable(s) and set up the equation for all examples, but to completely work through only one or two designated examples.

Mrs. Armstrong's approach might seem time-consuming and unnecessarily elaborate (unless you've recently read the results of the NAEP [National Assessment of Educational Progress] study on the ability of young adults to solve typical word problems—the findings were *far* from encouraging). Our own classroom experience has convinced us that the rote application of a standard tabular arrangement can be presented quickly and will yield fairly high scores on a quiz given the next day on exactly the same kind of problems. However, if the teacher slips in investment problems, dry food mixture problems, coin problems (all possessing the same structure as solution problems), the students will be lost and must be taught to memorize still more procedures for each of these "types." Mrs. Armstrong's initial investment of time will pay off dividends in meaningful learning and make her later attempts to transfer learning of this rule to all problems in the same family far more successful. We will pay particular attention to retention and transfer in a later section. For now we suggest that you begin to add a wide variety of relevant and interesting problems to appropriate folders in an algebra resource file.

Theorems in geometry Why are theorems being considered under the heading "Strategies for Rules/Principles"? Did you classify proof constructing as problem solving rather than rule learning? If so, then you're correctly focusing on one of the important processes of mathematics. However, theorems involve products of mathematics as well as processes. The learning outcomes are a mixture of both rules and novel problem solving, of both process and product.

Almost every theorem serves a dual role: (1) to be applied in computational questions and (2) to be applied to deduce other theorems. For example, the inscribed angle theorem is utilized as a formula to find the measure of an inscribed angle given its intercepted arc. It is also the key theorem used to obtain a proof of other angle/arc measure theorems. Both of these related, but distinct, rules must be specifically provided for in instruction—from the advance organizer to the model problem to appropriate practice. That sounds simple enough but in reality it is far from simple. In particular, the rule-use of deducing other theorems adds the ingredient of the axiomatic process to the mixture.

Each and every theorem is so classified because it is capable of being deduced from prior concepts, postulates, and/or theorems. And certainly the overall major objective of secondary-school geometry is to help students become capable proof constructors. But there are two stages in proof construction. The first, designated the *analysis* of the proof, is the process by which the student, through selective trial and error, conjecture, and testing, actually arrives at a solution to the deductive proof. The student's solution may or may not appear on paper. It may be symbolized by drawings and notes but, in any case, if questioned, the student would be able to outline the core of the solution. This analysis stage, *if left to the students,* is correctly characterized as Gagné's problem solving.

However, the axiomatic method systematized by the Greeks demands that the solution to a proof be presented in a prescribed manner. This presentation, usually written, must start with the hypotheses and then list new results deduced from the hypotheses and prior assumptions or theorems until the desired conclusion is reached. Each new assertion must be justified by a generalization that already exists in the axiomatic system. When asked to write a short, deductive proof, Juan wrote the following:

$\overline{AB} \cong \overline{CD}$ *was given, as was D as the midpoint of* $\overline{CE}$ *and therefore,*
$\overline{CD} \cong \overline{DE}$ *from the definition of a midpoint.*
Finally $\overline{AB} \cong \overline{DE}$ *since segments congruent to the same segment are congruent to each other.*

Whether paragraph style or column style is used, the requirement is the same. The written *synthesis* must repeatedly illustrate the logical relationship

$$p \text{ and } p \to q$$
$$\therefore q.$$

Thus, the prescribed character of the synthetic presentation is rule bound. And although the nature of this complex rule is constant, it is not a rule to be taught in a single lesson or two. Indeed, throughout the course, variations in synthetic arguments will occur (such as presentation of the indirect proof) and objectives, instructional strategies, and feedback strategies must be designed to account for these variations.

A single theorem, then, has the potential for yielding both a new numerical rule and a new axiomatic rule and, in the process of applying the latter, of involving the student in the problem solving of analysis and the rule learning of synthesis. Is it any wonder that mathematicians hailed Euclid as a mastermind and for centuries reviled any who criticized lapses in his work? Even the simplest part of the axiomatic system requires a thorough analysis of all these aspects of the content by the teacher if meaningful plans are to result.

But why should students be guided through the reconstruction of proofs which could be given to them as models for study? If the teacher and the class explore conjectures together (Textbooks should be closed! Students can seem like geometry wizards otherwise.), then the process has a purpose analogous to the presentation of type word problems. In the analysis stage, the teacher can help the class with key questions which they should ask when they try novel proofs. Guidelines and ways to test conjectures can be suggested.

Guideline 1: *If you think an auxiliary segment is needed, try to use points already in the diagram and draw as few segments as possible.*

Hans thinks: *"I need a central angle. I'll use radius* $\overline{OC}$ *and point A."*

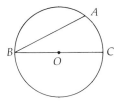

Guideline 2: *If you think a relationship exists, put in numbers and check it out.*

Royann thinks: *"I think m $\angle$ DCA = 2m $\angle$ ACB. I'll try 20 for m $\angle$ A and 70 for m $\angle$ B, then m $\angle$ ACB = ..."*

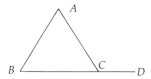

The teacher needs to characterize some hints as rules of thumb to be used with every proof of a certain type while others should be classified as more general cues to be tried when the proof cannot immediately be classified as to type. (Cues of the latter kind are sometimes called *heuristics*. We will be seeing more of them in the section on novel problem solving.) It's obvious that the teacher who shortcuts the analysis stage by forcing or given certain answers is wasting everybody's time.

Occasionally novice teachers erroneously assume that the only thing that really counts is a two-column proof. In their enthusiasm, they omit the analysis stage entirely and immediately ask the following sequence of questions: "What is step 1? Why? What is step 2? Why? What is step 3?" and so on. The students soon tire of the guessing game and begin reading the answers from their texts. Synthesis without analysis is *nonsense* mathematics; but analysis can live without routine subsequent synthesis. When the reconstruction of the synthetic argument will provide needed practice in writing proofs or when it includes a novel element, then the *students* (Notice. We said "the students.") should reconstruct the argument with gradually decreasing teacher assistance as the course proceeds. As always, the teacher must let the students in on the objectives of this part of the lesson, help them identify novel aspects of the argument, and clarify the use to be made of both the product of the proof and the process of the proof.

Planning for Problem Solving

Problem solving of the nontypical variety is really where it's at if you believe, as we do, that the ideal of education is to teach the students to think for themselves. In the section on theorems in geometry, we designated the analysis stage of proof constructing as problem solving if the analysis is the responsibility of the students rather than of the teacher. Similarly, in an algebra class, it would be a natural next step to move from the solution of type problems to the solution of diverse novel problems. However, in both cases, that shift of responsibility in student thinking has to be gradually fostered by the teacher. Thus, a teacher who wants to teach future problem-solving lessons should incorporate into rule lessons such behaviors as descriptions of the way the teacher had fumbled toward a solution, questions as to the advantages and disadvantages of various modes of attack, and hints as to ways to get feedback without recourse to the teacher.

Mr. Potter had been gradually moving his geometry class toward problem solving. He included all the above behaviors and, in addition, encouraged students to work in groups and to defend their work to one another. On the day of the first problem-solving lesson, here's what he told the class:

> We're going to try a little different approach today. You're going to behave like working geometers—conjecturing, deducing, and assessing different approaches. I've prepared a handout containing five geometry problems—all of which can be proved in more than one way. I've divided you into groups and assigned two of the five exercises to each group. When I give the signal, the groups will move to their assigned locations and the group leader will take over. The leader must see that all directions are followed and should try to give all students in the group an opportunity to participate. The group should find as many different proofs for each of the two exercises as possible, should work each out and then discuss the advantages and disadvantages of each. Is one proof more rigorous, easier to follow. . . ? Keep a record of your conclusions so that you can report on them when I ask you to stop the group work. Questions? [Two or three questions.] When I turn on the overhead projector, you can see the location where you are to sit and the exercises assigned to your group. The leader's name is starred [overhead projector turned on]. You may move into groups now.

Mr. Potter's success today will depend partly on his prior developmental work and partly on the nature of those geometry exercises. They must be above level III in the cognitive taxonomy with enough of the known so that the students can begin hypothesizing and with the right mix of novelty to insure the blocking of "pat" solutions. Mr. Potter's quoted preamble would constitute his introduction to the activity. You should be able to identify the next instructional strategy he would list in his plan and the associated feedback strategies. That's right. Next would come small-group discussions with Mr. Potter touring, listening for conjectures, trying to get students to help one another, praising an insightful idea, and so on. Take a few minutes now and sketch out the rest of the plan with appropriate time estimates. Don't forget to write the instructional objectives. After you have produced an outline, check it for correspondences with the instructional strategy pattern presented below.

Instructional Strategy Pattern for Problem Solving

Present problem; may question students to elicit alternative approaches and will emphasize the desirability of a variety of search strategies; arrange for individual work or small-group discussion (in the collection of data, the analysis of data, the making and testing of conjectures); reassemble class; ask students to weigh the advantages and disadvantages of the proposals (including processes) which resulted from the group discussions or individual work.

If your plan failed to match the pattern above, you should check out the discrepancies with other classmates. If you're still out in left field, see your instructor as soon as possible. Inability to repair weaknesses in planning at this stage can be fatal later.

We've already emphasized the need for a gradual move toward instruction in problem solving. These prior teacher-led explanations of "thinking through a problem" need to be continued and expanded during problem-solving lessons. Reread the final step in the strategy pattern. That step includes class consideration of the *processes* used by small groups and individuals during work on the problem. It is analogous to the summarizing step prior to practice in lessons on type word problems, but the difference resides in the focus of the summary. In type problem work, the teacher's aim is to clarify the structure that is reflected in all similar problems and, therefore, the typical steps to solution. However, in novel problem-solving lessons, the aim of this summarizing step is to clarify the tactic(s) used in this case and to consider the pros and cons of a more general use for these specific processes. Over time, the teacher should help the class attain a list of such tactics that *may* illuminate a problem but are not guaranteed to produce a solution. Such tactics are labeled *heuristic strategies* or simply *heuristics*. *Heuristic* is a word derived from the Greek *heuriskein*, which translates roughly "to find out." Of course, sometimes you *find out* that the particular heuristic you tried is of no help. We've compiled a list of potentially useful heuristics with some examples from the literature and our classroom experience. We recommend that you begin collecting additional examples of these heuristics as well as listing other heuristics you may identify.

Heuristics

1. Use analogy or contrast. ("This series is behaving *like* the coefficients of Pascal's triangle. Let's try....")

2. Change the scale (or frame of reference). (Shrink or stretch a shape.)

3. Use successive approximation; exploit errors. (Guess, try, correct, guess again.)

4. Exploit symmetry. (Use a folded paper disc and the radial symmetry of the circle to prove that the perpendicular bisector of any chord passes through the center of a circle.)

Match *A* to *B*

5. Ask "What if such and such were *not* true?"

6. Change the exercise to a problem you can do. (Polya [1962] describes a use of this heuristic when demonstrating the solution of the problem: Given

any triangle ABC, inscribe a square in $\triangle ABC$ so that one side of the square lies on $\overline{BC}$. Polya tells us to inscribe a small square, one vertex on $\overline{BA}$, and one side on $\overline{BC}$. He claims that the original problem has been solved. [Hint: Draw a line through P and parallel to $\overline{AC}$.])

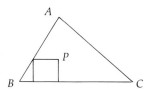

7. Look at the extremes. (In a computational situation, let the numbers get very tiny and very large.)

8. Focus attention on one part of the problem at a time.

9. Ask another student.

It's clear that some of these heuristics may not be useful tactics in the case of certain problems and students will need to be reminded that none of them guarantees success. A "guaranteed successful" stamp on a tactic is a sign that the tactic is really an algorithm and the problem, a type problem.

Are novel problems difficult to find? Not at all. There are numerous sources of novel problems, including professional journals, some curriculum projects, and texts devoted solely to them. But you do have to know what to look for. Here is a sampling of a few different kinds of novel problems.

1. Find the smallest number of persons a boat may carry so that n married couples may cross a river in such a way that no woman ever remains in the company of any man unless her husband is present. Also find the least number of passages needed from one bank to the other. Assume that the boat can be rowed by one person only. (A puzzle-problem attributed to E. Lucas, 1883.)

2. How can one get the greatest number of cars into a parking lot, permitting access and egress? If one needs to put in still more cars, how can we do it so as to minimize shuffling? (A real-world problem from *Goals for the Correlation of Elementary Science and Mathematics, 1969.*)

3. A teacher prepared a quiz on a stencil, but when she ran the stencil, part of one question did not show up on the copies. Complete the question in as many ways as possible and answer each such question:

Q5 Let a, b and c represent the sides of a triangle and let $a + b = 7$, $a + c = 9$, and $b + c = 8$. Determine _____ (a reconstruction of an otherwise typical problem).

4. In the product 9 · HATBOX = 4 · BOXHAT, find the 6-digit numbers
    HATBOX and BOXHAT. (A cryptoarithmetic example from *The Mathematics Student Journal*, November 1964).

Novel problems, then, may belong to recreational mathematics (puzzles, riddles), advanced mathematics (more complex problems on the topic being studied), or applied mathematics (problems from the sciences or social sciences). There is a time when each of these kinds of problems serves an instructional purpose. We encourage you to begin collecting such problems for appropriate folders of your resource file. Be sure to organize *within* folders so that novel problems are separated from relevant type problems. You will find selected references to use as a starting point at the end of this chapter and additional references in Chapters 9, 10, and 11.

Planning for Psychomotor Learning

The final type of plans relevant to junior/senior-high-school teaching is that aimed at instruction in psychomotor skills—the kind of learning which was identified in Chapter 6 as *chaining*. These motor chains are also sometimes called *manipulative skills*. Be alert, future mathematics teachers, to a second meaning assigned to the term *manipulative skills* by some mathematics textbook writers. Often they equate calculation abilities with manipulative skills. We will *never* use the term *manipulation* in that sense.

Refer back to Section 5.3 for a sample of the kinds of psychomotor objectives which might be included in a mathematics teacher's plan. This is also a good time to refresh your memory as to the possible errors made by students drawing circles with compasses for the first time. Now assume that the teacher of a junior-high class wants to design instructional strategies on the use of such compasses so as to avoid most of the gross student errors listed in Chapter 5. The teacher makes the following decisions:

1. All students must have identical sturdy compasses, sharpened pencils *not* in the compasses, pieces of cardboard, and looseleaf notebook paper.

2. The teacher will use a pair of demonstration compasses as much like the students' as possible rather than the large board compasses.

3. A demonstration will be performed on the lighted overhead stage, rather than at the desk or even on a sheet taped to the board.

Why are each of the above decisions likely to promote the learning of these particular kinds of psychomotor skills? If either decision 1 or 2 isn't clear, then you've forgotten scenes from your own classroom days. The pencils jammed way up in the holder, the holes in desks, the cheap compasses that would not fix at a given radius, the board compasses whose use differed in major respects from the normal instrument (that is, if you could even manage to draw a circle with it)—all of these deterrents to learning need to be avoided. The value of decision 3 may be unclear if you've never experienced the use of the overhead projector's stage to *project* enlarged images of objects held above or on the stage. There is one problem with this decision. Can you guess what it is? Think of the images of a pencil lying on the stage, then a pencil held so it is vertical to the stage, and, finally, a pencil tilted at an acute angle to the stage. The image of that vertical pencil is a dot, if you can get your hand out of the way. So the teacher will have to practice with the compasses until a reasonably good image is obtained. An eraser on the tip of the sharp point will keep the stage from being scratched.

With the above decisions in mind and your knowledge about the conditions for learning motor chains (If you need a reminder, check Section 6.3 as needed), you are ready to try your own hand at a draft of a plan on the topic of compass use. Remember, the students may never have seen compasses before this day. In any event, you should assume that they have not used them and may not even know for what they are used. When you have completed your rough draft, compare your instructional strategies with those outlined below.

Instructional Strategy Pattern for Psychomotor Skills

Demonstrate (show how) one step at a time; have students perform each step immediately after the teacher shows each; watch students perform each step and get and give feedback; then show several steps put together; have students do several steps put together; have students practice to get closer approximations of correct performance; then have them practice to increase speed.

How did your draft compare? Did you identify steps such as the way to physically find the correct location for the pencil, the proper placement of the hands on the instrument, the positioning of the other hand on the paper, the proper starting position of the hand, the movement of the hand on the instrument during the drawing? The students would have to be told that the teacher had already performed the first step for them—the sharpening of pencils. Have you figured out why? Just envision 30 seventh-graders up at the single pencil sharpener. Chaos! (Since the students will be expected to sharpen their own pencils on other days, the teacher had better devise a routine to avoid a similar mob scene.) The feedback strategies and the use of feedback elicited can make or break this lesson. One of our student teachers was sufficiently concerned about feedback to construct an audiotape to correspond to her planned demonstration. She felt that this freed her to observe student progress more completely. An experienced teacher made a home movie of the complete demonstration which he was able to show in a partially darkened room while he gave directions and toured systematically. Segments of the movie were repeated whenever he observed a common error.

The use of the compasses to generate complex designs may be motivation for some students. Others may get "turned on" by an introduction to the draftsman's trade. What did you employ as a tieback to previous lessons and how did you elect to present a rationale for learning this skill? Pictures, architect's plans, pop art charts, and excerpts from the history of mathematics are possible instructional aids you might use to achieve both a tieback and a look ahead. You should now understand the need for yet *another* kind of material to add to your resource file.

Putting It All Together

Just as single plans often contain a mix of concepts and rules, so psychomotor skills are frequently taught in conjunction with some rule. In other words, two or more of the instructional strategy patterns outlined earlier will typically be needed in a single lesson. The teacher must analyze the content to be taught and choose strategy patterns to match each content category. Our students have found it helpful to have all the basic strategy patterns in one place, so we have combined all into one table (Table 7.1) for easy reference.

Compare the strategy patterns with the kinds of objectives described in Chapter 5. Can you identify instructional strategies that could be matched to objectives at varied levels of each of the domains studied? If you're hesitating, you probably noticed the lack of specific reference to the affective domain. Is this domain being ignored? Reread the definitions of levels I and II of the affective domain and take a second look at the two sample plans. Even if all the planner was after was the promotion of level I (compliance), the demonstrations, laboratories, and references to the everyday world would be appropriate. Voluntary participation (level II) should ring a bell. Check the places on the plans where volunteers are specifically noted or where at least some students have a choice as to what they will work on next. Level III is not clearly identified on either plan, but then objectives at this level require attention over time and might not show up in isolated plans. In fact, affective objectives at all three levels are more appropriately written for a unit of work, rather than for a single day's lesson. Moreover, although Gagné did not specifically refer to the affective domain in his Hierarchy of Learning Types, he repeatedly emphasizes this domain in his work on the Events of Instruction. For more on the interconnections between attitudinal objectives and the lesson plan, refer to that part of Section 7.5 that focuses on *gaining and controlling attention*.

One final piece of advice before we leave the topic of daily lesson plans. Write post-mortem comments on each *immediately* after implementing them in an actual classroom situation. Revise time estimates, make notes on what worked particularly well, and earmark any section in need of repair while the experience is fresh in your mind. Our student teachers who have followed this practice report that it pays big dividends when they teach the same lessons to

Table 7.1
Instructional strategy patterns for mathematics content

Kinds of content outcomes	Examples	Instructional strategy patterns
Vocabulary	Polygon, prime, limit, i, Δ	T shows object (or exemplifies idea); then pronounces name and spells on board; students do likewise, repeat (practice), give name; T elicits statement of idea or description of object; T has Ss use in context.
Concepts	Number, prime, greater than, point, locus	T shows differentiated examples and non-examples in sequence; gets essential and nonessential characteristics identified; elicits operational definition; models by analogy; elicits generalization of the concept to a variety of specific instances not previously used.
Rules/principles	Square root algorithm, "Work" problems, Congruence proof writing, T.P. = $-$ b/2a	T gets Ss to recall/review prerequisites; indicates nature of expected terminal performance; cues (via questions, lab work, applications) Ss to find the pattern by chaining concepts; gets rule stated (by Ss if possible); has Ss demonstrate several instances of the rule in a variety of situations. *Fix and maintain skills by spaced and varied drill.
Novel problem solving	Analysis of novel proofs, puzzles, coding, or transportation problems	T presents problem; may Q/A Ss to elicit alternative approaches to solution; T emphasizes desirability of a variety of search strategies; individual work or small group discussions (data collecting, data analysis, making and testing conjectures); T asks Ss to weigh advantages and disadvantages of proposals (and processes) which arose from group discussions and / or individual work.
Psychomotor learning (manipulative skills)	Use of compasses, protractor, geoboard, clinometer, slide rule	T demonstrates (shows how) one step at a time; has Ss do each step immediately after T shows each; T watches Ss perform each step and gets and gives feedback; then T shows several steps put together; has Ss do several steps put together, practice to get closer approximations of correct performance, then practice to increase speed.

other classes of students. They have advised us to urge you to form this habit. Consider yourself urged.

7.3 LONG-RANGE PLANNING

As we have already seen, day-to-day planning is vital to success. But isn't long-range planning also a very important part of the teacher's task? It most certainly is if the daily lessons are to fit together into a meaningful whole. In fact, logic would seem to dictate blocking out plans for the entire year-long course first, next developing the more specific treatments to be given each of the major sequential subcomponents to fit within that overall frame of reference, and then finally working out the details of each daily lesson. Yet we will be directing your attention to these levels of planning in the reverse sequence. Why? Our experience has been that novices find long-range planning both a hopeless and meaningless task until they have first acquired some experience in designing and implementing daily lessons—another example of how learning most often follows psychological rather than logical principles.

Unit planning After a couple of weeks of experience with planning and implementing daily lessons, you should be ready to try your hand at the next step—blocking out plans for the next unit. By a *unit* we mean a *major topic that will occupy approximately two to four weeks of instructional time.* One full-year course is typically composed of a sequence of eight to twelve such units.

Recall the Resource File modular assignment we asked you to begin in Chapter 4. At that time we directed you to select one unit as a point of departure for building a file of teaching ideas that would serve as a base for future plans. Obviously, you need to fill other folders (as directed in that assignment) prior to starting plans for additional units because these materials will constitute the idea base for future instructional strategies. Also recall the Sequence of Content assignment given in Chapter 6 which provided experience in both (1) identifying major objectives expected of all students and (2) analyzing the interdependent relationships among parts of a major topic. Since there is not time to teach all possible aspects of a unit topic, selectivity is essential. In addition, prerequisite concepts must be introduced prior to the point at which related principles/rules and their applications are taught.

But how does a beginner get started with the actual task of unit planning? If a course syllabus or outline is available, we suggest you use that source as a point of departure for the initial development of a topic outline. If no existing syllabus can be located, consult relevant sections of two contemporary books which are widely used as student texts for a similar course. If you were able to obtain a syllabus to construct a tentative outline, modify that outline as needed after studying the treatment of the same topics in relevant student texts. Be especially alert to rethinking matters such as breadth and depth of treatment and potential sequencing schemes. We advocate treating the final topic outline as a guide and source of stimulation for your own thinking rather than as a prescription to be followed blindly. Experience convinces us that the final outline must make maximum sense to you, the teacher, if it is to serve as a frame of reference for daily lessons which will make sense to your students.

The second stage in moving from a topic outline to a unit plan involves allocating the available time and listing some initial planning ideas. You will need to block out projected time allotments for parts of the outline. Next, you should begin to list (by descriptive title) potentially useful labs, demos, films, field trips, homework assignments, bulletin board ideas, and the like from your resource file. If the "cupboard is bare" for some topics, you will have to dig into the sources suggested in Chapters 9, 10, and 11 to fill these voids. Otherwise, the unit plan will *not* serve its intended purpose of providing an adequate base for more detailed day-to-day planning. Projecting time estimates is always a problem for the novice, and there is no use in pretending that there is any substitute for the experience of having taught the unit several times to varying groups of students. However, the problem must be faced squarely by every novice, and we can make suggestions which will alleviate the task to some extent. Attempt to identify topics which are either extensions or modifications of ideas previously presented. These, along with content ideas familiar to students, as a result of extensive experience outside of school, can be assigned shorter instructional time than totally new work. Even longer time periods should be planned for new ideas which will provide a base for more sophisticated notions to occur in future units.

Arrange the unit plan in three columns as shown in the partially completed sample of a unit plan on Systems of Equations for Ninth-Grade Algebra (Fig. 7.7). Column 1 lists topics in the sequence determined by the teacher while column 2 contains the estimated number of instructional periods to be devoted to each topic. Column 3 contains the gold-mine entries—selected instructional resources from the teacher's resource file described in a phrase as well as by a coded entry. The phrase provides immediate information for the reader and the coded entry (for example, A-BB #1 for the first [Ath] folder of the unit, the first bulletin board item) shortens the retrieval time as the teacher moves from the unit plan to daily plans. You should provide your cooperating teacher with copies of (1) the unit plan and (2) all the related resources identified in the learning activities column. Schedule a conference after your cooperating teacher has had time to review these materials but far enough in advance of the time when you must begin developing the first few daily plans of the unit. At least several days of lead time are desirable. Heed the advice given and ask questions to help in thinking through various possible approaches. Examine further learning activities which may be suggested, talk out potential problems which may arise, and discuss details which might be added, or deleted, in case time requirements deviated from those projected. Having done all this, you are bound to feel much more confident in developing detailed day-to-day plans, and smooth transitions within the unit will be more easily achieved. As in the case of daily lesson plans, notes to yourself written on the unit plan as implementation proceeds will prove valuable in future years.

Unit 5: SYSTEMS OF EQUATIONS
NINTH-GRADE ALGEBRA

Topics	Time estimates in days	Learning activities
A. Algebraic solution of two first-degree equations		Bulletin Board—"Peanuts" problems (A-BB #1)
		Demo—box with red & blue marbles—problem situation (A-D #2)
1. Addition/subtraction	3	
2. Substitution	2	Demo—Dienes blocks in illus. +,− (A-D #31)
		Problem cards HW from Sci. & Bell's *Everyday World* (A-P #3-10)
B. Graphic solution of two first-degree equations and inequalities		Bulletin board of Howard Johnson problem (B-BB #2)
1. Equalities	2	Lab with springs and weights (B-L #2)
		Demo with acetates (B-OH #1-5)
2. Inequalities	2	Problems from newspapers (B-C #3)
C. Solution of two second or one second- and one first-degree equation		Filmstrip on fishery work (C-F #3)
1. _____	____	
2. _____	____	

Fig. 7.7   Part of a unit plan.

Course planning. Speaking of years, 365 days does seem like a lot of time. But is it, in terms of how many full periods of 40 to 50 minutes are actually available for instruction? State regulations, school calendars, and administrators frequently refer to a 180-day school year. (This figure takes into account weekends, various holidays, and summer vacation periods when school is not in session.) Does this mean one should plan a course based on the assumption of 180 full periods available for instruction in new content? Not quite. Further time reductions must be planned to account for the nasty realities categorized below:

1. There will be times when *all* students entirely miss one or several of your class sessions. The most obvious include the half-days of school that

precede Thanksgiving, Christmas, and Easter (these *were* counted as full days in arriving at the 180-day figure). The calendar may or may not have allowed for a few days off due to snow, power failure, or an energy emergency. "All sophomores will report to the auditorium at 9 a.m. this Thursday to take the SAT examinations," blares the P.A., and there goes another day of geometry instruction. Friday's daily bulletin carries a reminder that all seventh-grade students are to remain in their home-rooms on Monday for the administration of the Iowa Achievement Tests and thus another day of seventh-grade mathematics class has evaporated. Then, too, we must deduct for any midyear and final exam days which were counted among the 180 official school days. Neither can we forget any half-day faculty meetings scheduled during school hours. Indeed, some schools have been known to cancel all classes during the last day prior to final exam week so students can clean out their lockers.

2.   Sometimes you will find yourself meeting only about 50 percent of the class even though none of the previously mentioned events has occurred. Half are out with the flu on Tuesday and the other half are stricken on Wednesday. Since you are a mere mortal, chance favors you being sufficiently contaminated to require bed rest on Thursday (plus Friday if your resistance is a bit low). Just when all are healthy and present again, the daily bulletin brings the glad tidings that all varsity and junior varsity basketball players and cheerleaders will leave for the playoffs at the start of eighth period tomorrow and that all involved in the senior play will be excused from the morning classes to participate in the dress rehearsal. Similarly, don't expect very many able-bodied males to report for classes on either the first day of deer season or the first day of trout season if you are teaching in a rural or rural/suburban school. You should also count on one or more days of heavy absences due to all-day off-campus field trips sponsored by other departments.

3.   A variety of events cause students to miss *parts* of instructional periods. Among these are assembly programs, visits by college recruiters, schedules for yearbook photographs, and fire drills.

4.   Your own schedule of tests/quizzes and related activities will also deduct from time available for instruction in new content. It is reasonable to assume 8 to 10 full-period tests and 25 to 35 quizzes (10 to 20 minutes each) will be given during the year. These are essential encroachments on instructional time, but is it really defensible to add fuel to this fire by routinely coupling a full-period review and a half-period post-mortem to each of the big tests? Review the recommendations in Chapter 2 aimed at minimizing these problems.

So much for the myth of the 180-day school year. We have found 140 periods a much more realistic target when planning a year-long course. Eight units would thus *average* about 17 full class periods. Obviously some could take longer, but you would then have to compensate with lesser time for others in the long-range plans.

Is a course plan, then, merely a list of unit plans designed to fit within a 140-day school year? There are basic flaws in that approach even when each individual unit has been thoughtfully designed. Yes, sequencing may be out of whack, interconnections may be ignored, and mindless redundancy may take the place of meaningful spiraling. There's only one way to avoid these disasters and that way begins with a thorough study of all available course materials. We recommend that you commit to paper the five to ten *major* instructional objectives of the course. If your list exceeds ten statements, you may be including some less important objectives or ignoring the dependency of one objective on another. When you're satisfied that your list emphasizes the core of the course (Did you include processes as well as products of mathematics?), sketch in a simple learning hierarchy among just those five to ten major objectives. Don't worry if interconnections are rare. You may have identified several distinct and relatively independent terminal objectives. These statements, then, would become the uppermost objectives in a vast learning hierarchy. Next, match the topics, into which you had previously subdivided the course, with the objectives. If interconnections are missing here, go back to the drawing board. Finally, estimate the number of days to be assigned to each topic on the basis of (1) the relative contribution of each unit to the objectives, (2) the location of each unit in the sequence, and (3) the presumed range of student intellectual development. One student teacher, Ms. Aronowitz, attempted to develop this kind of course outline for tenth-grade geometry. Her planning culminated in the schema shown in Fig. 7.8, where the topics are matched to the five objectives listed below.

1. Construct a synthetic argument of a proof, given hypotheses from the major topic areas of plane geometry.

2. State generalizations in "If, then" form, given experiences reflecting a pattern.

3. Evaluate an argument (geometric or nongeometric) as to its match to deductive logic, its precision, and rigor.

4. Calculate measures (linear, angular, circular, area, volume) given data and relationships, diagrammatic or written, from which to obtain further data and/or rules.

5. Analyze the relationships between real-world phenomena and corresponding mathematical models from geometry.

This schema, or thought model, was designed to illustrate the major sequences among unit topics and the gradual development of the learnings

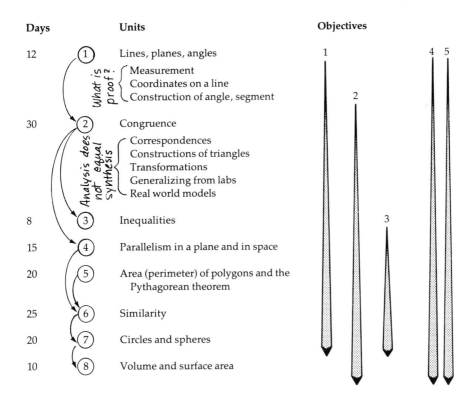

Days   Units   Objectives

12   (1)   Lines, planes, angles   1   4   5

              ⎧ Measurement
  What is   ⎨ Coordinates on a line
  proof?   ⎩ Construction of angle, segment   2

30   (2)   Congruence

              ⎧ Correspondences
  Analysis   ⎪ Constructions of triangles
  does not   ⎨ Transformations
  equal   ⎪ Generalizing from labs
  synthesis   ⎩ Real world models

8   (3)   Inequalities   3

15   (4)   Parallelism in a plane and in space

20   (5)   Area (perimeter) of polygons and the
                   Pythagorean theorem

25   (6)   Similarity

20   (7)   Circles and spheres

10   (8)   Volume and surface area

Fig. 7.8   A course planning schema for tenth-grade geometry.

needed to attain each objective over time. Eight unit topics, each with a distinctive contribution to the attainment of the major objectives, were arranged in the proposed teaching sequence. Where is *construction* or *coordinate geometry*? They aren't being ignored, but Ms. Aronowitz saw each of these topics as threads, or strands, which help to integrate geometry. Notice the subdivisions under "Lines, planes, angles" and "Congruence." Constructions are listed under both unit topics, and coordinate geometry under unit topic 1. Subdivisions yet to be added under the remaining units will include these and other strands. Such strands have great unifying potential *if* the teacher builds on the earlier work and helps the students sense the ever-widening scope of these themes. The first two units were further synthesized by the addition of topic sentences. The beginning of proof considerations—for example, distinctions between assumption and theorem, a geometric proof and a rationalization based on many experiences—will be the focus of unit 1, while serious work on the writing of synthetic arguments will be left for unit 2. But that's also depicted by the growing width *and* length of the arrow associated with objective 1. Why does that arrow stop short of topic 8? Ms. Aronowitz is following well-established practice in eliminating this topic from those in which students will have to be able to write proofs. That doesn't mean that they won't develop some proofs together as a class. Each of the other arrows representing objectives identified the units in which that objective will be given explicit attention. See if you can follow Ms. Aronowitz's thinking here. The curved arrows connecting unit topics are other indicators of sequence. However, these represent the axiomatic links inherent in the connected topics rather than strands or overarching themes. "Parallelism" and "Area of polygons" both contain essential prerequisites to the proofs of the basic similarity theorems in the sequence being followed by Ms. Aronowitz (the SMSG Geometry texts, 1971, are one source of this kind of axiomatic chain) so two curved arrows lead from unit topics 4 and 5 to topic 6. Each of the other curved links has a like reason for existing and, of course, multiple other links are possible. But this student teacher envisioned these as the major hierarchical connections which she should emphasize. A thorough grasp of the subject matter of geometry is clearly essential.

Now turn to the time allotments for each unit. Why were some planned for as little as 8 days while others were given as much as 30 days? Ms. Aronowitz identified "congruence" as the initial key to proof—the time to involve students in many concrete experiences, to build on the concepts and principles introduced in unit 1, and to develop the writing of simple synthetic arguments. Her cooperating teacher agreed and advised her to hold back on the writing of such arguments until this unit, despite the suggestions in various texts to include them in unit 1. On the other hand, "inequalities" would be built on prerequisites from other courses. The novel aspect would be the introduction of the indirect method of proof which would then become a strand throughout successive units. Other time allotments were arrived at by this kind of balancing of sequence, prerequisites, extent of novelty, and nature of objectives. There is nothing sacrosanct about these figures. They represent the best thinking of this student teacher after having discussed time estimates with her cooperating teacher. And, as always, feedback during instruction may result in some differences between the actual and estimated times. Such information should be recorded and appropriate modifications made for future years. But any change must continue to reflect the conceptual framework designed by the teacher.

Now reflect on your own reactions to this particular overall course-planning guide. Does it provide a frame of reference for planning the sequence of the units it subsumes? Does it represent an accurate view of contemporary geometry and yet have high potential relevance for tenth-grade students? Are interrelationships among ideas within and across units clarified? We answered yes to all these questions after studying the fully completed schema. Further, we judged it to have excellent potential for its intended purpose (after discussing it thoroughly with the young woman who created it). Does this mean that this specific scheme is *the correct one* to use for a tenth-grade geometry course? No, we would make no such claim—only that it ought to be a highly usable one for the person who created the schema and could implement it effectively. Personally, we would use the same major objectives but subdivide the units somewhat differently and use a variation of the proposed sequence. These changes wouldn't necessarily make our schema superior for *your* use, though it should for *our* use. Hopefully you would design a plan that might be somewhat different, but equally fruitful, for your own use.

7.4 RETENTION, TRANSFER, AND PRACTICE

We have already seen that humans are ill-equipped to behave like walking computers. People are just not capable of storing huge numbers of isolated information bits to be retrieved instantly upon demand. It is also true that much of what our students will need to know in their adult life 10 to 50 years hence is not now in the storehouse of human knowledge. Thus, formal schooling must educate the young to *transfer* learning—to develop both the *ability to perform new tasks at about the same level of difficulty as previously learned ones (lateral transfer) and also perform those of more complex difficulty by using the base of past learnings (vertical transfer)*. Students who have learned that corresponding altitudes, medians, and angle bisectors of similar triangles are proportional to any pair of corresponding sides should give evidence of lateral transfer by making like predictions for the ratios of corresponding diagonals, altitudes, and the like in the case of similar parallelograms. Further, we look for evidence of vertical transfer from students who, having mastered the algebraic solution of a system of two first-degree equations in two variables, later encounter problems involving three first-degree equations in three variables. How can we as teachers ensure that the maximum amount of transfer takes place? We can make certain that those factors known to promote transfer are accounted for in planning and implementing lessons.

Transfer, going beyond present learning, obviously depends upon a student's ability to *remember* or *recall previous learning (retention)*. There is firm evidence to support the assertion that the most important factor in promoting retention is the degree of meaningfulness of the initial learning. Here Gagné's task analysis technique and work on the conditions of learning, as well as Ausubel's conception of advance organizers to promote meaningful verbal learning, provide specific guidance for instruction. Review Sections 6.2, 6.3, and 6.4 at this point in case you run the risk of teaching lessons whose learning outcomes are retained at the levels illustrated in the classic "forgetting" curve (see Fig. 7.9).

Frightening, isn't it? Applying the findings of Gagné and Ausubel to the design of instructional strategies, as described in Section 7.2, will certainly help to avoid the situation depicted in Fig. 7.9. There are also other measures we can take to promote retention. The hierarchical structure of the subject matter makes it possible to teach retrieval strategies directly. Students who have learned the nature of the basic sine and cosine curves and the relation-

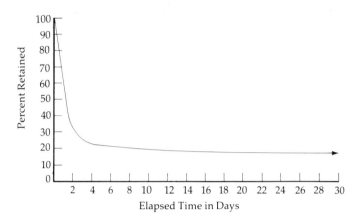

Fig. 7.9   Retention of nonsense learning.

ships of amplitude and period connecting graph to equation can be taught the value of recalling the picture of the basic sine or cosine curve first whenever questions concerning the amplitude, frequency, or value of $f(x)$ for (a sin bx), (c cos dx), or even (a sin bx ± c cos dx) are encountered. A few experiences are usually sufficient to convince students that one can employ the pictoral model to reconstruct curves such as $f(x) = 2$ sin $x/3$ or to retrieve the value of $f(\pi/4)$ where $f(x) = (½)$cos $2x$. Demonstrating the degree of utility of an understanding of these basic trigonometric functions also helps establish both the intent to remember and the students' confidence that they will be able to remember—two additional factors which promote retention. Skillful teachers have learned to stress at every opportunity the relatively small number of thought models which need to be mastered in order to give one command over the many principles and in turn, the even more numerous concepts which constitute the product aspect of mathematics. Of course, such instruction must be coupled with strategies aimed at getting, giving, and using feedback if it is to produce the desired effect.

What is the role of practice (drill) in promoting the retention so essential to transfer? Surely the time-honored tradition of drilling students on newly learned content must contribute heavily to later recall? Not necessarily. In fact, there is good evidence that practice *per se* has little effect on retention! The desired positive effects of practice can be enhanced by attending to four factors.

First, drill sessions should be kept short and spaced out over time. Massed sessions in which students "plug in" numbers for a, b, and c in a dozen or more Pythagorean theorem type problems are neither efficient nor effective. The equivalent amount of time divided into five- or ten-minute practice sessions distributed throughout a two-week period would yield higher returns for the time invested.

Second, practice should be provided on both the parts and the whole in the case of complex tasks. This has been found to be true for both cognitive and psychomotor learning. For example, it is important to devote some portions of practice to each of the separate subtasks of calculating the log of a product, interpolating to obtain a mantissa which can't be directly read from the log table, calculating the log of a root where the base is smaller than 1 (such as $\log \sqrt[3]{.045}$), obtaining an antilog, and all the other separate applications of logs. But it is also necessary to allow sufficient time to practice performing all of these parts within the total problem complex. Similarly, students need practice in manipulating the sliding scale of the slide rule and the plastic "finder" separately and in performing each manipulation as part of the total integrated technique. One caution! In the case of motor skills, the sequence of part-whole practice needs particular attention. Motor skill learning is enhanced when the whole skill is practiced first, followed by practice on each subskill and, finally, by practice on the subskills as part of an integrated whole (known as the whole→part→whole approach). Moreover, even in the cognitive domain it has been found effective to begin with practice on a simplified whole, then practice on gradually more complex applications of the parts, and finally practice on the integrated whole. In the example given, the teacher might begin by using logarithms to the base two and contriving examples in which all numbers were powers of two. The student is able to calculate $(128 \times 32^3)/64$ very quickly from a table and appreciates the calculation shortcut. After sufficient practice on the total process, the students already have ideational scaffolding on which to hang the learning associated with the parts in the usual logarithm to the base ten problems. Sound like an advance organizer? It is.

Third, practice is more effective when it is structured within a framework that the student would consider meaningful—such as real-world applications. For instance, students who have demonstrated success in finding the ratio from two pieces of data can be given a ten-speed bicycle and asked to find the ratio of the number of revolutions of pedals to the number of revolutions of the rear wheel for each shift position. It is clear that while the initial instruction may emphasize the solution of type problems (level II—cognitive taxonomy) that retention depends partly on practice which places the student in a novel situation (level III—cognitive taxonomy).

Last, but perhaps so logically obvious that we tend to forget it, practice must *actively* involve the learners. This does not imply that students must be engaged in some sort of obvious and vigorous physical activity. It does mean that repetitious and boring drill which encourages students to take a mental vacation must be avoided. How can anyone profit from a cognitive learning activity if the mind is not attending to the task at hand and if there is little effort or intent to learn? Because of past unpleasant experiences, just the use of the label *drill* is enough to turn off many students. Thus, we have learned not to use the word with junior- and senior-high-school students (ditto for *home-*

work). Remember the example of using the ten-speed bicycle to extend practice so as to include applications in context? Doesn't it at the same time include drill on the computation of ratio and on the way to obtain like quantities for substitution in a ratio expression? It certainly does, and students do become mentally active in working out these solutions. Section 7.5 will provide additional cues on ways to promote such mental activity.

If we have been able to design instruction to maximize retention, the task of promoting transfer is 90 percent accomplished! What constitutes the final 10 percent? Double-check that your plans incorporate a way to share with students the *expectation of transfer* to novel situations, as well as learning experiences commensurate with that goal. Such learning experiences should include ways to help students *verbalize the methods* of problem solution, in the case of the cognitive domain, or *comprehend the theory* which underlies each motor skill to be learned, in the case of the psychomotor domain. Yes, understanding the reasons for performing a manipulation in a certain way does promote transfer of that skill to other relevant situations—another reminder that the three domains of human learning are not like watertight compartments.

Above all, *never* think or behave as if transfer is automatic. Prior to the beginning of the twentieth century the human brain was commonly considered analogous to a muscle. One could certainly develop the biceps by repeated lifting of a blacksmith's hammer, and this increased the capability of that set of muscles to move any other heavy object. Thus, it seemed logical that exercising the memory portion of the brain by memorizing lengthy Latin poems ought to facilitate the memorization of mathematical formulas. However, countless research studies performed from 1890 through the present day continue to refute this notion. Muscle tissue does *not* equal nerve tissue! Analyzing the standard proofs in geometry does not automatically help students cope with data analysis in physics or vice versa. Transfer must be specifically taught if it is to be caught.

7.5 EVENTS OF INSTRUCTION REVISITED

In Chapter 2 we introduced you to the Events of Instruction in order to explore the feedback relationships implied in that sequence of occurrences. However, this same set of events should be considered a capsule treatment of daily, unit, and course plans. As you study the seven steps listed below, look for connections to planning strategies at *all* levels.

The Events of Instruction

1. Gaining and controlling attention.
2. Informing the student of expected outcomes.
3. Stimulating recall of prerequisites.

4. Presenting the new material.

5. Guiding the new learning.

6. Providing feedback.

7. Appraising performance.

Do events 2 and 3 sound familiar? They have been illustrated in each sample lesson plan and the strategy patterns for each kind of learning. Unit plans and course plans should also be designed in terms of the Events of Instruction. For example, initial lessons and initial units should be high in interest—"gaining and controlling attention"—and subsequent lessons and units dependent on earlier prerequisites can be planned to take less time—"stimulating recall of prerequisites." It should now be clear that the time spanned by each event varies widely. In a daily plan, "informing the student of expected outcomes" may be effectively done in less than two minutes, whereas "presenting the new material" might span 10 to 15 minutes, with part of that time taken up by "guiding the new learning." In the actual implementation of a plan, the events will flow smoothly into one another IF *the teacher is able to get the process started.* Gaining and controlling attention, thus, becomes an event of more than passing importance.

Gaining and controlling attention. Some authors refer to this event as motivating the students or turning them on to instruction. Whatever it's called, its presence is essential to meaningful learning. Experience convinces us that the teacher must *get attention on the lesson* within the first few minutes of the period—the sooner the better. We're not referring to commands to "Pay attention." The *repeated* occurrence of these are cues that the teacher has failed to get attention *on the lesson.* One of the easiest ways to "turn off kids" is to begin every class in the same way, to conduct routine homework post-mortems without concern for feedback, to announce the topic in "big-word mathematics" prior to any concrete experiences.

What does get attention? We've provided you with several examples thus far—the most recent being the demonstrations in the sample concept/vocabulary plan and the use of the overhead to seek patterns in the sample rule plan. Notice that each illustration is highly interwoven with the subsequent strategies and so immediately highlights the lesson, while having potential for sustaining attention beyond the first few minutes of the class. These are the two necessary characteristics of the opening gambit: (1) to get attention focused *on the lesson* and (2) to keep attention focused on the lesson. Can you see why the opening joke by the comedian teacher is likely to get attention, but *not on the lesson?* Teachers who use this ploy deliberately often create a monster they can only handle with authoritarian reprimands—not a very promising start to a class, is it? However, sometimes even apparently content-related material does not get attention on the lesson. For example, a junior-high teacher decided to teach set union and intersection by having the students place groups of jelly beans inside string loops. To his dismay, most student interest was focused on consumption of the jelly beans. Thus, the "on the lesson" aspect is critical—as a preventive of undesired student behavior; but more important, as an initiator of learning. Once attention has been focused on the content, keeping attention should follow automatically. Right? It should, but it often doesn't! If the class begins with the presentation of a relevant problem dealing with teenage drug addiction, the teacher is likely to get attention on the mathematics needed to solve the problem. However, if the problem is dropped and the mathematical rules are presented with no cross-references to the initial problem posed, the potential for keeping attention descreases rapidly.

What kinds of things do usually elicit student attention? The following categories have proved consistently useful:

1. An unfamiliar, puzzling or unusual event or idea.

2. A familiar object (circumstance) behaving in an unfamiliar way.

3. Materials or activities with high sensory appeal (the use of color, sound, or touch, for example, or a combination of these).

4. The use of materials or circumstances matched to the students' "here-and-now" interest.

4ʹ The use of materials or circumstances corresponding to applications in other school subjects or in the nonschool world, where these applications have potential for students' "here-and-now" interests.

In the preceding chapters, as well as in earlier sections of this one, each of the above has been illustrated. The use of labels from soup cans to introduce the surface area of a cylinder is another example of (1) a familiar object being used in an unfamiliar way and (2) a material with sensory appeal. We used 4' to indicate that this idea is a corollary to the previous one. However, if none of the students take shop, there's not much interest potential in an introductory problem based on that subject. Similarly, income tax forms are rarely of any concern to the adolescent who will not fill out a real one for several years.

What *are* some major "here-and-now" interests of students? The following list contains a sample of interests obtained from the only reliable source, junior/senior-high students themselves.

Students' here-and-now interests

Acceptance by peers	Body (*theirs*)	Music	Sports
Acne	Drugs	Pets	Ten-speed bikes
Alcohol	Hobbies	Sex (opposite)	Tobacco
Automobiles	Money		

Your own students will help you add to this list if you listen. And where can you find source materials related to these areas of interest? Check the local and school newspapers, *Popular Mechanics*, professional journals, and source books, plus all the many other resources identified in Chapters 9, 10, and 11.

7.6 CROSS-REFERENCES TO A SAMPLE CBTE EVALUATION INSTRUMENT

By this point in the text, the gradual spiraling built into the treatment of instructional design and objectives has crystallized in the shape of plans. The lesson plan is the subject of category 2 of the sample CBTE instrument. But read that descriptor carefully! The ability to write lesson plans is not assessed directly in this list of competencies. The rewards are attached to successful *implementation* of plans which are well-designed. In our experience well-written plans do not always result in effective implementation, but effective implementation is inextricably linked with thoughtful planning.

There is one competency which is operationally defined on the bases of the strategy patterns of this chapter. Competency 2.1 reflects not only the need to match taxonomic level of objectives with strategies, but also the need to match type of learning (concept, principle, and so on) with strategies. And since 2.1 and 2.2 are dependent on the use of modes (2.3), sequencing (2.4), and feedback getting and giving (2.5), these competencies are also reflections of aspects of the plan. Moreover, the plan should communicate the nature of the subject matter (all items of category 3), may get attention by relating to the students' "here-and-now" interests (2.62), and will thus utilize instructional resources (almost all items in category 4). It's all beginning to come together, isn't it? Planning for systematic instruction is a synthesis of the many facets of instruction which we've previously considered. The competencies are not distinct features to be attained in isolation, but complexes of capabilities. You should continue to find this kind of interweaving in Chapters 9, 10, and 11 where resources of varying kinds are identified. Meanwhile, in Chapter 8, we spiral back to feedback and add the aspect of judgment in the topic known as "evaluation."

7.7 SUMMARY AND SELF-CHECK

This synthesis of all the previous chapters presented strategy patterns for vocabulary, concept, rule/principle, and problem-solving learning, as well as for the learning of motor chains. Sample daily plans were analyzed both in terms of the related strategy patterns and the earlier work on objectives, modes, feedback, human intellectual development and the nature of the subject being considered. In addition, the unique characteristics and problems of lessons on type word problems and proof were outlined. Finally, the ways in which affective objectives might be incorporated in daily plans were outlined.

From daily planning, you were directed to guidelines for constructing unit and course plans. The pervasive threads of retention, drill, transfer, and attention getting were treated in the final sections of the chapter.

Now you should be able to:

1. Operationally define strategy, lateral and vertical transfer, retention, and practice.

2. Describe in sequence the major steps in the strategy patterns for vocabulary, concept, rule/principle, problem solving, and psychomotor instruction.

3. Describe in sequence the major steps to be included in the strategies for type word problems and for theorems.

4. Write a daily plan for a lesson in your major teaching field which corresponds in format to the plan outline, and in structure to the strategy patterns appropriate to the content as well as to all components of the systems analysis model treated in the first six chapters.

5. Critique both a unit plan and a course plan for correspondence to the characteristics of sequence, time estimates, and use of resources.

6. Write both a unit plan and a course plan which correspond to the sample plans in this chapter *if* you have taught, and justify the choices of topic, sequence, and timing made.

7. Construct or modify mathematics problems so that they will be suitable for a junior/senior-high mathematics lesson aimed at novel problem solving.

8. Cite incidents where the teacher appears to be aiming at retention and transfer, given a "live" or canned lesson, and justify your answer.

9. Cite evidence from a "live" or canned lesson of practice which is either meaningful or rote and defend your decisions.

10. Design practice sets for a specific rule/principle which have potential for retention and transfer.

11. Critique the opening five minutes in a "live" or canned lesson for its attention-getting and -keeping potential.

12. Design an "attention getter" for any given lesson and outline its projected use throughout the lesson.

The next section contains exercises aimed at some of the above objectives and simultaneously provides additional samples of instructional activities.

7.8 SIMULATION/PRACTICE ACTIVITIES

A. The Introductory Activity of this chapter included a partially inadequate response to a quiz question. You were asked to critique that response. Now compare your judgments against Tony's more adequate response to the same question. Tony's two objectives were just rephrasings of those written by Mary, but here is his sequence of modes:

Tony's response to the sequence of modes portion of the question

1. Do a demo of the ruler lab with the assistance of two Ss.
2. Divide Ss into groups and have them perform the lab.
3. Have small groups record data on board.
4. Have Ss in small groups discuss optimum way(s) to display total class data.
5. Ask group leaders to report on consensus and use Q/A and lecture to reach a decision.
6. Have each S graph data.

After identifying the favorable characteristics of Tony's response, reread the objectives and compare his response with the objectives. What else would you add in order to match the instruction to the objectives?

B. Construct a lesson plan designed to introduce junior- or senior-high mathematics students to one or more concept(s) and/or rule(s). Be sure to check your choice of topic with your instructor.

C. Why should a plan aimed at novel problem solving include either individual work or small-group discussion prior to question/answer while type problem lessons might begin with the use of developmental question/answer?

D. Choose a rule from the content of junior- or senior-high mathematics. Assume it has been introduced meaningfully. Now, a week later, you plan to assign spaced practice. Design a practice worksheet which has potential for optimizing retention and transfer.

E. Read the following article. Then construct a daily lesson plan for either junior- or senior-high students based on the suggestions in the article.

Ranucci, E. R. "Drawing for Mathematics Teachers Who Can't Draw." Pages 49-52 in *Updating Mathematics*, edited by F. J. Mueller. High School Teachers ed. New London: Croft Educational Services, 1964.

F. Design a problem suitable for initiating a novel problem-solving lesson based on the information below:

In Raleigh, N.C., rats have been found able to survive 2½ to 6 times the normal killing dose of a commonly used rat poison. Apparently a genetic trait is involved. In Scotland about half the farms have rats, and 40 percent of them are resistant to this same poison. (*The New York Times*, October 17, 1971.)

G. Outline way(s) in which each set of objects in column 2 might be employed to *get* attention on the corresponding topic in column 1.

Column 1	*Column 2*
(A) Fractions—Grade 7	(a) Sections of colored egg cartons and pebbles
(B) Sine curve—Algebra 2	(b) Tuning fork and clarinet
(C) Motion problems—Algebra 1	(c) Toy car, inclined plane, stopwatch
(D) Alternate interior angles, given parallel lines— Geometry	(d) Periscope

SUGGESTIONS FOR FURTHER STUDY

Bruner, J. S. *The process of education*. Cambridge, Mass.: Harvard University Press, 1962.

This little book has become a classic reference on the curriculum reform movement of the 1950s and 1960s. The second chapter, "The Importance of Structure," is an especially good follow-up to our emphasis on unifying strands or themes in the "Long-range Planning" section. In addition, Bruner illustrates ways in which the teaching of the structure of the subject matter will facilitate retention and transfer. The examples are often from secondary school science and mathematics and Bruner's style of presentation makes the text highly readable.

Fremont, H. *How to teach mathematics in secondary schools*. Philadelphia: Saunders, 1969.

We highly recommend Chapters 6-20 of this well-written text. These chapters are a rich source of ideas for getting and keeping attention as well as for introducing and developing concepts and rules in mathematics. Both experienced and novice secondary-school mathematics teachers have found this book to be a gold mine which seems to be replenished on return visits. What higher praise is there?!

Kane, R. B., M. A. Byrne, and M. A. Hater. *Helping children read mathematics*. New York: American Book Co., 1974.

The printed language of mathematics consists of words, symbols, and even schematic arrangements. The authors identify the distinct problems faced by the student reading a problem, a symbolic expression, and/or a graph. There are clear cues to the teacher as to the special components of vocabulary planning regardless of grade or ability level of the students. Chapters 1 and 4 are highly recommended as bases for planning.

NCTM. *The teaching of secondary school mathematics*. 33rd Yearbook. Washington, D.C.: NCTM, 1970.

Several chapters of this yearbook are especially useful in the planning of lessons. Chapter 10, "Generalizations," and Chapter 11, "Skills," include specific problem material and suggestions for use in lessons. In Chapter 10 there are multiple connections to our earlier work on the processes of mathematics and an excellent section on the use of target tasks. Chapter 11 includes several examples of practice sets designed to promote retention and transfer. In addition, you are directed to the material in Chapters 12-15 in which planning ideas for arithmetic, algebra, and geometry are illustrated. Above all, don't miss the superior classroom tactics described in Chapter 16, "An Example of Planning for Low Achievers."

Polya, G. *Mathematical discovery.* Vol. 1. New York: John Wiley and Sons, 1962.

The subtitle of this book is "On understanding, learning, and teaching problem solving," and that subtitle tells it all. Yes, Polya is referring to *novel* problem solving, albeit of the purely mathematical variety, rather than the real-world correspondences. Polya emphasizes seeking patterns, making conjectures, and testing these systematically. You must become a problem solver yourself as you work through his insightful and challenging questions. A note to the less confident: Answers are provided!

CHAPTER EIGHT

THE PROOF OF THE PUDDING
EVALUATION OF INSTRUCTION

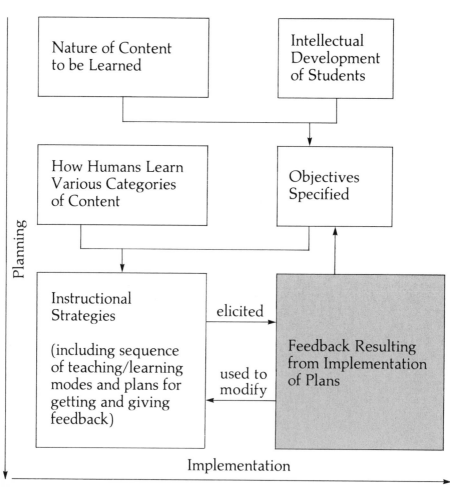

ADVANCE ORGANIZER

Yes, we are again back to the feedback box portion of the systems analysis model! Our previous treatment of this aspect in Chapter 2 focused on the minute-to-minute and day-to-day concerns of systematic instruction. The emphasis was on getting data on which to base immediate decisions such as whether to speed up, slow down, stop, or repeat the presentation with more or varying examples. Similarly, we explored ways and means of providing learners with information on their progress toward attainment of the objectives set for today's lesson, last night's assignment, or yesterday's laboratory exercise. Much dependence was placed on quick and informal means to these ends—spreading oral responses among a wide sampling of students, taking straw votes (followed by verbal expression of the "hows" and "whys"), observing student written work in progress, and giving verbal praise for correct answers and productive approaches to problems. All of this background is prerequisite to the longer-range view and the more formal means of assessment to which we now turn our attention.

Note the two feedback loops depicted on the schematic. One leads to and from the key component "Instructional Strategies," which is the summation of the planning dimension of the instructional model. If indeed "The proof of the pudding is in the eating," then crucial decisions as to what worked and why, and what didn't work and why must be made intelligently on the basis of evaluation of results. Then and only then are we in a position to redesign instructional inputs with realistic hopes of improved success. Likewise, the feedback loop to the "Objectives Specified" component points to the need for a firm data base upon which to judge the degree to which important objectives have been realized. Initial objectives may need to be revised in the light of this experience.

But haven't all of you been evaluated throughout your lives both within the school setting and in the broader context of all life's activities? After all, you have spent a good bit of time on the receiving end during some 16 years of formal schooling. Shouldn't this experience equip you to deal with the topic on the sending end of the process? Without a doubt, you do bring some background to the task. Our students can quickly reel off a list of ineffective techniques and practices. However, they are frequently at a loss when asked to provide the remedy: examples of good models.

And there is no sense trying to wish the problems away. Within a week to ten days of student teaching the novice typically constructs, administers, and scores one or two short quizzes. By the end of three or four weeks of student teaching a full-period test will have to be prepared, duplicated, administered, and scored. These tasks are formidable enough, but soon come even tougher jobs—interpreting results and making a host of decisions regarding future instruction, midmarking-period warning reports, remedial instruction, and marking-period grades—all of which have to be defensible in the eyes of students, parents, school administrators, and, most important, yourself! You will hear claims that the test wasn't "objective." A parent who is an exteacher may question its "validity." These and other bits of jargon are abused in both everyday speech and in dictionary definitions. Yet there is no hope of communicating feedback to students, parents, or administrators without some clear understanding of the language associated with the topic of evaluation. The next section provides you with the background for the understanding of several relevant terms. It is up to you to communicate that understanding to those who receive the results of evaluation.

8.1 OPERATIONAL DEFINITIONS OF SELECTED KEY TERMS

Evaluation and *testing* are often used synonymously but, in fact, they refer to closely related but distinct processes. The root word *value* is clearly central to the concept of evaluation. Thus by *evaluation* we mean *any process of making judgments against selected criteria as to the worth of things or ideas, based on relevant data.* Relevant data? That's where the notion of testing comes in. *Testing* will be used to label *any procedure designed to collect evidence that indicates degree of attainment of objectives.* Note that we do not limit testing to paper-and-pencil instruments and that the results of testing are not viewed as evaluations but rather as data upon which judgments (evaluations) will be based. For example, Jill's record shows percentage scores of 85, 99, 80, 76, and 92 on a set of written tests/quizzes over a ten-week period, but the 85 refers to a pretest given prior to instruction on the unit. As a result of that score, Jill was placed in the group of students who were to work on individualized learning activity packages. All of the remaining scores were attained on quizzes or tests administered after certain periods of instruction. How should her teacher evaluate this data for purposes of deciding her report card grade? If your response is, "I can't tell because you still haven't told me all of the criteria against which to judge this data," you are catching a good part of the idea.

Jill's first test is an example of a test given for purposes of *diagnostic evaluation*. *Diagnostic evaluation* is *characterized by one of two purposes: (1) to place a student at the proper instructional starting point or (2) to find out the causes of instructional defects that have been isolated during instruction.* So Jill's score on the pretest would not properly be included in the data her teacher is evaluating for purposes of reporting a grade. Jill's teacher had administered the second type of diagnostic quiz to a group in the class who had been making repeated errors on three of the rules in the unit. By including diverse examples at varied levels of rule application, the teacher had tried to pinpoint the *causes* of the errors. Should these scores be included in the report card evaluation? Absolutely not! However, each of Jill's other scores will be included in some fashion (more on that in the next section!) in the final evaluation. Thus, all of these test scores will serve the purpose of *summative evaluation, the use of data to make judgments as to the extent to which instructional objectives have been achieved by the students.* All final course exams fall in the category of testing solely for the purpose of summative evaluation. Regardless of error patterns now identified by the teacher, instruction is over and this data cannot be used to affect the instruction of that group of students (see Fig. 8.1). In contrast, such data can and should be used to plan future instruction in the case of short quizzes and, to some extent, even in the case of unit tests. Thus, a third kind of evaluation, *formative evaluation*, is regularly used in systematic instruction. The chief purpose of *formative evaluation* is *to assess the extent to which progress is being made on the instructional objectives and to identify those objectives that need further instructional attention*. Note that data from a short quiz serves two purposes: summative and formative evaluation. So does data from a unit test, but the emphasis from such data is now primarily on summative evaluation. The circle graphs in Fig. 8.1 depict this changed evaluative emphasis as one moves from final exam to unit test to quiz. (The sectors in these graphs should not be interpreted as exact matches to the evaluation emphases in any particular test/quiz. They should be thought of as representing a ball-park figure.) Notice the small sectors designated as diagnostic evaluation in both the quiz and unit test circle graphs. Although these tests are not *constructed* so as to discover causes of known errors, data from them sometimes yield this kind of information. You will be making judgments based on data gathered for all three types of evaluation. Knowing the what, when, and why of the process is a paramount aspect of those judgments.

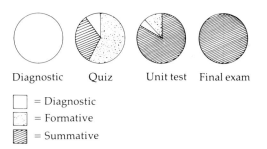

Diagnostic   Quiz   Unit test   Final exam

☐ = Diagnostic
▦ = Formative
▨ = Summative

Fig. 8.1   Relative emphasis on the three kinds of evaluation in various types of tests.

Another commonly expressed concern is to what extent the data gathered on the written tests were objective as opposed to subjective. *Objective* literally means that *no judgment entered into the process*. We take the position that no quiz or test could possibly meet this definition! Why? Didn't someone have to *decide* (judge) such matters as (1) which questions to include, (2) how to phrase the questions, (3) the answer or range of acceptable responses, (4) the credit value of each item, and (5) the time limit for responding to the test? Thus we will not speak of objective tests or of objective test items. It seems more honest to refer to *objectively scored test items* as *those which would be scored identically by anyone using the same answer key*. Similarly, *subjectively scored test items are those which allow judgment of the rater to enter into the scoring process.* A word of caution is in order at this point. Do not automatically equate "objectively scored" with "good" testing technique nor "subjectively scored" with "poor" testing technique. The nature of the subject matter at stake and the particular kinds of objectives being assessed help determine which approach makes the most sense.

But don't objectively scored items lead to increased test reliability? Generally. However, let's not fall into the trap of confusing reliability with validity. *Reliability is the extent to which a test yields repeatable results* (tells a consistent story) whereas *validity is the extent to which a test measures what it is supposed to measure* (tells the truth). Assume that a manufacturer inadvertently produced a batch of metric rulers 11 cm long which were each calibrated into 100 equal segments labeled mm and 10 equal segments marked cm. Skillful students using these devices would get consistent (reliable) but incorrect (invalid) measurements! Obviously we want both valid *and* reliable tests so that the data they produce can be trusted as a base for educational decision making. Fortunately for us, it turns out that highly valid tests typically turn out to yield respectable reliability. (Warning: the reverse is *not* true!) How does one insure validity? Obtaining a close match of well-structured test items to carefully specified objectives is the best way to achieve content validity. It is with this last task in mind that we direct your attention to the Introductory Activity that follows.

8.2 INTRODUCTORY ACTIVITY

Step right up and take a chance!

Test taking shouldn't resemble a carnival game of chance but, sadly, it sometimes does. Very often, this is due to poorly constructed test items. This activity places you in the role of a critic of a set of such test items. Assume that clear directions on the recording of responses have been given but that each item in the following set is defective in some way—for example, might be incorrectly answered by capable students or might be correctly answered by poor students who guess well. For each item, identify the defect(s) and rewrite the item so as to eliminate the problem(s).

Selected items from a Math 8 test

Completion

1. A line segment joining two points on a circle is _____.
2. _____ circles are those with the same _____ and unequal _____.

Multiple choice

11. The area of a triangle with a base of 4 and an altitude to the base of 2 is
    (1) 16   (2) 2   (3) 4   (4) 8

12. A polygon with all sides congruent is called

    (1) equiangular (2) equilateral
    (3) regular   (4) a square

True-false

16. π is equal to 3.14.

17. Certain line segments containing the center of a circle sometimes divide a circle into two semicircles.

After you have completed your analysis of these items, compare your corrected versions with those produced by another student. Then check on your thinking as you interact with the material in the next section.

8.3 ASSESSMENT OF COGNITIVE OBJECTIVES

The first objectives you will encounter in assessment situations in student teaching will be those belonging to the cognitive domain. And, for better or worse, this kind of outcome is what schools emphasize above all others. Further, the means of assessing these are often limited almost exclusively to paper-and-pencil instruments called quizzes and tests. This doesn't need to be the case, since there are several other ways to gather evidence of achievement of this kind of objective. Oral reports of library research can be presented and assessed. Written reports on take-home labs and in-class projects can be assigned and analyzed for evidence of learning. Moreover, don't the actual products produced in laboratory work and bulletin board design sessions yield relevant data? Also consider the potential use of anecdotal records made by the teacher as students participate in question/answer, small-group discussion, supervised practice, and laboratory work. Some illustrative examples of such non-paper-and-pencil tasks were provided in Chapter 1; for further help on ways of deriving assessment data from such tasks, see the Suggestions for Further Study at the end of this chapter. Our brief treatment here will focus on making the best possible use of the common paper-and-pencil approaches.

Types of Paper-and-Pencil Items

In the following section, we consider each of five types of short response items and the essay, or long response, item-type. In each case a set of clear directions to students must be our first concern, since we have to be sure that the answers reflect an understanding of the content being tested rather than the students' skill in guessing the nature of the task. Ambiguous directions not only promote undesirable guessing, but also frequently confound the teacher's work when the completed tests are being scored. Yet one of the major advantages of short response items is supposed to be ease and speed of scoring. Therefore, we will begin our consideration of each item-type by providing you with good models of directions. We suggest that you type each model direction on a separate card for inclusion in your resource file. Second, we will list a set of sample items, but these will be a mixed bag of acceptable and defective items that you will be asked to analyze with the help of some cues. Advantages and disadvantages inherent in the item-type under consideration will be treated next. Finally, some "rules of thumb" to follow when constructing the item-type will be suggested.

True-false Most test construction experts classify this item-type as one of the most abused. Let's see why as we analyze the sample items that follow. (Remember to type the model directions on a card for your resource file.)

Directions: Circle "T" if the statement is true. Circle "F" if the statement is false.

1. The altitudes of a triangle meet in an interior point,
   called the orthocenter.   1. T   F

2. π is equal to 3.14.   2. T   F

3. The lowest common multiple of 5 and 12 is 60.   3. T   F

What answers would you accept for these? The point where the altitudes meet *is* called the "orthocenter." However, it is not an interior point in the case of the right triangle or the obtuse triangle. How, then, would you interpret the instructional implications of a response of "false"? Would you infer that (1) the student knows about the varying location of that intersection point or (2) the student believes that the name of that point is something other than "orthocenter"? Also consider how to interpret a response of "true" to example 2. Would this mean the respondent knows that the value given is correct to the nearest hundredth, or that the respondent thinks that 3.14 is the exact numerical value of pi? The problem of interpretation of true-false results is further complicated by the fact that the probability of guessing the correct response for each item is 50 percent.

There are other problems, too. Try writing a number of statements that are unequivocally true or false, and you will soon discover that your efforts fall into one or both of two categories. The items either test *only* recall of specific facts and/or contain key words which clue students to the correct responses. Test-wise students have learned that words such as *may, generally, usually, sometimes,* and *probably* correlate with true statements while *never, none, always,* and *all* usually indicate false statements about mathematics.

Is there *any* advantage to be claimed for using this type of test item? The one most frequently cited is that many items can be answered in a short time, thus increasing the sampling of knowledge. However, this feature quickly loses its attractiveness in view of the fact that what is typically sampled is simple recall of factual information. In our view, the true-false item-type is better suited for review/discussion purposes than it is for testing use. Thus, no rules of thumb for constructing this type are included.

Modified true-false This item-type is sometimes considered an improvement over true-false items, since the student must do more than circle *T* or *F*. How much better is this item-type for sampling student achievement? Consider the set of items that follows.

Directions: Circle "T" if the statement is true. If the statement is false, fill in the blank with the word or number that would make the statement true when substituted for the <u>underlined</u> term.

1. The value of the 6 in 635 is <u>two</u> times the value of the 3.   1. T _____

2. Trapezoids are <u>parallelograms</u>.   2. T _____

3. As x increases from $3\pi/2$ to 2π radians, sin x <u>increases</u>.   3. T _____

Item 1 is a good example of this type of item. The underlined word identifies a frequent misconception with respect to place value and the statement clearly calls for the substitution of the correct value. Now contrast this with the second example. Start listing the range of literally correct responses which could be made by the students who have been taught that trapezoids are not parallelograms. How would you interpret responses such as "polygons," "simple closed curves," "four-sided," "interesting," and "important"? Another kind of problem is built into the third example. Did you spot the fact that it really describes a two-tailed situation? Obviously, the value of sin x is likely to either increase or decrease (a constant value is an unlikely guess), thus making this item a true-false type in disguise! This example would be much improved by amending it to read" . . . increases from −1 to 0."

Good modified true-false items are not easy to construct, even when aimed at lower levels of the cognitive taxonomy. Another limitation inherent in modified true-false is student lack of experience with this type of item. For this reason we suggest using modified true-false items on homework or worksheets prior to employing them on tests. Students have to be taught that no credit will be given for (1) substitutions of equally correct terms in true statements or (2) simply writing *false* after incorrect statements.

Two advantages of the modified true-false over straight true-false can be claimed. The guessing factor is reduced substantially from 50 percent if the items are well structured and if approximately half of the statements are false. Also, it is more nearly possible to construct items that measure learning at levels II and III of the cognitive taxonomy. You may find this advantage to be more theoretical than practical because novices find such items very difficult to construct.

Rules of thumb for constructing modified true-false items include:

1. Construct "true" statements that are absolutely correct.

2. Use the same number of true as false statements.

3. Underscore a word that limits the range of substitutions in false statements.

4. Avoid where possible the use of specific determiners.

5. Avoid negative statements.

6. Keep all true and false statements approximately the same length.

Completion or fill-in-the-blank This item-type often brings out unsuspected creativity in students, much to the teacher's chagrin. Try your own hand at the items in the sample set.

> Directions: Read the sentences below and determine the word(s) or number(s) that would have to be substituted for the blank to make each sentence complete and correct. Then write these words or numbers in the correspondingly numbered blanks on the right.
>
> 1. The man who wrote *The Elements* was ____(1)____.   1. _____
>
> 2. Lines which intersect a circle in exactly one point are called ____(2)____.   2. _____
>
> 3. The ____(3)____ of a number, N, is the exponent indicating the   3. _____
>   ____(4)____ to which it is necessary to raise a given ____(5)____   4. _____
>   to produce N.   5. _____
>
> 6. The positive value of tan (arc cos $\sqrt{2}/2$) is ____(6)____.   6. _____

Do you want to demoralize a class and start a near riot? A surefire way to accomplish both is to include items such as 1 and 3 on a quiz. Then go into class the next day and tell the students you marked their plausible, but nonkeyed, answers as incorrect because they were not "what I wanted." Students who completed item 1 with an appropriate descriptor of the man's age, occupation, nationality, or mental ability cannot reasonably be marked wrong, since the question does not specify that his name is the desired response. Note how examples 2 and 6 avoid this common error, which shows up even in many published texts and workbooks! Can you guess the process that generates such monsters as that illustrated by item 3? Typically such ill-conceived guessing games result from teachers "lifting" a sentence (out of context) from the textbook and compounding the felony by replacing several key words with blanks. Fortunately this type of gross error is becoming rare in published textbook and workbook examples, but watch out for it in exercise material you assign. Yes, a serious limitation of the fill-in-the-blank item-type is the difficulty of constructing unambiguous items. It *seems* so easy that novices can hardly resist trying. We have nicknamed these *suicide items* for good reason! You will also find it difficult to write items of this type that go beyond level II of the cognitive taxonomy—another serious limitation.

Advantages for this type of item are few and far between. Some kinds of mathematics content, such as recalling concept labels and formulas and solving type problems, do lend themselves to this format. In fact, many mathematics tests contain a number of items such as item 6 in order to test ability to respond accurately to basic rules and principles. However, consider the limited interpretive value in computational problems of knowing *only* the answer a student arrived at and not seeing steps in the student's work which led to the response. Thus it makes sense to use completion for single-step computational problems during the early stages of formative evaluation and to test multistep problems by other means. Subsequently, mathematics teachers may wish to include some slightly more complicated examples as completion items, especially those where many of the steps are based on skills learned much earlier. An item like the one which follows, worth two credits with no partial credit allowed, falls into this category for intermediate algebra students.

7.   Solve for x: $2^{x+5} = 8$.   7. _____

At a later stage of formative evaluation, the teacher may not be seeking feedback on process errors, but instead may be emphasizing accuracy via this item-type with its "all or nothing" credit format. Mathematics teachers would be well advised to combine this valid emphasis on accuracy with an equal (or greater) emphasis on process which must be achieved by means of other item-types.

Some useful rules of thumb follow:

1.   Use only *one* blank per item.

2.   Put all blanks at the end of the sentence.

3.   Make all blanks the same length.

4.   Be sure the statement clearly delimits the range of potential responses (include degree of specificity of numerical answers).

5.   Avoid the temptation to copy out-of-context sentences from texts and then substitute a blank for a key term within the sentence.

6.   In the case of computational problems, restrict use to the single-step type during the early stages of formative evaluation.

Multiple choice   This type of short response item is *the* one most favored by professional test writers. They find that well-structured multiple-choice items can more effectively assess many of the lower level outcomes often measured by other short response items. Additionally, this item-type can also measure a variety of objectives at levels III and above of the cognitive domain. As is the case with all other item-types, good items require careful attention to construction details. However, the inherent advantages of multiple-choice items make for higher payoff in terms of time invested. Now check out each of the following examples for structural soundness and make mental note of any changes you might propose.

Directions: Write in the blank provided the capital letter preceding the expression that best completes *each* statement or answers each question.

1. The sequence $3, 6, 12, \ldots, 3(2)^{n-1}, \ldots$ is an example of a(n)

    A.   arithmetic progression
    B.   finite sequence
    C.   geometric progression
    D.   harmonic progression   1. _____

2. In triangle ABC, if $m \angle A = 30°$, $a = 15$ and $b = 12$, then triangle ABC must be

    A.   acute
    B.   isosceles
    C.   obtuse
    D.   right   2. _____

3. If the set of points represented by $\{(x,y) \mid |x| = 3\}$ is translated two units to the right, the new set is represented by $\{(x,y) \mid$

    A. $|x + 2| = 3\}$
    B. $|x - 2| = 3\}$
    C. $|x| = 5\}$
    D. $|x| = 1\}$ 3. _____

4. Rhombi

    A.   have unequal sides
    B.   have area equal to the product of the diagonals
    C.   belong to the class of parallelograms
    D.   can be separated by one diagonal into two scalene
          triangles   4. _____

5. If in triangle ABC $m \angle B = 60°$ and $AB > AC$, then which relationship must be true?

    A. $m \angle C > m \angle A$
    B. $m \angle C < m \angle A$
    C. $m \angle C = m \angle A$
    D. $m \angle C < m \angle B$
    E. $m \angle C = m \angle B$ 5. _____

Did you spot one weak alternative among the four supplied in item 1? Finite sequence is a poor distractor, both because it clearly limits the number of terms and because it does not refer to the dependence of terms on one another. A sound item always includes only plausible alternatives (distractors) plus one correct response (keyed answer) among the alternatives. The net effect of one poor distractor in a list of four alternatives is to increase the guessing factor from 25 percent to 33⅓ percent. Note that none of the other sample items suffer from this type of error. In fact, you may have wondered if item 5 would be considered to be superior on this account since it contains five plausible alternatives. The five-alternative type does reduce the guessing factor 5 percent (from 25 percent to 20 percent) but the trade-off is both (1) the difficulty of finding that extra distractor and (2) increased reading time, thus reducing the number of questions students can answer in the time available. For these reasons test experts recommend use of the four-alternative type. You must have noticed that item 4 looks different from the others because the alternatives are all long in comparison to the one-word stem of the question— just the opposite of what is considered sound structure. Each of the other sample items contains a stem that is either a clear question or a completion-type statement followed by shorter alternatives parallel to each other in con-struction. This practice makes the task more explicit to students—always a desirable feature of a test.

Are there any other limitations in addition to the care needed in item construction? Yes. Multiple-choice items emphasize recognition of the correct response. Thus this item-type is a weak match for objectives that specify outcomes such as recall or synthesizing an original response. We must employ other means of assessment to match these objectives.

Obvious advantages inherent in use of multiple choice include the sampling of many objectives in a short time, speed and ease of scoring, and reduction of the guessing factor to 25 percent. But don't overlook additional good features that may not be so obvious. The fact that a student has been attracted to a particular distractor has diagnostic value and also tells the teacher something about the functioning of the questions included in the test. Both of these desirable features are explored in detail in a later section on item analysis. Items 3 and 5 illustrate the potential of assessing level III and above objectives with multiple-choice questions.

Rules of thumb to follow in constructing multiple-choice questions in-clude these:

1. Write the stem so that it presents a single, specific problem in either question or completion statement form.

2. If the stem incorporates an exception, emphasize it by solid caps and underscoring (EXCEPT).

3. Include in the stem all words that would otherwise be common to all alternative phrases.

4. Make the alternatives as brief as possible and in no case any longer than the stem.

5. Keep alternatives within an item parallel in form and grammatically consistent with the stem.

6. Employ distractors that are plausible and incorporate typical errors and misconceptions.

7. Sequence alternatives in alphabetical or numerical order.

8.   Place alternatives in a vertical column, not in a horizontal row.

Matching In our view, this is really a special case of multiple choice. Two columns are presented, and the task is to match items from one column with items from the other column—multiple stems to match with multiple alternatives. Thus you should apply what you just learned about multiple choice as you analyze the examples that follow.

1.   Directions: On the line to the right of each quantity in Column B, write the letter of the phrase in column A that corresponds.

Column A		Column B	
A.	$\sin^2 \theta + \cos^2 \theta$	$\cos \theta$	_____
B.	$\dfrac{\sin \theta}{\cos \theta}$	$\cos (A - B)$	_____
C.	$2 \sin \theta \cos \theta$	$\cot \theta$	_____
D.	$\tan (270° + \theta)$	$\cot^2 \theta$	_____
E.	$\dfrac{1}{\sec \theta}$	$\sin \theta$	_____
F.	$\cos A \cos B + \sin A \sin B$	$\sin (A + B)$	_____
G.	$\tan (90° - \theta)$	1	_____
H.	$\sin A \cos B + \cos A \sin B$		
I.	$\csc^2 \theta - 1$		
J.	$\cos \theta \tan \theta$		

2.   Directions: On the line to the right of each expression in Column B, write the letter of the expression in Column A that fits it. Each expression in Column A may be used <u>only ONCE</u>.

Column A		Column B	
A.	$y = 4x + 3$	circle	_____
B.	parabola	$y^2 + 4x^2 = 3$	_____
C.	$y^2 + x^2 = 3$	conic	_____
D.	ellipse	first-degree equation	_____
E.	$y^2 - 4x^2 = 3$	$y = 4x^2 + 3$	_____

No doubt you have surmised that one of these two examples is intended as an illustration of correct performance, while the other suffers from several defects. How many defects did you spot in example 2? Did you catch on to the fact that the use of short and equal columns guarantees that achievement of three correct matches will assure the fourth match (since each Column A expresssion can be used only once)? If the directions hadn't included this restriction, the teacher would have even a bigger mess to contend with. Conic is correctly associated with all but one of the expressions in Column A! Note that example 1 avoids these kinds of problems in a number of ways. Recommended practice was followed in that (1) unequal length columns were used, (2) the shorter column lists five to seven items, and (3) the longer column includes a maximum of ten alternatives. Yes, it is difficult to construct items that meet these criteria, and that is an important limitation of matching questions. This limitation becomes even more severe when you attempt to avoid the problem of multiple combinations of matches illustrated by example 2. Homogeneity of the type of expression within each column is the best way to eliminate this particular difficulty. For instance, Column A could be limited to characteristics of various geometric shapes with Column B listing the names of those shapes. But doesn't this process then restrict matching questions to the assessment of fairly low-level objectives? Yes, it frequently does just that—another limitation of this item type.

Aside from quick and easy scoring, it is difficult to conceive of other advantages of matching questions. Even the advantage of quick and easy scoring pales in view of the fact that other types of short-response items will accomplish that and more. No doubt this explains why matching questions are rarely used by professional testmakers.

Rules of thumb for matching questions include the following:

1. Include, in the directions, the basis for matching and state whether or not responses may be used more than once (if either is needed for clarity).

2. Place the briefer expressions in the right-hand column (easy to scan) and in alphabetical or numeric order.

3. Use unequal length columns with a range of five to seven items in the shorter, right-hand column and seven to ten in the longer, left-hand column.

4. Make the material within each list homogeneous. For example:

Column A	Column B
Achievements	Names of persons
Events	Dates
Definitions	Concept names
Graphs	Equations
Concepts or rules	Symbols or formulas
Geometric shapes	Classification categories

Long response This item-type is also often referred to as essay, although some teachers use the term *essay* to indicate that the response desired is in the form of one or more paragraphs. Long-response items of various kinds are needed in order to test important objectives that are difficult or impossible to assess by means of short-response items. They are also required in order to get feedback on students' ability to function at the highest levels of the cognitive domain.

Just as is the case with all other item-types, considerable care in construction is essential if such questions are to serve their intended purposes. Study the following examples and decide if each is likely to achieve its purpose.

1. a. Solve for tan θ to the *nearest tenth*:

    $$2 \tan^2 \theta - 5 \tan \theta + 1 = 0$$

    (Show all work.)

    b. How many different *acute* angles are there which satisfy the equation in part a? Justify your answer.

2. Given: $\angle A \cong \angle B$, $\angle C \cong \angle COD$ in the figure $ABOC$.

    Prove: $\angle OEA$ is a right angle.

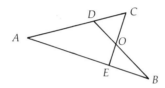

3. Show that the equation $x^2 + xy + y^2 = 3$ represents an ellipse.

4. Solve graphically the system of equations below:

    $$3x + 4y = 5$$
    $$x - 8y = 4$$

    Be sure to indicate the solution set.

5. Can you explain why we need three different measures of central tendency (mean, median, and mode)?

6. Write an essay on the importance of geometry to everyday life.

7. Black Bart guided his drag racer through the standing one-quarter-mile strip in 5.95 seconds. However, Wonderful Willie beat him by .025 seconds. How much faster was the average acceleration of Willie's car in ft/sec²? Show all work.

8. Two separate streams of lava have flowed down a mountainside and are slowly approaching a fence. The first gets closer and closer to the fence but never goes beyond. The second moves closer and closer to the fence, and, at last, flows one-eighth inch beyond. Could the fence be described as the limit of either flow? Of both? Explain.

9. Write out the "multiplication" table for the integers 0, 1, 2 under addition, mod 3. Compare this table with the table of rotations of an equilateral triangle. In what way(s) are they similar, different?

Would you have a difficult time grading responses to item 6? We would! This question as stated is so broad as to provide for a virtually limitless range of student responses. The item could be vastly improved by indicating a narrower frame of reference that would focus responses more directly on a cluster of specific objectives the teacher wishes to assess. Unless you can list the main points or key ideas that should be included in a complete and accurate answer, the question requires reworking. Otherwise that item will have very low reliability. Did you spot the error in item 5, or is it so simple a fault that you overloooked it? The answer to that question as phrased is a simple yes or no (another true-false item in disguise). Eliminating the first two words and replacing the question mark with a period would repair the item quite well. Care in item construction is needed in order to overcome the two big potential limitations of long-response items—low validity and low reliability.

The other seven examples are intended to illustrate well-constructed items which assess objectives that cannot be assessed by short-response items. Outcomes that have to do with abilities such as solving equations, writing proofs, applying a rule in a slightly novel way, constructing graphs, analyzing the structure of mathematics, and using logic, evidence, and basic assumptions to justify answers all require long responses if the assessment means are to match objectives. But why not cast item 7 into a completion or multiple-choice format? The cue here is in the direction to "show all work." The long-response format of this question allows the teacher to find out more than just who was able to obtain the correct numerical answer and to assign partial credit to responses that indicate varying degrees of comprehension.

Rules of thumb for constructing/scoring long-response items include:

1. Use long-response questions to assess *only* those outcomes that cannot be satisfactorily measured by short-response item-types.

2. Limit the scope so that the task is clear and the time required to respond is reasonable in terms of both credit and time to be allotted to each question.

3. Prepare a key that includes all major points of information, ideas, and steps that should appear in correct and complete responses. Be sure that the part of the response involving major principles and/or concepts is assigned more credit than simple computation aspects.

4. Plan the credit distribution within each question (in terms of item 3) prior to administering the test.

5. Do *not* use "optional" questions. (Otherwise, not all students have taken the same test.)

The Unit Test

Within a week you will be completing instruction in a major unit in trigonometry, the solution of problems involving the oblique and/or right triangle. Feedback obtained during group problem-solving sessions, super-vised practice, lab work, and on the results of a few short quizzes is convincing evidence of the class progress to date. But can they put it all together? It's time to draft the BIG ONE!

Designing the unit test Unfortunately, some unit tests are patchwork quilts of earlier quiz or textbook questions with no apparent parallel to instructional emphasis or sequence. If you've been victimized by such unit tests, you probably learned to play the game of school in order to reap the rewards of second-guessing the teacher. But a comprehensive assessment of objectives ought not to be a capricious collection of items, but a thoughtful, integrated part of systematic instruction. And that can only happen as a result of thorough planning.

What are you assessing? Of course, it should be the extent to which the students can meet the objectives of the unit. If your instruction was directed toward the mastery of objectives at levels I and II of the cognitive domain, then your *test results* should *report whether students did or didn't achieve each objective (criterion-referenced testing)*. If you included low-level objectives that all or most students should master as well as higher-level objectives that fewer students might attain, you would be more likely to report *test results in terms of each student's relative position in the group (norm-referenced testing)*. Each of these types of testing is characterized by specific test construction principles. However, since the junior/senior-high classroom teacher's test more often is modeled after norm-referenced instruments, we have chosen to examine testing from that vantage point. In either case, test planning begins with the consideration of the selected objectives.

Mr. Zidonis, a twelfth-grade teacher, began his test planning by com-pleting the margins of what is called a *table of specifications*. He listed major content topics of the unit in the horizontal margin and the taxonomic levels of the cognitive objectives in the vertical margin (see Fig. 8.2).

Why not list each cognitive objective? Just imagine the size of the result-ing table! Such a table of specifications would become an unmanageable monster rather than a guide to test design. Mr. Zidonis apparently included all levels of the cognitive taxonomy in his instruction, and he intends to sample the effectiveness of that instruction at each level. But levels of the affective domain and objectives in the psychomotor domain are not included. Did he ignore these domains in his instruction? Definitely not, but he chose to assess them by means of long-term projects and anecdotal records during lab work rather than as part of the unit test. (Other ways of assessing objectives in these domains are treated in later sections of this chapter.) Decisions! decisions! There's no getting around it. The decisions which result in the completed margins of the table of specifications have already framed in the future test. But the professional testmaker wouldn't stop there. Next each box of the table would be filled in with a decimal representing the instructional emphasis

UNIT ON TRIANGULATION—MATH 12

Objectives by level	Units of content					
	Law of sines	Law of cosines	Law of tan	Mixed law problems	Bearing problems	Derivation problems
I. Knowledge						
II. Comprehension						
III. Novel application						
Above III						

Fig. 8.2   A partially completed table of specifications.

given the topic and the matching objectives. "You must be kidding!" Well, *they* aren't, but then their purpose and skills differ from those of the beginning teacher. Your next step is considerably different. You begin selecting and/or generating items.

You already know quite a bit about the need to match items to objectives, the pitfalls inherent in some kinds of items, the advantages of others, the scoring of varied types of items. Where do you begin? The guidelines below, have been found useful by both novice and experienced classroom teachers.

1. Review all classroom quiz items, key homework, lab, and project questions. Review the text, source books, and items you had collected earlier and stored in your resource file on this unit.

2. On five-by-eight-inch cards, write individual items which are related to the objectives and major content topics in your table of specifications. We'll also be using these cards after the test. Answers should be written on the back of the card. Number the cards *in pencil* (the final ordering of the items comes later), and place in parentheses the credit you will assign to a correct response.

3. As each card is completed, record *in pencil* the number of the item and the credit assigned to that item in the appropriate box of your table of specifications. For example, item 4, worth 2 credits, which requires typical use of the Law of Cosines in a computational problem, would be entered in the second row and second column of Mr. Zidonis's table as $4^{(2)}$.

4. As the entries begin to fill the table, stop and "eyeball" both the number of items in each box and the percent of the total credit thus far assigned to each box. You may need to begin writing items in different areas of content or for neglected levels of objectives. The test emphasis represented by credit assignment may need adjusting if the test emphasis is to match the instructional emphasis. A pencil with an eraser will come in handy!

5. When, in your judgment, the completed matrix seems satisfactory, divide the item cards into clusters on the basis of item-type—that is, multiple choice, true-false, and so on. You may need to rewrite items or alter the number of items of a particular kind to improve the validity of the total test. This is the time to review the section on different kinds of test items. When you are satisfied with the particular items, turn your attention to the ordering of the clusters and of the items within each cluster. There are no hard and fast rules to follow. In general, short-response items are positioned ahead of essay-type questions. Most test-writers suggest that the order within a cluster of multiple-choice items, for example, should be decided by considering both adjacent items and the array of responses. The answer to question 4 should not "give away" the answer to question 5. Moreover, if the teacher has unintentionally built in a pattern in which the correct response is always *A* or *B*, the test-conscious student who is sure of *some* of the answers has a good chance of guessing correctly all the rest. Since the order of choices for each item should have already been established as per the suggestions early in this section, the teacher must rearrange items, *not* choices within an item.

6. At last the clusters are ordered, the items are ordered and renumbered, and appropriate changes have been made in the table of specifications for later reference. Mr. Zidonis's completed table of specifications is reproduced in Fig. 8.3. Notice that individual items are worth varying amounts of credit. Superscripts enclosed in parentheses have been used to indicate any items or parts of items *worth other than two credits* Each part of item 20 deals with the same unit of content at varying levels and their placement on the grid reveals these emphases. The table indicates a test which (1) loads almost equally on all units of content with a to-be-expected heavy loading on Mixed Law problems, (2) provides minimal rewards for straight recall, and (3) gives equal and heavy weighting to level II and to levels III and above combined. This picture of the test would satisfy Mr. Zidonis if it matched closely the objectives and instructional emphasis of the unit. But wait. The total credits don't add up to 100! There is no reason, other than superstition, to suppose that it should. What sense would there be in juggling credits to obtain this magic number? This might destroy the match of test-to-objectives emphasis he had worked hard to achieve. It would be far better to use a

UNIT ON TRIANGULATION—MATH 12

Unit of content

Objectives by level	Law of sines	Law of cosines	Law of tan	Mixed law problems	Bearing problems	Derivation problems	
I. Knowledge	3	12	7				6
II. Comprehension	1,6, 17[3]	4, 11[3]	2, 14[3]	9,16[3] 19,[3]20a[5]	5,8		34
III. Novel Applicaltion		13[3]	15[3]	20b[5]	21[10]		21
Above III	10[3]					18,[3] 22[10]	16
Credit totals	12	10	10	18	14	13	77

Fig. 8.3 A completed table of specifications.

conversion table, slide rule, or calculator to translate raw scores into percentages if that is desirable in terms of local custom or school policy.

7. Your next-to-last step is to insert the direction cards from your resource file with credit noted before each cluster of item types. Add the heading for student name, date, and class section, and the test is ready for the typist. Since that typist will probably be you, remain alert to the need to leave space for hand printing of any required symbols. Follow the earlier suggestions on format to promote ease of scoring. Recall that short-response items can be scored more easily if they are written in blanks in a right marginal column or a separate response sheet. (Do number these blanks to correspond to the item! See why?)

Do these guidelines seem to represent a lot of work? They do, but if valid feedback on instruction is your chief objective, then there's no way to avoid careful planning of a major test. As you work on varied stages of test preparation, be sure to share your attempts with your cooperating teacher. Modifications based on thoughtful experience will simplify your job.

To review or not to review? Since you decided, on the basis of positive feedback, that the time for a comprehensive test had arrived, the students must be "ready" for the test. So why review? If your feedback sampling has been adequate, the answer is that typical review days are often a waste of valuable instructional time. Remember those mythical 180 days and the ways the total gets reduced? Suppose your feedback sampling was spotty. In that case the chances are that a review day will become a hurried and ineffective attempt to repair several weeks of instruction.

But don't students need some guidance in how to study for a test? Yes, and they also need information on the parameters of the test you have designed. After all, you made selective decisions as to the scope of the objectives to be tested. You may have decided to extend the test over two class periods, to include laboratory work as well as paper-and-pencil items. These decisions should not be secrets. The students should know what will be expected of them and the nature of the testing conditions.

At least four to five days prior to the test date, announce the scheduled test. *If* your instruction has consistently employed spiraling of objectives, your class will profit from a structured working session where you help them highlight major concepts. That *if* is not to be passed over lightly. You must teach students how to approach a mixed collection of content topics—no easy task, since most textbooks encourage students to expect that all items on the same page will deal with the same task. Supervised practice may have given students the false impression that they are ready to "ace" a test. What a surprise when the teacher refuses to answer questions during the test! Test preparation throughout the unit should include some simulated test sessions where no teacher or student-to-student help is given. This kind of preparation cannot be left until the day before the unit test, or it is doomed to failure. However, a summing up of all earlier practice sessions, a planned review work session focusing on areas the teacher has diagnosed as needing special attention, and a summary lecture in which the teacher outlines the format of the test, its scope, any special conditions—all these are not only desirable, but necessary. But watch out for poorly planned review sessions that depend heavily on student-initiated questions. This frequently degenerates into a "fishing expedition" as students cast about blindly trying to locate specific items that are to appear on the test. Some characteristics of pretest work sessions which teachers and students have found successful include:

1. Short periods of supervised practice interspersed with student demonstrations, board work, or quick recall Q/A.

2. Carefully sequenced sets of written questions, where the concepts, rules, and problems most in need of review are placed first in the set, where there are more questions than can be completed in class, where separate answer keys are distributed at the end of the session to those who wish to complete the set.

3. Tactics that allow more capable students to assist the teacher or to spend this time on a project of their choice.

Some teachers use review games very successfully. Others warn that games in which chance dictates the order and kind of question to be answered may entrap the class into experiencing a pleasurable today and a painful tomorrow when test questions fail to match game questions.

Administration of the test Because testing periods which are badly managed lead to control problems, the basic routines for administering a test are outlined in Section 12.3 under "Effective and Efficient Handling of Routines." You may wish to look ahead at that short section right now. That section alludes to the problem of student cheating. Why do some students cheat? You know, from your own experience, that external pressures of varying kinds tempt even some bright, but tense, students to engage in dishonest practices. What can you, the teacher, do to help students resist yielding to temptation? For despite meaningful instruction, a variety of assessment procedures, well-designed test preparation, and careful student guidance, we know of no magic way to eliminate the very real pressures of college admissions, parental expectations, peer mores, and the like. There are, however, some specific suggestions which we can make that have proved useful in (1) convincing students that their chances of "getting away with it" are minimal and (2) demonstrating that built-in self-administered penalties are likely to result for culprits.

Copying from an adjacent classmate is probably the simplest and most frequently used tactic of cheaters, and even the most vigilant teacher may fail to spot occurrences in a crowded classroom. In Section 12.5, "A Pound of Cure," we describe a classroom where the teacher suspects copying has occurred during a test. Three courses of follow-up behavior are outlined, and you are asked to react to the relative effectiveness of each. Take the time to read ahead in that brief section now. Of course, we'd all prefer to avoid discipline problems so that a "pound of cure" wouldn't be needed. We'd recommend as an "ounce of prevention" in test administration that the first unit test be prepared in two parallel forms and distributed to the class in accordance with the directions in Section 12.5. There are several ways to concoct question papers that look alike when one glances over a shoulder. You might (1) scramble the order of the choices for each multiple-choice question (as suggested in Section 12.5), (2) reorder individual questions within each cluster of items, (3) include items that are open-ended and require a variety of responses, and (4) alter the details of explanations students are to give in the open-response items. Be sure to read the details of distributing papers, proctoring, scoring, and returning papers in Section 12.5. If you have more than one class section taking a test on the unit, you will have to construct parallel forms unless you are willing to give an automatic handicap to the first class. But if you have a single section taking the test, must you always construct parallel forms? Those teachers who have used unannounced parallel forms for the first unit test find that only sporadic subsequent use is necessary. The students are now never sure whether their neighbor's test is the same form as theirs. An effective preventive? You bet it is!

But what can a teacher do to prevent the successful use of "crib" notes? The best defense against crib notes is the construction of test items that cannot be answered solely on the basis of any kind of notes. Questions above level II are guaranteed to frustrate the student who is depending on hidden notes. The next level of defense is constant vigilance on the part of the teacher and adherence to the routines recommended in Section 12.3.

Remember, the whole point of constructing a test is to sample what students have learned about the unit, not to assess their eyesight. You should be basing future instruction on the analysis of these results. If the results are a mixed bag of half-truths, who suffers most? The successful cheater now may be tempted to cheat even more on the next unit test.

Assessing test results   Before any student's paper is scored, a key should be prepared for each form of the test. The form of the key depends on the form of the test, but it should be constructed with efficiency of scoring built in. If the students have been instructed to write answers or the capital letter of the keyed choice in a space in the right margin, the teacher's key can be constructed by correctly filling in all spaces on a copy of the test. It also helps to write the total possible credit for each cluster of items in the right margin of the key at the end of each cluster. With the key beside a student's paper, the teacher can locate and mark errors and begin the recording of credit obtained by that student. Figure 8.4 illustrates a portion of such a key—one student's responses to the first two sets of short-response items and the teacher's scoring notations thus far.

Since Sam received 16 credits for his responses to the multiple-choice items, he must have correctly answered items 4-9. Notice how the teacher lined up the key on top of Sam's paper so that corresponding responses could be quickly checked, marked only the wrong responses, and indicated the credit accumulated for each section of the test above the slash line and the total credit obtained below the slash line.

It is good policy to score all short-response answers on each paper before addressing student efforts on open-ended questions. However, it is *not* good policy to score all open-ended responses on one paper before going on. Do you see why? Let's assume that the Locus test, alluded to earlier, required students to write two proofs. Even though your prepared key would include the optimum responses to each question, together with the credit to be assigned to varied portions of a response, you will find enough variability among responses to make judgments about scoring difficult. After carefully reading and assessing Sam's response to the first proof, you will compound the problem of keeping the scoring reliable if you now move on to a different item. That makes sense, doesn't it? What makes even more sense is the advantage gained in scoring efficiency. By the time the fifth student's response to the same essay item has been read, particular scoring decisions begin to repeat and thus can be made more easily.

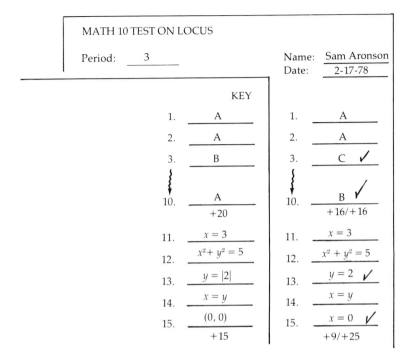

Fig. 8.4 A partially scored unit test.

At last all papers are scored and you are ready to record the grades in your grade book. An additional step may be needed at this point in the case of schools that require the reporting of letter grades. Your cooperating teacher may have guidelines that affect the meaning attached to certain scores. Remember, the raw scores represent *data* that will be used in *evaluating* student achievement. Whether raw scores alone are reported, or whether you need to report letter grade plus raw score or percentage plus raw score, you must make a judgment as to the meaning of the reported grade. Follow your cooperating teacher's suggestions if it is necessary to translate raw scores into letter grades. Consistency in the reporting of scores is important. Many faculty handbooks give only verbal descriptions of each letter grade: A—excellent, B—very good, C—adequate or acceptable, D—marginal or minimal, E—unsatisfactory. In the absence of clear written or unwritten guidance as to the quantitative and qualitative match of raw scores to these letters, we recommend that you determine a cutoff score to represent the lower limit of the Ds. This is a judgment based on the total possible credit assigned to the test and the nature of the content topics and objectives it was designed to sample. If you haven't already done so, arrange the papers in rank order and write all total scores in order in a vertical array. A consideration of clusters of scores will help you make further decisions about cutoff scores in the case of A, B, C, and D papers.

Notice that we haven't suggested "curving" the grades! Why not? What does "curving" the grades mean? Every one of you is familiar enough with statistics to know that "curve" in this case relates to the normal curve. If the population being tested is normally distributed—an assumption hard to defend with class-size groups—then the mean and standard deviation can be calculated and grades of A, B, C, D, and E can be awarded on the basis of the number of deviations of a raw score from the mean. That means that a little over 2 percent of the class must be assigned an A, but another 2 percent must be assigned an E. Is that what happens when your professors tell you they've "curved" the grades? We'd bet it usually doesn't. What does happen? We'd bet that variable numbers of points are added to raw scores to make the total picture congruent with what the professor wants. A more honest term for this procedure would be "fudging" the grades. The correct use of the standard deviation to transform raw scores is described in any standard test and measurement text and will not be treated here.

Why do teachers resort to "fudging" the grades? They may have found themselves in the predicament faced by Mrs. Rely. Her scores on a unit test consisted of a few high grades, but most were dismayingly low. What should she do? Teachers have responded to similar situations in *each* of the following ways: (1) tell the class that most of the scores were so bad that the test won't count and a makeup test will be given to all students, (2) "fudge" the grades by adding an increment to the low scores, (3) tell the class the test is over, the scores are in the grade book, and that's all there is to it, or (4) carefully analyze details of test results to identify the possible source of poor student achievement and then share this information and the decisions based on it with the class. Which of these approaches is likely to increase student confidence in the evaluation process and the teacher? To us, the fourth choice is the only reasonable alternative, particularly if the teacher has followed earlier suggestions on test preparation. But after carefully constructing the test and accurately scoring all the papers, what else is left to analyze? That's exactly what we need to consider next.

Item analysis   Even the trained and experienced test constructors at the Educational Testing Service (ETS) regularly reject items after analyzing the scores of a sample population. Why would they reject well-written items that match the objectives in terms of *logical* analysis? *Logical* does not equal *psychological* here any more than it does in the areas of intellectual development and human learning. Students may "read into" an item a meaning not intended by the teacher. If this happens in a random way, the teacher might well discount it. But if a number of students misread an item in exactly the same way, the item needs to be studied for the source of the error. Suppose a number of high-scoring students got a particular item wrong, while a similar number of low-scoring students got the same item correct? You'd check that item very carefully. Would you be suspicious of an item's wording on the basis of the score of one student? Two students? More than half the class? How do you decide? Here we are guided by the professional test constructor who employs, among other things, two indices: item difficulty and item discrimination.

The usual unit test (unless all the unit objectives are to be mastered by all students) is designed to differentiate the more capable students from the less capable. The teacher intends to include items of varying degrees of difficulty. But is that what happened? It's easy to ascertain. The *item difficulty* (or *difficulty index*) is defined as the *percentage of the class who succeeded on that item*. The computation of the difficulty index for two different scoring cases is illustrated in examples 1 and 2 below.

Example 1: Item 7 is a multiple-choice item worth one credit. No partial credit is given. Twenty-four students in a class of 36 students answered item 7 correctly.

$$D_f = \frac{Ns}{T} = \frac{24}{36} = .67 \text{ or } 67\%$$

(D_f = difficulty index; Ns = number of successful students; T = total number of students taking the test.)

Example 2: Item 20 is worth 10 credits, but partial credit is given. In the same class of 36 students, some students received 10 credits; some 8 credits; some 5 credits; and so on. The sum of the credits obtained by the entire class on the item was 183 credits.

$$D_f = \frac{\Sigma_c}{T \times I_c} = \frac{183}{36 \times 10} = \frac{183}{360} = .51 \text{ or } 51\%$$

(Σ_c = sum of the credits and I_c = credit for the item.)

The two items in examples 1 and 2 would be considered moderately difficult. In general, you should expect the difficulty index of main items to range between 60 percent and 70 percent. It is often appropriate for a teacher-made test to include a number of items with higher or lower indices. The teacher is the final judge of this, and that judgment depends in part on the original objectives set for the unit. In any case, test designers would recommend a long, hard look at items where $D_f \geq 90\%$ or $D_f \leq 30\%$. Notice the apparent anomaly. Professional test constructors call this index *item difficulty*, even though the higher percents represent easier items. As you read further, remember a D_f of 92 percent represents a very easy item for that group of students, while a D_f of 21 percent represents an item to which few students were able to respond successfully.

We also have need of another kind of index to help with analysis of items. Why? What else might we want to know? Study the information given in example 3 and try to identify one further area of potential concern.

Example 3: Item 7 described in example 1 was successfully answered by 6 of the 12 students in the top third of the class on this test, by 8 of the middle group of 12, and by 10 of the 12 lowest scoring students.

What did you infer from the data in example 3? If you red-flagged the scoring results of the highest and lowest students on this item, you're on the right track. A test is supposed to differentiate those who have met the instructional objectives from those who haven't. Yet in this case half of the top scorers, the "haves," got the item wrong while more than three-fourths of the low scorers, the "have nots," got the item right. Why did this happen? The wording of the stem may be at fault, or the nature of the distractors may suggest erroneous information to the brighter student. So even though this item seemed to be acceptable from the point of view of item difficulty (D_f=67 percent), it is providing false information as to the "haves" versus the "have nots." The data provided in example 3 is that used in determining the second index, *item discrimination*. Example 4 illustrates the computation of item discrimination for the item referred to in examples 1 and 3.

Example 4: In a class of 36, $\frac{1}{3} \times 36 = 12$. Therefore, there are 12 highest scores (H) and 12 lowest scores (L) on the total test.

$$D_c = H_s = L_s = 6 - 10 = -4$$

(D_c = discrimination index; H_s=number of H students who successfully answered this item; L_s = number of L students who successfully answered this item.)

The negative number is an immediate signal that this item is discriminating in the reverse direction from that desired. The use of the top third and bottom third of the class is a modification of the 27 percent used by professional testwriters—a modification that works fairly well for a class size of 20 or more. If the class size is less than 20, the top and bottom halves of the class should be used with the middle scorer ignored in the case of an odd number. How is the discrimination index computed for an open-ended item worth partial credit? Example 5 illustrates the computation for item 20 referred to in example 2.

Example 5: Students could obtain 1 to 10 credits on item 20. The H group (12 students) in this class of 36 accumulated 96 credits, while the L group (12 students) accumulated 36 credits.

$$D_c = (H_\Sigma - L_\Sigma)/I_c = (96 - 36)/10 = 60/10 = 6.$$

(H_Σ = sum of credits obtained by H group; L_Σ = sum of the credits obtained by the L group; I_c = total possible credit value of the item, as before.)

Notice that the formula has been adjusted in example 5 to obtain, as before, an integer. In the case of item 20, where item difficulty was 52 percent, this positive discrimination index characterizes this item as discriminating in favor of the top scorers. On the basis of these two indices, item 20 appears to be doing the job it was designed to accomplish.

Figure 8.5 depicts the presentation of test data in an item analysis sheet prepared by Mrs. Bowden. Mrs. Bowden uses sheets of this form for each of her unit tests, so she has prepared copies with sufficient columns to list members of each of her classes and sufficient rows for tests of varying number of items. She followed these steps in recording the data:

1. Arranged the scored test papers in rank order.

2. For each paper, recorded student name and score on each item to be given part credit (as in the case of 30_a and 30_b), distractor selected for multiple-choice items, $(+,-)$ for true-false or other short-response items. "O" is recorded wherever the student left a blank on the answer sheet.

3. Completed the N_s (or Σ_c) column.

4. Computed $H = L$ and used colored ink or squiggle lines to visually separate those groups on the table.

5. Computed D_f and D_c.

Mrs. Bowden has completed most of the table in accordance with these five steps. On item 2, for example, blanks represent correct responses; O, omitted responses; and other letters, selected distractors. Six students got this item correct. Therefore, $D_f = 6/26 = .23 = 23\%$. Item 2 has a discrimination index of 4 since $H_s - L_s = 4 - 0 = 4$. Now it's time to check your understanding of the two indices. First, check Mrs. Bowden's figures where given. Then complete the rest of the table. You'll find our results in Simulation Exercise D.

Does the work Mrs. Bowden must have done after the test seem like "closing the barn door *after* the horse has wandered away"? It will be just that if she files the analysis sheets neatly in a file drawer and does nothing further with them. But she had much more in mind. Think back to the test post-mortem suggestions of Chapter 2. *Before* Mrs. Bowden returns these tests, she will have completed the item analysis in order to (1) identify error patterns and (2) identify students who need individual help or a differentiated follow-up assignment. The test post-mortem will be planned on the basis of this information and structured according to suggestions given in Section 2.3, Chapter 2. If Mrs. Bowden has isolated some "poor" items, identified items that many students got wrong or omitted, or identified items in which one distractor was chosen above others, she can share that information with the entire class. As a result, her questions can focus on these distractors and items. There's nothing more deadly than a test post-mortem in which the teacher or a student recites

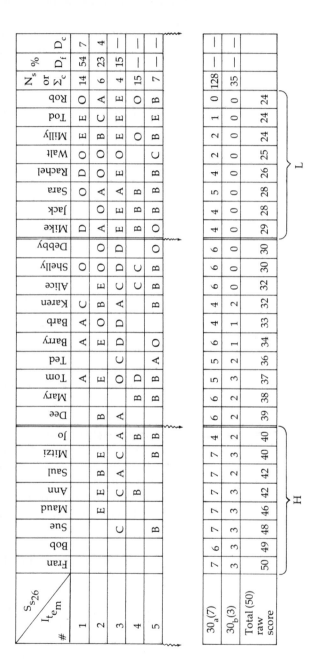

Fig. 8.5   A partially completed item analysis sheet.

each and every answer and tells why it's correct, item by item. The "haves" are bored by the repetition, and the "have nots" are discouraged. Often the "have nots" had some reason for making an error and they are likely to repeat the same error if the limitations of that reason are never explored. However, if Mrs. Bowden draws attention to item 5 on the test by presenting the data below, the entire class can be meaningfully absorbed in questioning the results.

Item 5.   If f and y are the measures, in feet and yards respectively, of a given distance, find k in $f = ky$.

   A. 1/12
   B. 1/3
   C. 1
   D. 3
   E. 12

Results

Choices	A	B	C	D*	E	Omit
# of Ss	1	13	1	7	1	3

Question/answer and straw poll techniques can then be used to get students thinking about *why* half of the class members were misled by distractor B while so few selected A, C, or E. She might also profitably explore the reasons for the three omitted answers. Such an analysis not only has attention-getting and -keeping potential, but helps to convince the students that the test and the test post-mortem are parts of systematic instruction. Mrs. Bowden may subsequently decide that the multiple-choice format is not the best way to test this objective. And if Mrs. Bowden wants to avoid reinventing the wheel each year, she'll record that kind of change when she adds information on these test items to her resource file.

Storage and retrieval of test items Busy teachers need a system that allows quick location and organization of a large number of test items. A useful system is one that facilitates adding, deleting, and refiling individual questions—both from past classroom tests and from published test sources. The system described here can be constructed at nominal expense. Obtain a supply of five-by-eight-inch index cards, a knitting needle (preferably steel), a pair of scissors, and two pieces of scrap plywood. Cut both pieces of plywood to five-by-eight-inch dimensions and trace a pattern of holes similar to that shown in Fig. 8.6. Clamp a two-inch thick stack of index cards between the two pieces of scrap plywood (pattern piece on the top) and drill the series of holes. The plywood serves to stop the cards from twisting as the drill mounted in the drill press is depressed through the stack. Select a drill size slightly larger than the diameter of the knitting needle to be used. (If you are not mechanically inclined, this is a good time to enlist the support of the industrial

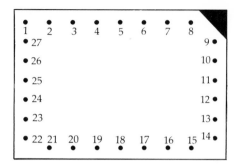

Fig. 8.6   Pattern for drilling holes.

arts teacher.) Work slowly to ensure that heat dissipates rapidly enough to prevent the cards from scorching. Cut off the upper right-hand corner of the stack with a table or band saw before unclamping to ensure proper orientation of each card. Teachers with more than one course preparation will avoid confusion by making stacks of differently colored cards. For example, white cards might be used for algebra and green cards for eighth-grade mathematics.

The area remaining inside the border of holes should be approximately four by seven inches, a convenient size for pasting questions cut from dupli-cated tests of other teachers or from xeroxed pages of other noncopyrighted material. Remember the five-by-eight-inch cards we recommended for use in your own item construction? If you wrote your items on cards drilled as described here, you are now ready to begin coding information on each item you intend to store for possible future use. The various categories of infor-mation about each test item are encoded and stored by using scissors to notch out the appropriate hole, as shown in Fig. 8.7. It is also imperative to keep the cutoff corner in the upper right-hand position while notching, since that is to be its position during later retrieval. The rear of each card is employed to record the answer or answers for which credit is allowed, let the point value of

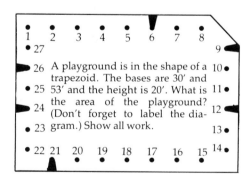

Fig. 8.7   Sample front side of item storage card.

the answer or each of its parts, and any other pertinent information desired by that system user.

Obviously, a master or key card must be designed and its schema must be followed exactly in the encoding process. A suggested code for junior-high mathematics is illustrated in Table 8.1.

Table 8.1
A code for junior-high mathematics test item storage

Hole #	Examples of encoded information	Hole #	Examples of encoded information
	Unit		*Question Type*
1	Whole numbers	15	True-false
2	Fractions	16	Modified true-false
3	Decimals	17	Completion
4	Integers	18	Multiple choice (including matching)
5	Open sentences		
6	Geometry	19	Graph/construction
7	Graphs	20	Original problem (essay)
8	Measurement	21	Other essay
		22	Laboratory performance
	Domain		*Difficulty Level*
9	Cognitive	23	Easy ($\geq 85\%$)
10	Psychomotor	24	Moderate
		25	Hard ($\leq 30\%$)
	Cognitive Level		*Discrimination*
11	Knowledge—I	26	Positive
12	Comprehension—II	27	Zero
13	Novel application—III		
14	Above III		

All but two of the categories of information to be stored can be determined and notched into the card before an item has been administered to students. The two remaining categories, difficulty level and discrimination, must await scoring of actual test results and calculating those indices. But why is affective domain omitted from the code? That's right—the measurement of stable and consistent attitudes requires assessment forms not reduced meaningfully to single items.

Once a pool of test items is begun, additional cards can be easily added at any time. There is no need to be concerned with placing new cards at any particular place in the deck or stack. The only relevant point with regard to inserting cards is to keep the cutoff corner in the proper (upper right-hand) position. It is obviously a simple matter to remove and discard cards for items made obsolete by curricular changes.

Retrieval is where the investment of the small amount of time and effort required to start and maintain this system pays high dividends. Teachers who do not use this or similar systems typically save copies of tests given in previous years. Two or three days prior to a scheduled test date they search through sheets of test pages trying to recall which items proved to be ambiguous, too difficult, or too easy. More often than not, such a procedure yields disappointing results because one finds many items of a single type focusing on one particular aspect of the subject matter and not enough of other types (usually the more difficult to construct) dealing with other aspects of the subject matter. Hours or days later, the net result is many sheets of old test items with cross-outs, red arrows, and lines sandwiched between sheets of hastily written (or uncritically plagiarized) items. How good are the chances that an ordinarily cooperative spouse, friend, or school secretary can produce the desired stencil or master given this sort of raw material?

Consider the procedure employed if you start with the system described here. If your task is to construct a major test on similarity and indirect measurements, you begin by placing the knitting needle through hole 6 and lift up. All cards containing items on similarity fall out and are stacked in one pile. The needle is then inserted in hole 8 of the cards remaining in the master deck to sort out cards relevant to indirect measurement. The master deck is returned to its place and further attention is given to the two subdecks in question. Similar successive uses of the needle on these subdecks will quickly sort out the sampling desired in terms of the cognitive level, as well as the demonstrated difficulty and discrimination indices. If more items than needed are available, the decision of which to use is made and the needle quickly sorts the selected items by question type. A rubber band is used to secure each group by question type with a card of directions for answering each type on top of the proper pile. The piles are arranged in order of desired appearance on the test and then the typist is approached with the usual request for assistance. Some claim this is more effective than a box of candy. After the typing is completed it is a simple matter to return the cards to the master deck. As long as the cutoff corner is in the upper right-hand position, they can be retrieved at any future time. Cards containing items not used previously should be held out until after the test is scored and the results tabulated. Then the difficulty level can be notched before filing these particular cards. The entire system is once more ready for the next major test, quiz, or even review session.

Report cards and final grades Scores on paper-and-pencil tests of cognitive objectives are by far the most commonly used ingredient in arriving at report card grades. Such grades, which are intended to represent student achievement over a marking period or entire course, are indicators of summative evaluation. But shouldn't degree of attainment of affective and psychomotor objectives also be incorporated into such grades? That depends partly on the overall objectives of the course and partly on the definition of the report card grade used by a particular school. As a student teacher, be sure to check both written policy and unwritten tradition on these matters with your cooperating teacher. Also find out if class participation and work on out-of-class projects is to be quantified and, if so, how. Suppose some diagnostic quizzes were given solely to determine the future course of instruction. Clearly these should be omitted from the calculation of a composite score. Were the unit tests designed to assess units of varied importance? If so, these scores cannot be treated as if each unit was of equal worth. Figure 8.8 depicts part of a page from Mr. Means's grade book. Since Mr. Means has recorded raw scores, he may simply obtain the sum of the raw scores of all quizzes and tests, except that given on October 31 (we hope you noticed that). For the four weeks depicted here, he would obtain 141 for K. Abrams. If a percentage grade is to be reported, that would be calculated by dividing each such total by 150. Notice the difference in total possible scores on the congruence test and the inequality test. If Mr. Means averaged 88 and 35 (K. Abrams's scores), K. Abrams would be assessed as achieving 62 percent of the work on these two units. If K. Abrams is as alert as her actual score on these two tests (123/130=95%) indicates, she'd be complaining loud and clear! And she should! But what if Mr. Means had converted individual raw scores into percentages—a common practice—and then averaged all unit tests? Would the students have been treated fairly? Let's see what happens to K. Abrams. She'd receive 98 percent (88/90) on the first

GEOMETRY—PERIOD 4 CLASS

Week 6

		10/10-10/14					10/17-10/21					10/24-10/28					10/31-11/4				
		M	T	W	Th	F	M	T	W	Th	F	M	T	W	Th	F	M	T	W	Th	F
1.	Abrams, K.	8				88					10					35	15				
2.	Anthony, R.	7				78					6					30	15				
3.	Butler, A.	10				85					9					38	18				
23.	Zeh, T.	2				55					5					25	10				

Quiz—Congruence (10) • Test—Congruence (90) • Quiz—Inequalities (40) • Test—Inequalities (40) • Diag. quiz—Parallels (20)

Fig. 8.8 Sample entries from a grade book.

test and 88 percent (35/40) on the second. Her average on the two would be 93 percent. A fluke? Well, teachers-to-be, you know how to find out. Is

$$[(a/b) + (c/d)] \div 2 = [(a + c)/(b + d)] \div 2?$$

You'd be amazed at the number of teachers who behave as if they are! Of course, all this concern for conversions which maintain the initial emphasis of instruction and related assessment is based on the assumption that the teacher took this into consideration in planning and test construction. Moreover, if Mr. Means had identified ambiguous items after item analysis—items which substantially altered final test scores—and did nothing to indicate the lesser degree of validity of that test, then all bets are off.

Mr. Means has a number of other decisions to make. Should a composite score for short quizzes be calculated separately and weighted less than a composite score for unit tests? Mr. Means also required a construction/design project of each student. How should this assessment count? In addition, several students volunteered to construct a large demonstration model after engaging in the required project. If Mr. Means's school reporting system includes a space for "effort" or "attitude," he has a specific way to report such results. Otherwise, what does he do? We'll take another look at effort assessment in the next section, since this area is clearly intended to emphasize the affective domain.

There are no easy answers to the questions posed here. However, it is important to keep three principles up front:

1. Report card grades always represent a *value* judgment. (Don't let the use of numbers fool you or anyone else into believing they are objective and quantitative results.)

2. Decisions as to weighting of varied projects, quizzes, tests, activities, and so on, must reflect instructional emphasis. (The die is cast after you report results on the first project—perhaps during construction of the first test.)

3. Overall principles guiding the decisions you've made must be shared with the students as early in the game as possible. (Past experience may cause them to mistreat assessment.)

The last principle is not as easy as it sounds. A good practice is to introduce the students to the teacher's task in marking report cards. Have each student keep a record of his or her individual scores and let each use a calculator during a class period to compute composite quiz and test scores according to the principles you've decided to follow. Sammy, who had a string of 8s on seven of nine 10-credit quizzes and a pair of 2s begins to see what those low scores do to a composite score. You can explain how projects, laboratory work, and the like are included and then have each student submit (1) the individual computations, (2) the final grade deserved (in his or her judgment), and (3) the reasons why that grade is deserved. Teachers who employ the above practice are often pleasantly surprised by the students' discerning judgment. But the success of the venture depends largely on the respect the teacher has engendered thus far in evaluation and the realization that while the final *judgment (evaluation)* is the teacher's responsibility, any new data will be studied carefully before that decision is made. Respect, values, perception as to results—sounds like we've spilled over into the affective domain. Let's take a direct look at assessment in this area.

8.4 ASSESSMENT OF AFFECTIVE OBJECTIVES

Complying, choosing to participate, exhibiting stable and consistent attitudes—these were the three levels of the affective domain described in Chapter 5. How does the teacher get at the evaluation of such objectives? There

> Directions: This exercise will measure what students think about mathematics. The exercise will *not* be used as a part of your grade because there are no correct or incorrect answers.
>
> 1. *Do Not* put your name on this exercise.
> 2. *Do* answer all the questions on this exercise.
> 3. *Do* answer the questions on the basis of what you have been taught in this course.
> 4. When you are finished with the exercise, turn your paper over on your desk. You will pass them to the end of the row when everyone has finished.

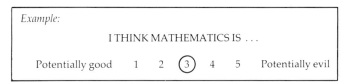

If you feel mathematics contains all potential for good and none for evil, circle 1.

If you feel mathematics contains mostly potential for good, circle 2.

If you feel mathematics contains about as much potential for good as for evil, circle 3.

If you feel mathematics contains mostly potential for evil, circle 4.

If you feel mathematics contains all potential for evil and none for good, circle 5.

In this example the student felt mathematics contained about as much potential for good as for evil. Number 3 has been circled.

DO NOT BEGIN UNTIL YOU ARE TOLD TO, then turn the page and begin work.

Fig. 8.9 Directions for semantic differential scale.

are two chief types of assessment that teachers have found useful: (1) self-evaluation forms (that is, student response forms intended to reflect their beliefs, attitudes, opinions) and (2) observation instruments.

One kind of student self-evaluation was referred to in the last section when students were to judge the composite grade deserved. Student self-evaluations that are honestly completed represent a student's *perceptions* of likes, dislikes, ability, and beliefs. It is well to keep in mind that how a person perceives he or she will behave is frequently inconsistent with actual behavior. Yet, taken as a whole, the direction and intensity of likes and dislikes are a key to attitudes and so, to behavior. For this reason, it is considered more useful to sample aspects of the same attitude by means of several items on an inventory. Students are sometimes asked to react to a collection of statements by checking or circling a numeral representing the extent to which one of two polar words represents their attitude. Such a scale consisting of polar descriptors listed in two columns is called *semantic differential*. The scale depicted in Fig. 8.9 requires the student to identify an overall attitude toward mathematics.

A scale such as this is easily constructed. All you do is list polar words or phrases characterizing the extremes of the attitude being assessed in two columns. One caution! All the verbal cues that tend in the same attitudinal direction should *not* be listed in the same column. For example, "Valuable to Me" and "Valuable to Society" are listed in opposite columns in Fig. 8.10. This construction feature will tend to identify the careless reader who circles numerals without reading each pair of descriptors. Why are the students directed *not* to put their names on the response sheet? Anonymous responses are more likely to be honest—especially when attitudes perceived as undesirable from the teacher's point of view could be revealed. Student teachers have used this scale early in their teaching experiences to assess student attitudes toward algebra and geometry as well as toward the broader area of mathematics in general. The results can be helpful to the teacher. For example, if most of the students' responses indicate little or no correlation between

I THINK MATHEMATICS IS . . .

Simple	1 2 3 4 5	Complex
Easy to learn	1 2 3 4 5	Difficult to learn
Boring	1 2 3 4 5	Interesting
Impractical	1 2 3 4 5	Practical
For everyone	1 2 3 4 5	For scholars only
About real things	1 2 3 4 5	About theories
Valuable to society	1 2 3 4 5	Worthless to society
Worthless to me	1 2 3 4 5	Valuable to me
Related to biology	1 2 3 4 5	Unrelated to biology
Related to physics	1 2 3 4 5	Unrelated to physics
Unrelated to art	1 2 3 4 5	Related to art

Fig. 8.10   Sample semantic differential scale.

mathematics and biology, the mathematics teacher can begin to work cooperatively with the biology teacher to emphasize such interconnections as the kinds of symmetry exhibited by living things or the surface area versus volume relationships so important to cell size and function. A readministration of the scale at the end of the course is a way of assessing changes in attitude. But don't expect a *massive* shift in attitude despite your concentrated efforts! Experience and research results indicate that attitudes are not easily changed and are sometimes held despite controverting data—witness the long and continuing struggle against racial prejudice. Yet, attitudes are affected by long-term, heavy doses of exposure to data nonsupportive of the initial attitude. *At the least,* let's not have students exiting from a course with less positive attitudes about the content than they possessed at entry level.

It is a truism that teachers unintentionally affect attitudes by their verbal and nonverbal behavior. If the teacher verbalizes about the importance of mathematics in real life but responds in an impatient manner to students who volunteer examples, then student faith in that teacher's attitude is shaken. On the other hand, a teacher may praise such class participation but structure all assignments and tests so that real-life applications are ignored. It is not easy for the teacher to identify these and other forms of unintentional affective instruction without systematic feedback from the students. One way of obtaining such feedback is by means of a short teacher-constructed inventory which focuses on what the teacher hoped was included in instruction. These instructional features, often in the form of questions, might be rated on a three-point scale (agree, disagree, or uncertain) or on a five-point scale. Such scales are fashioned after that developed by Likert and are thus called Likert-type scales. Figure 8.11 illustrates one such scale.

Answer each of the following questions by circling one of the numerals after each question.

	Definitely Not	I Don't Think So	I Can't Decide	I Think So	Definitely Yes
A. Do you understand most of the mathematics in the unit just completed?	1	2	3	4	5
B. Do you see any way to apply the content in this unit to real-life situations?	1	2	3	4	5
C. Were the homework assignments usually too long?	1	2	3	4	5
D. Were the homework assignments usually interesting?	1	2	3	4	5
E. Was the mathematics presented in a way which interested you?	1	2	3	4	5
F. Do you think too much was expect of you in this unit?	1	2	3	4	5
G. Would you like to study further on this topic?	1	2	3	4	5

Fig. 8.11   Sample Likert-type scale.

Scales such as this one can be modified by providing a "comments" space after each item. For instance, a student who circled 1 or 2 on item A might then list areas or topics that seemed most difficult. Notice the use of the pronoun *you*. It is important to emphasize that each student should reflect personal reactions rather than what "a lot of kids say." It's strange how "a lot of kids" boils down to two or three individuals when all students have a chance to express themselves without fear of peer pressure. On the other side of the coin, it's well to avoid the pronoun *I*. Item F, for example, was designed to avoid this problem. The reader often reacts differently to the same item if phrased: "Do you think I expected too much of you?" When should this kind of inventory be administered? Certainly not after the students have been given test grades. Can you imagine the effect on responses of the student who was coasting along and got a failing grade in this test? Generally, such inventories prove most useful if administered after the students have taken the test, but before the test grades are reported. Could the inventory be administered on the test day after students have handed in the completed test? If *all* students have the five to eight minutes needed to complete the inventory, this is an effective use of time. Otherwise, it can be administered at the beginning of the next class. In any event, inventories should not be completed by some students on one day and by others at another time, and *never* ask students to complete these outside of class time. In both cases, responses are more likely to reflect pressure group comments than individual perceptions, although forms taken out of class may just disappear—to be found in the cafeteria, on the school bus, or tucked in a library book. A carefully thought-out inventory that is administered with due regard for these cautions is more apt to be taken seriously by the students, especially if feedback is shared with them later. But overkill can destroy the impact! An inventory administered after *every* test becomes a routine.

Another type of student self-evaluation of attitudes involves the writing of open-ended essays. The students may be asked to write an essay on the topic "What Mathematics Means to Me" or to expand on a topic sentence or question, such as: "A mathematician, like a painter or a poet, is a maker of patterns." Again students need to be told why the teacher is asking them to write these essays and what use will be made of them. Feedback on areas of major agreement and disagreement should be shared with the class. This particular device has enormous possibilities for identifying and affecting attitudes as to the nature of mathematics. (Remember the "invented versus discovery" question raised in Chapter 4?)

Essay statements might be utilized in structured small-group discussions or expanded on by the teacher, who could present data from the history of mathematics or about the nature of mathematical processes and products. The teacher who has specified objectives dealing with attitudes on the structure of mathematics can use phrases from those objectives as topics, topic sentences, or questions. The essay device used in this way becomes a powerful technique for getting feedback and for using that feedback in subsequent instruction. This is another instance in which the cognitive domain is closely meshed with the affective domain. An essay on the real-life applicability of mathematics may disclose strong biases based on erroneous or completely missing information.

The second of the assessment types listed at the beginning of this section—observation—does input individual data. Moreover, systematic observation shares with the essay technique the ability to focus on higher-level affective objectives. However, observation has an advantage over all types of student self-analysis because actual behavior, rather than perceptions as to likely behavior, can be identified and recorded. The form shown in Fig. 8.12 was designed by a teacher who had specified seven affective objectives to be aimed at during a lab-oriented approach to a unit on similarity. Observation must be conducted on a scheduled basis so that no students are inadvertently ignored. Hours after the event, selective recall takes over and much relevant data is lost.

Students

Behavior	*Paul*	*Barry*	*Lisa*	*Allan*	→
1. Works cooperatively with other students.					
2. Treats equipment with care.					
3. Tries some of the "extra" challenge problems.					
4. Volunteers to add real-life pictures, clippings, etc., to display on similarity.					
5. Works extensively on labs when given a choice between these and free time.					
6. Asks for reference books on surveying, art or photography.					
7. Chooses to do a science-related project.					

Fig. 8.12   A sample observation record form.

Observational data may be used to obtain a summative evaluation of the students' progress toward affective objectives. Indeed, some systematic collection of appropriate data must be employed if the teacher is to eventually record a grade in an "effort" category on the report card. As in all other school reporting systems, the operational definition of "effort" must be gleaned from faculty handbooks and tradition. The particular interpretation placed on that definition by any teacher should be communicated to students early in the year. This category of reporting is perhaps the most abused in the evaluation process. It is not unheard of for Mike to receive an A in the achievement column for algebra and a 1 (inadequate effort) in the effort column for that subject. Discerning parents will rightly question that teacher's understanding of evaluation and the objectives of instruction. Mike will undoubtedly continue sliding through algebra class when he can meet the objectives set by the teacher at the A-level without half-trying. In the same class, Eileen received an E in achievement and a 5 (top of the scale) in effort. Do you see any reason for a parent-teacher-guidance counselor conference before sending such a report home? Considerable care is always called for in reporting affective assessments.

8.5 ASSESSMENT OF PSYCHOMOTOR OBJECTIVES

A quick review of the psychomotor objectives listed in Section 5.3 of Chapter 5 will convince you that there is *one* valid way of assessing such objectives: systematic observation. The teacher must observe each individual perform the motor skill. An observational recording sheet analogous to that illustrated in Section 8.4 can be constructed for this purpose. Then the teacher can set up a lab practicum test in such a way that small numbers of students perform a task at lab stations with the teacher observing and recording behavior while the rest of the students work at pencil-and-paper tasks at their seats.

Although it is a rarity to find a schoolwide reporting system that includes specific attention to the assessment of psychomotor objectives, the teacher should not conclude that assessment in this domain is unimportant. Whenever psychomotor objectives are part of instruction, their assessment should be part of evaluation. In most cases, the report card grade referred to as *achievement* includes assessment of all instructional objectives. Once more the teacher has to make a policy decision. How much shall performance on these related psychomotor tasks count? The answer, as always, is based on the overall set of objectives and the emphasis of instruction.

8.6 CROSS-REFERENCES TO A SAMPLE CBTE EVALUATION INSTRUMENT

Evaluation is easily matched to a major cluster of competencies on the sample CBTE instrument. Category 5 contains direct references to test and item construction, to item analysis and its use, to written tests, and to other evaluative techniques. Notice how this cluster of competencies depends upon earlier items. For example, in order to effectively perform competency 5.2, the teacher not only has to construct well-written test and quiz items, but these must match the instructional objectives originally selected. If the level III cognitive objectives were focused on during instruction, evaluation should include attention to these objectives.

We begin to identify a variety of interfaces among competencies. The concept of formative evaluation demands that feedback obtained from varied forms of evaluation be utilized to remediate, to diagnose, to extend knowledge, perhaps to counsel students into other curricular choices. In category 5, competencies 5.3, 5.4, and 5.6 speak directly to this issue. But what other competencies might the teacher be demonstrating? Did you key in on the need to provide for "individuals"? The use of small groups in test post-mortems, of structured varied assignments, or even the parent-teacher-guidance counselor conference—all were suggestions focused on individual strengths and/or weaknesses. Did you identify competencies 2.64, 2.65, 2.66, and 2.67 as those related to these issues of diagnosis, remediation, and enrichment? We find that these competencies are among the most important ones for continued professional growth. Throughout Chapters 9, 10, and 11 we will continue to give attention to resource materials that have potential for enriching instruction, for remediating, or for capturing student interest, and thus preventing learning problems.

8.7 SUMMARY AND SELF-CHECK

Repeatedly throughout this text, the importance of feedback getting, giving, and using has been emphasized. In this chapter, the formal feedback process known as evaluation was explored. The reader was introduced to a variety of assessment techniques, including the observational scale and the student checklist, for each of the domains. For the cognitive domain, special attention was given to the advantages and disadvantages of various types of paper-and-pencil test items. The classroom test process was outlined from table of specifications through to the reporting of grades.

In focusing on evaluation, we are once again emphasizing the need to continually and systematically assess your teaching. All the popularity contests may be ignored if the "most pleasing" teacher never achieves the objectives of instruction.

After interacting with the material in this chapter, you should be able to:

1. Operationally define evaluation, testing, formative, summative and diagnostic evaluation, objectively scored items, subjectively scored items, norm-referenced testing, and criterion-referenced testing.

2. Identify the common defects in test items and revise such items in accordance with the rules of thumb in this chapter.

3. Construct test items that match specified objectives and correspond to the rules of thumb in this chapter.

4. Critique a completed table of specifications for match of items to the indicators in each margin, given a copy of the corresponding test.

5. Construct a table of specifications given a unit test.

6. Complete an item analysis sheet (including the calculation of item discrimination and item difficulty), given a set of scored test papers.

7. Critique individual test items, identify error patterns, and outline the major components to be included in a test post-mortem, given a copy of the test, the corresponding table of specifications, and the completed item analysis sheet.

8. Convert a set of raw scores into letter grades and justify your evaluation decisions.

9. Construct a key for a given test item and indicate how credit will be distributed on the basis of the guidelines in this chapter.

10. Design an inventory checklist or observational scale to assess given attitudinal objectives.

11. Design an observational scale to assess given psychomotor objectives.

Check your own mastery of these objectives as you respond to the exercises that follow. Both student teaching and regular contractual teaching will offer repeated opportunities to further develop your evaluation skills.

8.8 SIMULATION/PRACTICE ACTIVITIES

A. Each of the following questions is defective in some respect. Identify the defect and revise the question so as to correspond to the rules of thumb in this chapter.

Group 1: Completion items

1. A ___(1)___ is a parallelogram with one right angle.   1. _____

2. The length of a side of a square is 5 inches. Its area is ___(2)___.   2. _____

3. A parallepiped with rectangular faces is called a(n) ___(3)___ ___(3)___.   3. _____

Group 2: Multiple-choice

4. How many faces does a cube have?

 A. 6
 B. 8
 C. 4
 D. 5   4. _____

5. *ABCD* is an equilateral quadrilateral. *ABCD* is a

 A. parallelogram
 B. rectangle
 C. rhombus
 D. square   5. _____

6. Money paid for the use of money is called

    A. rate
    B. interest
    C. principal
    D. check   6. _____

B. Exercise F of Chapter 6 referred to the Sequence of Content Modular assignment in the Appendices. For each competency included in your response to that assignment, construct a matching assessment item.

C. A student teacher prepared a table of specifications which indicated that items 3, 10, and 12 were at level III of the cognitive domain. The cooperating teacher agreed with the classification of item 12 but questioned the designation of the other two items as level III. Help the student teacher reclassify items 3 and 10 and justify your decisions in terms of the criteria for each appropriate level of the taxonomy. *(No log table was given to students.)

    3. If the number 0.00000084 is expressed in the form 8.4×10^{a}, what is the value of a?   3. _____

    10. Express in simplest form: $\dfrac{(b/a) - (a/b)}{(1/a) - (1/b)}$ 10. _____

    12. If $\log_{10} 2 = 0.3010$, then $\log_{10} 5 =$

        A. 2.5×0.3010
        B. $0.3010/2.5$
        C. $2.5 + 0.3010$
        D. $1 - 0.3010$
        E. none of these   12. _____

D. In Section 1.3 you were asked to compute the values of the indices missing from Mrs. Bowden's Item Analysis sheet. The correct results are listed here.

Item #	$D_f(\%)$	D_c
3	15	3
4	58	3
5	27	5
30a	70	4
30b	45	7

With this information, the data on choices from the Item Analysis sheet, and the items themselves, Mrs. Bowden can now decide whether particular items need restructuring before storing them in her test file and whether certain error patterns provide an answer to learning problems.

    Item 5 has already been subjected to this kind of analysis. The remaining short-response questions are reproduced here.

1. A line segment $\overline{OP}$ is drawn from the point (0,0) to the point (6,4). What are the coordinates of the midpoint of $\overline{OP}$?

    A. (2,3)
    B. (3,2)
    C. (3,4)
    D. (6,2)
    E. (12,8)

2. If x, y, z, and w are all real numbers and none of them is zero, which of the following expressions can equal zero?

    I. $x + y + z + w$
    II. $x^2 + y^2 + z^2 + w^2$
    III. $x^3 + y^3 + z^3 + w^3$
    IV. $x^4 + y^4 + z^4 + w^4$

    A. I only
    B. III only
    C. II and IV only
    D. I and III only
    E. I, II, III, and IV

3. If $x(x - y) = 0$ and if y does not equal zero, which of the following is true?

    A. $x = 0$
    B. Either $x = 0$ or $x = y$
    C. $x = y$
    D. $x^2 = y$
    E. Both $x = 0$ and $x - y = 0$

4. The town of Mason lies on Eagle Lake. The town of Canton is west of Mason. Sinclair is east of Canton but west of Mason. Dexter is east of Richmond but west of Sinclair and Canton. Which town is farthest west?

    A. Mason
    B. Dexter
    C. Canton
    D. Sinclair
    E. Richmond

For items 1-4:

a) Identify those with distractors which contributed little or nothing to the item. Rewrite these.

b) Analyze the major error patterns and conjecture the probable learning problem that led to those errors. Suggest way(s) of remediating this learning problem.

E. On a unit test with a total possible raw score of 105, the following raw scores were obtained for a class of 31 students. Assume no problems with defective test items or test administration.

105—1	84—2	73—1
101—2	83—3	72—1
100—1	82—1	71—1
92—2	81—1	69—1
91—1	80—2	68—1
90—1	76—3	59—1
87—1	75—1	57—1
86—1	74—1	

Your school has a fixed scale to be used when assigning letter grades to test scores. Raw scores must first be converted to percentages.

Percentages

Scale		
	A:	100—91
	B:	90—83
	C:	82—71
	D:	70—65
	E:	Below 65

1. Assign letter grades to the above set of scores by using this fixed scale.

2. Find the median score (the score above or below which one-half of all scores lie). Assume that the median score should be assigned a C and assign letter grades to other scores based on "natural" clusters.

3. Contrast the results of the two methods outlined here. On what assumptions are each of these methods based? What are the advantages/disadvantages of each?

F. On an intermediate algebra test students were asked to find to the *nearest tenth* the roots of the equation $12 - 5x = x^2$ (10 credits). Patti's work follows.

$$x^2 + 5x - 12 = 0 \qquad x = 13.5/12$$
$$x = \frac{b \pm \sqrt{b^2 - 4ac}}{2a} \qquad\qquad\qquad \Big\} \ \text{Ans.}$$
$$\frac{5 \pm \sqrt{73}}{12} \qquad\qquad x = -3.5/12$$

1. Prepare a key for this item that includes the steps you would expect in a correct response and the points you would assign to parts of that response. Check your judgments against the guidelines provided in this chapter.

2. Using your key, ascertain the credits, if any, to be assigned to Patti's response.

G. Another student teacher has asked you to observe his classes in order to help gather feedback on student attitudes—from cooperation to subject-directed interest. You decide to construct an observation form of potential behaviors you might observe. Prepare such a form so as to include a range of attitudinal indicators and a recording system. What is the minimum number of observations you should make to collect meaningful data? To what extent would kind of modes being employed affect the behavior you expect to observe?

SUGGESTIONS FOR FURTHER STUDY

Bloom, B. S., J. T. Hastings, and G. F. Madaus. *Handbook on formative and summative evaluation*. New York: McGraw-Hill, 1971.

This handbook was written for use by classroom teachers and is one of the major references in the field. We recommend Chapters 3-6, 10, and 19 as ones which extend topics considered in this chapter and are particularly suitable for the novice. The mastery learning principles described in Chapter 3 provide another view of education and the related evaluation which has direct effects on testing and grading decisions. Chapters 4, 5, and 6 are devoted to summative, formative, and diagnostic evaluation. Illustrations of testing for each kind of evaluation are included. Chapter 10 provides further samples of assessment procedures for the affective domain. Finally, Chapter 19 (written by J. Wilson) concentrates on evaluation in secondary-school mathematics. Sample items for varied levels of the cognitive taxonomy are provided, as are items related to the affective domain.

Braswell, J. S. *Mathematics tests available in the United States*. (enl.) Washington, D.C.: NCTM, 1976.

This pamphlet lists tests by author, grade level (primarily K-14), and forms. It includes information on availability of norms, the name and address of the publisher and reference location in the *Mental Measurements Yearbook*.

Gronlund, N. E. *Preparing criterion-referenced tests for classroom instruction*. New York: Macmillan, 1973.

As the title indicates, this little paperback focuses on an increasingly popular direction in educational testing—tests designed to measure specified behavioral objectives. The author outlines the assumptions behind criterion-referenced testing and leads the reader through all phases of test construction—from the table of specifications through the storage and retrieval phases. This presentation provides an excellent supplement to the test construction, scoring, and analysis sections of Chapter 8 in which illustrations of the treatment of norm-referenced tests were considered. This book is highly recommended to prospective teachers.

Selecting an achievement test: principles and procedures. 2nd ed. Princeton, N.J.: Educational Testing Service, 1969.

In its continuing attempt to provide the teacher with simple but accurate guidelines to improve evaluation procedures, ETS has produced this small pamphlet on achievement test selection. The reader is introduced to the various kinds of validity that may be claimed in a test manual as well as some common misconceptions of the generalizability of test scores. This excellent source is a prime source for any teacher involved in standardized achievement test selection.

Sund, R. B., and A. J. Picard. *Behavioral objectives and educational measures: science and mathematics*. Columbus, Ohio: Charles E. Merrill, 1972.

This paperback book was listed in the references at the end of Chapter 5. Then you were directed to it as a source of objective statements in both the cognitive and the affective domain. Now we encourage you to use this text as a source of sample test items and assessment tasks in the cognitive domain and of various assessment techniques for objectives in the affective and psychomotor domains. For some unknown reason, the authors chose not to highlight the psychomotor domain, but careful readers will find illustrative assessment items hidden in observational record sheets, such as that found on page 119.

Teachers will find many useful models on which to base their own items. Indeed, mathematics teachers should not "pass over" assessment schemas labeled *science*. Many include mathematics items, or are easily adapted to mathematics objectives. One warning! Readers wishing to learn more about the preparation of a table of specifications, the item analysis technique or how to make best use of various types of item format should consult other references.

Wilson, N. *Objective tests and mathematical learning*. Hawthorn, Victoria: Australian Council for Educational Research, 1970.

No, the author of this text doesn't really believe that some tests are objective and others subjective. The title refers to ways to increase objectivity from item construction to scoring. The reader will find that the topics presented in this little paperback supplement those handled in Chapter 8 by Farrell and Farmer. Highlights include a host of diverse item types, two sample diagnostic tests, attention to sampling and assignment of letter grades, and a transcript of the conversation of a group of teachers constructing a common test. The reader will notice that the cognitive taxonomy is sliced somewhat differently from that used by your text and that the discrimination index is computed as a decimal. However, these minor differences should cause no problems in view of the wealth of helpful material contained herein.

CHAPTER NINE

WHICH WAY IS UP?
CONTEMPORARY MATHEMATICS
CURRICULUM OPTIONS

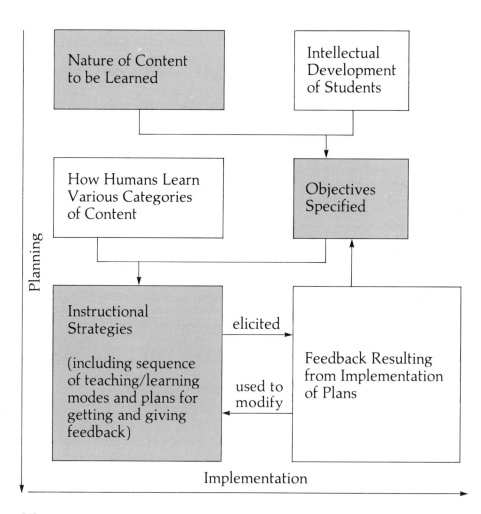

ADVANCE ORGANIZER

Why should you, the beginning teacher, be concerned about contemporary curriculum options? After all, it is highly unlikely that you will have much say about the "curriculum package" which will be adopted for use in your classes—if indeed any is to be selected. What, then, is your stake in the matter?

It is possible that you will be assigned to implement one of the national curriculum projects already in use at the school. In that case, understanding something about its characteristics and knowing the sources of further information will help you implement it effectively. Moreover, any other curriculum—whether commercially produced or designed by teachers locally—is almost certain to be some sort of variant of one or more of the national (and/or international) projects. Thus, regardless of the label affixed to the curriculum, you will find that effective implementation depends upon your understanding of the structure and emphases of the project prototypes. In addition, your teaching will be enhanced by selective adaptation of ideas from one or more of the national (international) curriculum projects. They are a *rich* source of resource ideas!

Before going any further it is important to understand the sense in which the term *curriculum* is used in this text. By *curriculum*, we mean *a course of study which prescribes a set of objectives and a systematic plan for achieving those objectives.* (Unfortunately it is the rare national curriculum project which spelled out its specific objectives in behavioral terms. However, these can be inferred from the topical treatment in the text, the exercise material provided and, of course, the tests.) A curriculum, then, cannot help but emphasize certain aspects of the nature of mathematics content, objectives students are expected to attain, and relevant instructional strategies to be employed, as shown by the shaded boxes in the systems analysis model. A contemporary curriculum package usually includes a wide variety of materials such as a teacher's guide, tests, laboratory exercises, special equipment, sets of audiovisual aids, and supplementary reading materials for students, in addition to a text. Thus they have the potential for being used as major sources of instructional ideas for the busy classroom teacher.

But wait a minute. The news media tells us that the new generation, *your* generation, is deficient in basic skills. "What has happened to the three R's?" they ask. And in the next breath, the reporter on national television asserts that the "new math" is the major culprit in the alleged decline in arithmetic skills. There is no doubt that some youngsters—too many!—are not able to perform the simplest of calculations. Moreover, an increasing number of college-bound students are less proficient in both verbal and quantitative tasks than those students of the pre-1960s. Then why draw water from a tainted well? Obviously we do *not* think that these projects, *per se*, are either the prime, or the sole, cause of the problem. Further, the tremendous amounts of time, talent, and money poured into these efforts produced many ideas and materials which have met the test of time in classroom use—when *adapted* to local needs by *talented* teachers.

A brief study of the trends of the past which have culminated in issues of the present will help provide a frame of reference for the contemporary scene and also give some direction to your own search for worthwhile ideas. In the following section you will be engaged in an activity designed to start you thinking along these lines.

9.1   INTRODUCTORY ACTIVITY

Those who fail to study the past are condemned to repeat it.

For this activity, you and a partner will need to obtain four texts. These should include:

1. The Milne *Algebra* text or any other algebra text from the first 20 years of the twentieth century,

2. an algebra text published during 1930-1940,

3. the School Mathematics Study Group *Algebra* texts—Books 1 and 2 (or any other algebra text of the "Curricular Reform Era" suggested by your instructor), and

4. an algebra text published after 1970.

Each person should analyze *two* of the texts. Study the contents of the two texts you selected for answers to the following questions:

1. Which text(s) employs vocabulary such as *commutative, associative, identity property?* What differences in symbolism are found? (Check for set symbolism and symbols used to illustrate square root, ratio.)

2. Which text(s) employ the axiomatic process to explain results or deduce new properties? (How, for example, is the product of two negatives explained?)

3. What real-life areas are utilized in the problem material? Which text(s), if any, use situations meaningful to the teenager?

4. Check through the exercise sets. Are these chiefly routine practice or is any novelty included?

After you've responded to all four questions, compare results with that obtained by your partner. Using the texts from the two most recent time periods try to identify those characteristics which might be aspects of "new math." Most parents believe that new math replaced an "old math." What, if anything, seems to have been replaced?

If you're beginning to believe that many of the new-math critics don't know the facts, that parts of the older texts are still very contemporary and that the newest versions seem to be barely a hair removed from the almost 20-year-old SMSG text (or parallel one you analyzed), then you're beginning to be properly skeptical about curricular bandwagons. Four texts hardly constitute an exhaustive sample, but they *are* a beginning in our search for data. As you read the following section, rethink your responses to this activity in the light of *events which occurred in* the time periods outlined.

9.2 FROM THE THREE R'S TO REVOLUTION AND REACTION

Although colonial America began to place some emphasis on the importance of formal education in the middle of the seventeenth century, mathematics was a stepchild that was rarely, if ever, mentioned. Massachusetts, for example, passed two education laws during the seventeenth century. The law of 1642 mandated instruction in reading and the catechism, as well as apprenticeship in a trade, while the law of 1647 required that each town of 50 families provide an elementary-school teacher and each town of 100 families provide a Latin grammar school. Reading, but not mathematics, was required for admission to the Latin grammar school. These Latin grammar schools were modeled on those of England. The typical seven-year course included *no* specific attention to mathematics, history, or natural science. Instead the students read the classics in Latin and presumably learned to speak Latin—accomplishments considered useful in a variety of occupations and required for admission to college.

At what point was some kind of mathematics also required for admission to college? Arithmetic was made an entrance requirement at Yale in 1745, at Princeton in 1760, and at Harvard in 1807. In 1820 Harvard became the first college to require algebra of its entering students. If these constituted the entrance requirements, then what mathematics was offered by the colleges? By the middle of the eighteenth century a program at Yale might have included arithmetic, algebra, trigonometry, and surveying. Some institutions included astronomy with mathematics; some added a course in geometry. Recall that a fraction of a percent of the population attended college and barely more enrolled in the preparatory schools, and you'll begin to realize that mathematics instruction, as we know it, was nonexistent. Add to this the dependence on Europe for texts. The first mathematics text written and published in the United States was Isaac Greenwood's *Arithmetick, Vulgar and Decimal*, printed in Boston in 1729. Greenwood was a professor of mathematics at Harvard from 1728-1738. Thus, his text was probably used in his college classes, but college students were not laboring over exactly the same calculation problems as the present primary-school children. Arithmetic texts were constructed to cover the waterfront from early work with vulgar (common) fractions to complicated applications to trade and commerce. Still, this brief background illustrates that the adolescent and the preadolescent in eighteenth-century America had little acquaintance with the mathematics you will be teaching.

When did that situation change and why? More states enacted laws requiring towns to establish schools and the Latin grammar school gave way first to academies and later the English high school. Arithmetic, algebra, and geometry began appearing in the offerings of these schools. By mid-nineteenth century, Horace Mann had started America on the road to free public education, which was to alter the character of the schools, increase the need for teachers, and raise unending questions as to the kind of instruction needed by this vast and diverse citizenry. None of this happened overnight. At the beginning of the twentieth century, men such as Joseph Mayer Rice, a pediatrician with an interest in education, wrote articles deploring the state of the schools, the plight of illiterate child laborers, and the lack of concern over the education of the increasing number of non-English-speaking immigrant children. Even compulsory education laws—the last in 1918 in Mississippi (since repealed)—did not dramatically increase the number of students who went to high school. Children simply continued to attend the grade school or rural one-room school until they reached the legal "leaving" age.

For all practical purposes, then, the bulk of our heritage in mathematics curriculum and instruction can be found by reviewing the texts, reports, and history of the nineteenth and twentieth centuries. This is what we propose to outline as we seek answers to the questions: Where have we been? Where are we now? and Where are we headed?

From the Three R's

The early colonial records of instruction in simple arithmetic describe a classroom scene in which pupils painstakingly copied exercises from the master's text or a slate. Later instruction followed a rule-recitation method. Without explanation, the teacher would recite or write the rule; the students were expected to commit it to memory and apply it to one or two typical exercises. This mechanistic approach to the teaching of arithmetic was applied to algebra and geometry through much of the nineteenth century (Some would add, even today in some classrooms!) despite the early, well-founded criticism of Warren Colburn. Colburn's 1830 lecture entitled "Teaching of Arithmetic" is a classic with applicability today. His exhortations to teachers to use practical examples, to guide the student through the process rather than provide solutions to be memorized, and to analyze the student's thinking processes are just as important today as they were in 1830.

Yet the history of the teaching of mathematics continued to be sprinkled with reformers' complaints about mindless drill, dull classroom teaching, and little transfer of learning. Recall the references in Chapter 6 to treating the mind as if it were a muscle and to subdividing content into tiny bits. Despite such emphasis on the "basics" (an emphasis presently being touted), testing of draftees in World War I identified a shocking portion of the citizenry who were deficient in the three R's of reading, [w]riting, and [a]rithmetic. During the next 30 years, some teachers did try to make mathematics relevant, interesting, and meaningful. There are reports of teachers who incorporated mathematics into interdisciplinary projects, of attempts to correlate mathematics and science, and of stress on insightful approaches to problem solving and proof. However, there were also reports that some elementary-school teachers ignored the teaching of arithmetic whenever possible and that the number of students taking geometry had reached a new low. At the beginning of World War II, Admiral Rickover, among others, roundly criticized the teaching of mathematics and science. Not only were large numbers of recruits deficient in basic skills, but that segment of the population who had been educated in high-school and even college mathematics were generally incapable of applying what they had learned in the various technological fields important to the war effort. Other critics deplored the outdated nature of school mathematics content.

The situation described in the preceding section appalled a good many people, but it was not news to many mathematics educators. For decades, journal articles, lectures at professional meetings, and committee activity of organizations such as the National Council of Teachers of Mathematics (NCTM) and the Mathematical Association of America (MAA) had identified the shortcomings of the system and recommended changes. The relatively few changes which did occur took place in an extraordinarily leisurely fashion. In the long run, change would only occur when implemented by teachers, and most teachers were not educated for change. Did World War II act as a catalyst? Not really. It took a Russian spaceship called Sputnik to jolt the nation.

The Revolution

In the Cold War era of the late 1950s, an advance like Sputnik I frightened the nation into pushing for massive strides in science and mathematics education. The National Defense Education Act (NDEA) of 1958 made it possible for agencies such as the National Science Foundation (NSF) to distribute large sums of money to implement changes in curriculum. The number of curriculum-writing projects mushroomed and the scope of their activities increased tremendously. Institutes, both summer and academic year, aimed to update the content background of tens of thousands of teachers. Demonstration classes, often involving talented high-school students, were incorporated into some of these institutes. Mathematicians from both industry and universities figured heavily in all projects. Mathematics educators and mathematics classroom teachers were involved on a somewhat less critical basis. Above all, the content was to be updated. And it was!

Plane Euclidean geometry texts at last corrected some of the problems inherent in Euclid's treatment—the erroneous characterization of axioms as "self-evident truths," the use of superposition in proof, failure to deal with betweenness, and so on. An attempt was made to treat algebra on an axiomatic basis rather than merely as generalized arithmetic. Topics from probability, abstract algebra, and even analysis found their way into precollege mathematics. Language, both symbolic and nonsymbolic, was given special attention. Precision and correctness were to be characteristics of such language from the earliest years. As a result, the symbolism associated with set operations appeared in texts since set was to be treated as a unifying theme. (Instead, it often appeared to be another attempt to mystify the common folk.)

These content characteristics, along with the introduction of topics such as number bases other than ten, unfortunately became known as the "new math." Some parents (and teachers) thought that *everything* unfamiliar to them must be newly invented mathematics. Labels such as *commutative*, which had appeared in the report of the Committee of Ten in 1895 (Bidwell and Clason, 1970, p. 135) and in algebra texts of the pre-Sputnik era, were assumed to be recent additions to mathematical language. No matter that mathematicians reacted against this misconception. "The arbitrary nature of axioms was established with the work of Bolyai, Lobachevsky, and Riemann over 100 years ago!" "The basic ideas of probability date back to at least the seventeenth-century work of Pascal." Few listened. Wasn't the subject called "new"? Some of it certainly looked new. In its anxiety to catch up with Russia, the nation accepted the superficial novelty and never understood the intended real novelty. According to recent studies (NACOME, 1975), most elementary teachers added the symbolism and language to their lessons, but not the meaningful patterns of mathematics. There's some evidence which suggests that many secondary mathematics classrooms reflected a similar concern for "grinding out" the products of mathematics with little or no concern for the processes. Too many teachers of mathematics, like the rest of the nation, just didn't understand the goals of this revolution and were unable to implement the new curriculum projects effectively.

The picture as of 1970 was not a success story, but neither was it a complete failure. For the first time texts contained "better" mathematics and reflected a clearer picture of the nature of mathematics. Curriculum writers, realizing the needs of teachers, began adding tests, background reading for the teacher, masters from which acetates or multiple copies could be made, texts on varied reading levels, and a wide variety of extratext resources. Mathematics educators, such as W. W. Sawyer, Robert Davis, Ernest Ranucci, and Max Sobel, wrote articles and lectured to teachers on creative ideas and meaningful approaches to mathematics. Mathematician Zoltan Dienes took Piagetian theory into the mathematics classroom. Even Morris Kline, a constant critic of the curriculum projects of the 1960s, spurred others to begin to interweave applications into mathematics texts. Thus, in addition to the potential resource ideas in curriculum projects, writers/teachers developed a wealth of curricular resource ideas in articles written for professional journals, in softbound "idea" books and in the yearbooks of the National Council of Teachers of Mathematics.

Curriculum projects By 1970, 32 American curriculum projects in mathematics and 49 international mathematics-related projects had developed sufficiently to be outlined in the *Seventh Report of the International Clearinghouse on*

Science and *Mathematics Curricular Developments*. Of these, we have selected 6 as representative of the curricular trends of the time.

The University of Illinois Committee on School Mathematics (UICSM) predated Sputnik, for it was initiated in 1951 under the leadership of Max Beberman. In an attempt to improve the mathematics taught to the college bound, UICSM writers and teachers attempted to produce high-school texts that would reflect the structure of mathematics as a unified whole rather than as disembodied branches. One of the then-novel aspects of text presentation was the incorporation of a guided discovery approach to the concepts and principles of mathematics. In order to train teachers in the modes of instruction deemed critical to the project, summer and academic year institutes for teachers included courses in mathematics and pedagogy, as well as demonstration classes. High-school texts, examinations for some of the units, self-instructional texts, and teacher-training films were produced by the project staff. With the completion of the high-school text project in 1962, UICSM staff turned its energies toward the development of innovative approaches to topics in junior-high-school mathematics—approaches appropriate for culturally disadvantaged students in urban areas. Reports of the successful reception of the resulting texts (such as *The Theory of Fractions* [Books 1-4]) in inner-city schools attest to the quality of these materials.

The School Mathematics Study Group (SMSG), which was organized in March 1958 at Yale University, is probably the best-known curriculum project of the revolution. Under the leadership of E. G. Begle, professor of mathematics, SMSG produced texts and teacher's guides for kindergarten through twelfth grade. Like UICSM, SMSG writers worked at incorporating precise language and careful logical deduction into the texts. However, SMSG texts (especially the 9-12 texts) seem to reflect the presentation of mathematics via thoughtful exposition rather than guided discovery. In addition to a set of core texts for each grade, SMSG produced alternate texts for various courses (such as *Geometry with Coordinates* as well as *Geometry* for grade ten), programmed texts, enrichment texts for teachers and/or students, teacher training texts, reports of research, and summaries of a longitudinal study of achievement. By 1970, 229 items were listed under the SMSG title. These included reprints of classics in mathematics and Spanish translations of many of its texts. With its emphasis on the structure of mathematics, SMSG's first secondary-school texts were designed for the college-bound student and that student had to be a good reader! Subsequently, alternate prealgebra texts were written at a slightly lower reading level. Where were the concrete examples, the practical illustrations? They were few and far between.

The Madison Project was initiated in 1957 as a supplementary mathematics program with stress on creative learning experiences. Based at Syracuse University, Syracuse, New York, and then at Webster College, St. Louis, Missouri, the writers and teachers produced supplementary texts and teacher's guides for grades 4-9 on topics in algebra, coordinate geometry, science applications, and even matrix algebra. In addition, project staff produced teacher-training films and audio recordings portraying actual classroom lessons. One of the most novel aspects of the Madison Project is the set of "Shoebox" packages for physical experiments. The impact of Piagetian theory is reflected in the nature of these laboratory exercises.

The School Mathematics Project (SMP) organized in 1961 in Southampton, England, is of particular interest to American mathematics teachers. Student texts and teacher's guides for courses at three levels were developed and used by over 100,000 students ages 11-18 in Great Britain. SMP texts A-H and X, Y, and Z reflect a spirit of guided discovery, the spiraling of concepts, and a wealth of concrete beginnings (labs, visualization, real situations). Practical examples are used to stimulate curiosity and to provide a firm base for abstract concepts. The advanced SMP texts—Books 1-4—present sophisticated mathematics with its applications to science. In addition to these texts, Activity-Investigation cards have been designed to supplement the work of Books A-H at three levels—remedial, mainstream, and enrichment.

The Secondary School Mathematics Curriculum Improvement Study (SSMCIS), 1965, was developed by Howard Fehr of Columbia University. SSMCIS texts contain a secondary-school (grades 7-12) course appropriate for the talented mathematics student. Emphasis is on the axiomatic structure of mathematics and the group concept is utilized as a unifying theme. Not only is the instructional treatment of topics at a sophisticated level, but the topics themselves include many usually taught at the post-secondary level.

The final project outlined here is the Comprehensive School Mathematics Program (CSMP) directed by Burt Kaufman since 1965. The purpose of CSMP was to implement the major recommendations of the 1963 Cambridge Conference. Individualization of the K-12 mathematics curriculum and presentation of sound mathematical content are the two priorities of CSMP staff. As a result, mathematicians are important contributors to text materials. At present, CSMP has developed a set of texts for highly verbal, well-motivated junior/senior-high-school students and activity packages using a variety of media for elementary-school children. Since all instructional materials incorporate considerable independent student work, the teacher must operate on a consultative, supportive basis. Hence teacher training in summer workshops or on a yearly inservice basis continues to be an important part of CSMP work.

Did you notice the diversity as well as the commonality in the six projects just described? There was continual emphasis on "good" mathematics—correct, precise, and up-to-date. There was also much emphasis on the college-bound student with an occasional interlude in favor of the slow learner, and then a swing to the talented student. Some projects seem to be constructed on the assumption that the college-bound student is a good reader and needs little in the way of concrete underpinnings while a few others, especially the Madison Project and the School Mathematics Project, lean heavily on labs, visual illustrations, real-life examples, and the like.

Recall that the impetus for the curriculum revolution was the space race that began with Sputnik. The emphasis of most of the curriculum projects produced in the following decade clearly reflects attention to producing talented mathematicians, engineers, and scientists. Yet by the middle of the decade, the attention of the American public was diverted from the space race to riots in the cities and school desegregation issues. Consequently, the limelight began to shift to the low achiever and the disadvantaged student. Federal funds were no longer available for any and all curriculum projects. The American taxpayer began demanding that schools be held accountable for teaching *all* the students.

The Reaction

There are more chapters to the story of the reaction than can be told in a few short paragraphs. Of importance to you are four curricular trends that were affected by the reaction against excessive symbolism, lack of real-world relevance of mathematics, and excessive formalism.

First, the *continued* low level of achievement in computation (Many writers acted as if pre-Sputnik children were whiz kids in addition!) received national attention. As a result, projects that promoted the *basic skills* were in. By the late 1970s, programmed texts, tape and slide kits, and criterion-referenced instructional packages were finding their way into most junior high schools and some senior-high-school classes. A *few* of these basic skills kits extended the area of basics to important concepts and processes and even a small number included instructional strategies *a la* Piaget.

Second, the calculator and its sophisticated distant relative, the computer, were finding a place in the mathematics classroom. By 1975, hand-held calculators were inexpensive enough to be owned by many high-school students. Whether the mathematics teacher approved or not (and some did not!), the calculator was being used to complete calculations. However, contrary to predictions of the 1960s, computer technology was still far too expensive to intervene widely into mathematics instruction. Nonetheless an increasing number of schools participated in time-sharing systems at nearby universities, industries, or regional educational centers. Programming in the BASIC language and problem solving via computer programs became the focus of an elective computer science course in some schools.

Third, the gap between the understanding of mathematics and the ability to use, or even appreciate the use of, mathematics led to an interest in teaching the nature of mathematical models. The mathematization of the social sciences, medicine, and business as well as the traditional natural sciences resulted in a spurt of college-level courses on mathematical models. At the other end of the spectrum, mathematics educators such as Max Bell saw in mathematical models a link to the everyday world of students. An increasing number of articles on ecology, wind chill, sports car rally directions, and even scuba diving were evidence of increasing interest in ways to implement instruction in mathematical models.

Finally, there has once again been a surge of interest in interrelating mathematics and science in schools. In particular on the junior-high level, some popular science curricula (such as ISCS—the Intermediate Science Curriculum Study) interweave mathematics skills and concepts throughout the chapters. Several SMSG texts that use science laboratory experiences to introduce mathematical concepts and rules are offered as supplementary mathematics curricular material. Chapter 10 is devoted to the wealth of available resources in this area.

9.3 IMPLICATIONS FOR THE MATHEMATICS TEACHER

Think back to the Introductory Activity. You were asked to compare a recent algebra text with a text characteristic of the revolution. You should have found the threads of emphasis on axiomatics, structure, precise language, and so on, throughout the newer text. We'd be willing to bet that the preface or foreword by the author(s) cited one or more projects from the revolution and/or recommendations from the Cambridge Conference as major sources for that treatment of algebra. The effect of the revolution on producers of commercial texts, films, tests, and other curricular materials can be similarly traced in every other mathematics area in the curriculum. More recently, curricular materials are being modified in response to one or more of the reactions referred to earlier. Thus, the first major implication of the curriculum reforms of recent decades resides in their impact on all curricular materials produced since 1970. There is no turning back, despite the cry of "Back to Basics," but modifications of the emphases of varying projects is possible and already in full swing.

This means that the texts you may be expected to use should correspond to a contemporary view of mathematics—a view your own mathematical background ought to reflect. Hence, you'll understand the basic mathematics well enough to reinterpret it for the students you teach. Right? Wrong! Under the best of circumstances, the *learning* of college mathematics does not require that the student translate mathematical concepts and principles into everyday situations, rephrase definitions in his or her own words, or explain the reasonableness of a definition or assumption. Under the worst of circumstances, the college student often succeeds in mathematics by dutifully grinding out responses and regurgitating proofs or problem solutions previously demonstrated in class. In the latter case, the college student suddenly (often painfully) realizes that he or she just doesn't undestand great chunks of mathematics. "Ask me to differentiate anything! I'm a whiz! But don't ask me why it works or what it's good for or what it depends on . . . ." If this is your plight, you might well question the mathematics instruction which allowed you to be stamped as "passed." More to the point, you need to start reeducating yourself, and some of the best source books for that purpose are curriculum projects such as SMSG or UICSM. Read the text material, work out the exercises, and constantly seek answers to "why?" as well as "how?" Moreover, even those luckier mathematics majors who did learn much of their mathematics meaningfully need to begin to work at the problem of communicating mathematics without distorting it. The teacher's commentaries accompanying each SMSG text are one source of potential help. For each topic in the student text, the teacher's commentary includes a parallel consideration but adds background material, gives a justification for the mode or sequence of presentation, and may refer the reader to other in-depth sources. Each of the other curriculum projects described earlier are similar sources of background material for the prospective teacher.

Finally, while novice teachers are rarely in a position to select new texts for a class, they should be constantly seeking resource ideas to add life and meaning to lessons. The curricular projects of the revolution are a gold mine of such ideas—to be selectively extracted from the project and modified as necessary. This third and final implication is of such importance to all mathematics teachers that we have selected a few representative examples to point the way.

1. You are looking for ideas which will improve your sequencing of content and thus clarify transitions.

All the curriculum projects have considerable strength in this area. Consider one example from SMSG's *Intermediate Mathematics* text. After a presentation of logarithms that depends on the areas under a curve (Fig. 9.1), the exponential function is defined as the inverse of the logarithmic function (Fig. 9.2).

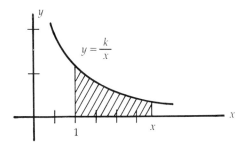

Fig. 9.1 The shaded area is used to define the log function.

The properties of the exponential function and the usual "laws of exponents" are then developed as a natural outgrowth of the work with logarithms. For example, from the graph the value of $E_a(0) = 1$; while $E_a(1) = a$ since $y = E_a(x)$ is the inverse of $y = \log_a x$. We recommend this atypical approach to the logarithmic function *providing that* the teacher give the students some reason for the sudden interest in areas under curves and a justification for tagging a

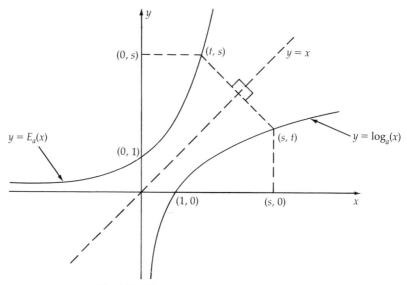

Fig. 9.2   A log function and its inverse.

label like "logarithm" on certain areas under the curve $y = k/x$. The SMSG text omits this entirely but your own memory of integral calculus should provide a clue. Yes, it is up to you to make intelligent selections and to modify where needed.

2.   You want to help students reach generalizations by developmental questioning or independent work on a set of exercises, and need sequences of examples.

Check the UICSM texts for your topic. The emphasis on patterns makes them a prime source for gradually differentiated examples but *watch out* for unnecessary and confusing-to-the-student symbolism. (SMSG shares in this excessive use of symbolism and of formal language.)

3.   You'd like to improve the spiraling betwen lessons and topics by incorporating an introduction to future tasks in homework assignments.

Here is one illustration from SMSG *Geometry* (Fig. 9.3) that appears in the congruence unit.

   If you look ahead to the parallelism unit, you'll recognize that the diagram in Fig. 9.3 looks like a twin of the usual diagram for the theorem: The line segment joining the midpoints of two sides of a triangle is parallel to the third side and equal in length to one-half the length of the third side.

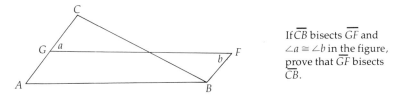

If $\overline{CB}$ bisects $\overline{GF}$ and $\angle a \cong \angle b$ in the figure, prove that $\overline{GF}$ bisects $\overline{CB}$.

Fig. 9.3 Congruence problem.

4. You need ideas for laboratory activities and demonstrations.

The SMP materials are rich in concrete illustrations of many kinds. Here is just one example.

From Book H, a spinner (Fig. 9.4) is used to initiate questions about probability.

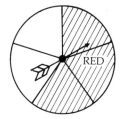

Fig. 9.4 Spinner with three red sectors.

1. What is the probability that the arrow will stop on a red sector?

2. If we spin the arrow 1000 times and succeed in obtaining "red" 609 times, what is our success fraction?

3. What is the probability the arrow will stop on the same color in two consecutive spins? in four consecutive spins?

In order to answer the more complicated questions, tree diagrams are utilized. The tree diagram showing possible outcomes for two consecutive spins on the red-white spinner is depicted in Fig. 9.5.

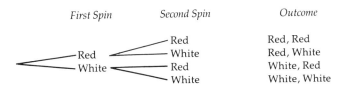

Fig. 9.5 Tree diagram for red-white spinner.

One goes from tree diagrams to mathematical models in one easy step, or so it seems, as the SMP authors model a traffic situation by means of a specially designed spinner (Fig. 9.6).

Number of commercial vehicles	In k sets of 10 vehicles
0	8
1	9
2	14
3	7
4	2
5	0

Fig. 9.6   Data and model for traffic survey.

The table represents data obtained by counting the number of commercial vehicles (for example, trucks and buses) in every set of 10 vehicles traveling on a particular highway. In all, 40 sets of 10 vehicles (400 vehicles) were counted. The table indicates that 8 sets included *no* commercial vehicles while each of 9 sets contained 1 commercial vehicle. The appropriate spinner can be designed by constructing the sectors so that the probability that the arrow will stop at 0 is 8/40, at 1 is 9/40, and so on. The students are asked to verify the 72° angle in the diagram, to find the other angles, to construct such a spinner and use it to simulate the actual traffic count.

The previous illustration is just one of many in the SMP texts, but don't stop there in your search for laboratory/demonstration/visual/everyday ideas. Browse through the junior-high texts produced by UICSM, the Madison Project materials, and articles produced by the CSMP staff (their texts are not available in libraries as of this date). Don't except to find a section labeled "Labs" in any of these materials. In most cases you have to read with care and extract laboratory and demonstration ideas from the texts. Even the spinner illustration from the SMP text does not specifically say: "*Show* the students a large spinner. *Have* individuals *spin* the arrow." It is up to the teacher to realize that here is an idea for both a demonstration and a laboratory activity and that one or both modes are essential *before* asking the students to predict results.

As a result of increased attention to the implications of Piaget's data, a large number of paperback pamphlets, hardcover texts, and articles in professional journals have focused on activities suitable for laboratory activities or demonstrations. These sources are outgrowths of the revolution and are treasure houses of instructional ideas. Some, like *Geoboard Geometry* (Farrell, 1971), give examples of the use of one manipulative device—in this case, a 5 x 5 pegboard—at several grade levels and within several topics.

Others, like *Vision in Elementary Mathematics* (Sawyer, 1964), illustrate the introduction and development of concepts and rules from pictoral and concrete examples. There are a wealth of examples in this little book, and all have

Problem: How many *different-sized* squares can you outline with rubber bands on this geoboard? (Hint: There are more than 6!)

Fig. 9.7   Geoboard problem.

pictoral dimensions. Don't be misled by the title. All these ideas are appro-priate for the secondary-school mathematics classroom. It is clear that such presentations help promote a meaningful sequence and that take-home labor-atory activities such as the geoboard problem could be used to encourage independent thinking. That leads us to the next potential area of resource needs.

5. You need challenge problems, in-depth development of topics, motiva-tional ideas for the capable mathematics student.

SMSG, UICSM, SSMCIS, the advanced SMP texts, and the CSMP materials contain problems on a number of different levels and thus are prime sources. In addition, journal articles, books of all kinds, and "activity" cards, written as outgrowths of the revolution, focus on this need. The National Council of Teachers of Mathematics' 27th yearbook, *Enrichment Mathematics for the Grades*, and 28th yearbook, *Enrichment Mathematics for High School*, represent just one kind of source book written during this era. They contain problems for the talented student as well as an introduction to atypical secondary-school topics (such as Gaussian integers and Farey sequences in yearbook 28 and topology and logic in yearbook 27). "Activity" cards containing a range of interesting problems—some using laboratory materials and some limited to pencil-and-paper questions—were designed by both curriculum project writers and commercial companies. The enrichment cards produced by the SMP writers noted in Section 9.2 are one excellent source. A British project not outlined earlier, the Nuffield Project, developed three sets of cards—a Green, a Purple, and a Red Set. Card 49 from the Green Set is reproduced in Fig. 9.8.

How many numbers greater than 0 and less than 1000 can you put into the counter register of a desk calculator by turning the handle twice?

Fig. 9.8   Card 49 from Nuffield Project's Green set.

The teacher's booklet provides not only an answer but also an outline of some problem-solving strategies and suggested ways to go beyond the stated problem. Catalogues from major distributors of mathematics-related instructional materials list still more potential sources of activity-card problems and questions. The uses range from independent projects, to level III and above test items, to discussion group problems, to introductory challenge questions.

6.   You are looking for novel remedial material for low achievers.

The UICSM junior-high texts are an excellent source for unusual approaches to required topics. However, most of the other major curriculum projects of the revolution were not addressed to this audience. With the reaction, there have been multiple commercial texts for teacher and students, articles, activity cards, filmstrips, workbook materials, games, and tests which are useful for low-achieving students. Sobel and Maletsky (1969) authored a three-volume set, *Essentials of Mathematics*, for low-achieving middle- or junior-high-school students. (However, many of the units are just as applicable to the average younger junior-high student.) Buckeye has developed a set of activity cards called *No Read Math Activities* which present the student with a pictoral representation of a problem and the teacher with directions. The cards are punched and coded to allow for ease of retrieval (after the fashion of the test item system described in Chapter 8). More of these sources will be listed in Chapter 11.

7.   You are incorporating machine-age concepts into one or more courses.

The resources, in this case, are far-reaching. Bell Laboratory mathematicians developed CARDIAC, a cardboard aid to computation. Class-size sets of CARDIAC(s) and an acetate copy for the teacher, together with a teacher resource text, introduce basic computer concepts in the absence of expensive hardware. A set of problems for computer solution and their solution in BASIC and FORTRAN (Rogowski, 1976) are offered for use in any computer-oriented high-school or junior-college course. The emphasis on flowcharting in programming courses has been utilized by others as an ideal way to teach everything from the solution of simple equations to the importance of clear directions. Figure 9.9 shows an example from *Mathematics One* (Suppes *et al.*, 1974, p. 409).

In addition to computer-related sources, there has been a flurry of articles on ways to use the hand-held calculator to solve problems, to introduce rules, and to provide immediate feedback.

If it seems as if the storehouse of resource ideas provided by the curriculum projects and the wealth of materials spawned by them are a bottomless pit, you have caught the idea. The usefulness of the materials is limited only by the constraints of time, energy, and creative adaptation you bring to task.

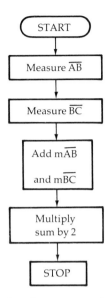

Fig. 9.9   Flowchart for perimeter of rectangle ABCD.

9.4 CROSS-REFERENCES TO A SAMPLE CBTE EVALUATION INSTRUMENT

There's little room for argument here. Competency 4.4 is a direct match to the background presented in this chapter. Moreover, in order to effectively perform that competency, the teacher will probably be working at most, if not all, of category 4 (Selected and Utilized Instructional Resources). What other competencies might the teacher be demonstrating? If you selected 2.37, 2.42, 2.64, as well as many category 3 items, you're in agreement with our thinking. Effort spent on mining the gold from curricular resources has a payoff in multiple facets of teaching.

9.5 SUMMARY AND SELF-CHECK

In this chapter we identified the major milestones in the history of school mathematics, the forces behind some events, and the issues which continue to affect mathematics teaching. The characteristics of the curriculum revolution of the 1960s were delineated and six projects of that era were outlined. Finally we introduced you to the central issues and forces which evolved after the revolution.

How can the beginning teacher of mathematics benefit from the efforts of the project writers and teachers of the revolution and the reaction? Answers to this question were explored by providing you with illustrations from selected sources.

At this point you should be able to:

1. Operationally define *curriculum*.

2. Describe the major characteristics of *at least four* major projects of the revolution.

3. Describe the two-pronged nature of the "new" math as seen by the curriculum writers/teachers of the 1960s.

4. Locate resources for the average, the slow, the talented in curriculum projects and related materials which grew out of the curriculum revolution.

5. Select ideas from the curriculum projects of the revolution which would be appropriate supplements to any curriculum which you are assigned to teach.

9.6 SIMULATION/PRACTICE ACTIVITIES

A. Choose a topic from secondary (7-12) mathematics. Research the approach to that topic in *one* of the curriculum projects described in Section 9.2. Then answer the following questions:
   1. What prerequisites do the authors of the chosen text assume as a basis for the selected topic?
   2. In what ways and to what extent are the processes and products of mathematics illustrated by the text presentation?

B. Teachers are expected to know where and how to locate curriculum project material. One excellent source is the *Report(s) of the International Clearinghouse on Science and Mathematics Curricular Developments*, which is annotated in the Suggestions for Further Study section of this chapter. Look up *two* mathematics projects not described in your text and briefly describe the characteristics of each.

C. The 1963 Cambridge Conference report, *Goals for School Mathematics*, outlines a possible secondary-school curriculum for 1990. Read the outline of that 7-12 curriculum. Would your mathematics background be sufficient preparation if you were to to teach such a program? Why or why not? How does the proposed curriculum differ from your secondary-school (7-12) mathematics program?

D. One of the most persistent critics of the curriculum projects of the revolution has been Professor Morris Kline of New York University. Read his views in "The ancients versus the moderns, a new battle of the books," *The Mathematics Teacher* 51 (October 1958): 418-427. Then read "Analysis of the Innovations," pages 14-22 of the NACOME report. (See the Suggestions for Further Study section.) Summarize the issues raised by Kline and the rationale advanced by the writers of the NACOME report.

SUGGESTIONS FOR FURTHER STUDY

Davis, R. B. *The changing curriculum: mathematics*. Washington, D.C.: Association for Supervision and Curriculum Development, NEA, 1967.

Davis describes the second decade of "Modern Mathematics Curricula"—the decade we've called "the reaction"—in a realistic, uncompromising manner. Of particular interest are the descriptions of less conventional approaches to mathematics instruction, such as the use of mirror cards, computer-assisted learning, and mathematics laboratory. Davis not only describes these approaches and others, but critically analyzes their assets and debits. This excellent paperback is worth reading from cover to cover.

Lockhard, J. D., ed. *Seventh report of the international clearinghouse on science and mathematics curricular developments*. College Park, Md.: Science Teaching Center, University of Maryland, 1976.

The materials produced by numerous science and mathematics curriculum projects throughout the world are described in this report. A few pages of highly compacted information are provided on each project, and the addresses needed to secure materials and/or further information are included. This report is updated every two years and it is the prime source for those seeking current information on K-12 curriculum projects in our field.

National Advisory Committee on Mathematical Education. *Overview and analysis of school mathematics grades K-12*. Washington, D.C.: Conference Board of the Mathematical Sciences, 1975.

This report (called the NACOME report, after its authors) is a direct response to "the reaction." Its writers considered issues in instruction, curriculum, teacher education, and evaluation, reported relevant data, and made policy recommendations. In a succinct and convincing manner, the NACOME report dismisses attempts to blame "new math" for all the ills of mathematics education and provides a balanced view of "the revolution."

National Council of Teachers of Mathematics. *An analysis of new mathematics programs*. Washington, D.C.: NCTM, 1963.

This pamphlet contains an evaluation of eight projects whose directors had responded to an open invitation to participate in an external analysis. Subcommittees of NCTM members from all levels of mathematics education reviewed text materials, read the director's report, and sometimes visited sites where materials were being used. Their assessment included a concern for the issues of social applications, placement of topics, the study of structure of mathematics, the kind of vocabulary, methods, relative emphasis on concepts and skills, attention to proof, and provision for evaluation. The novice teacher will find this kind of analysis helpful in identifying potential sources of useful ideas.

CHAPTER TEN

ISOLATION AND FRUSTRATION *OR* COMMUNICATION AND COOPERATION?

MATHEMATICS, SCIENCE, AND THE EVERYDAY WORLD

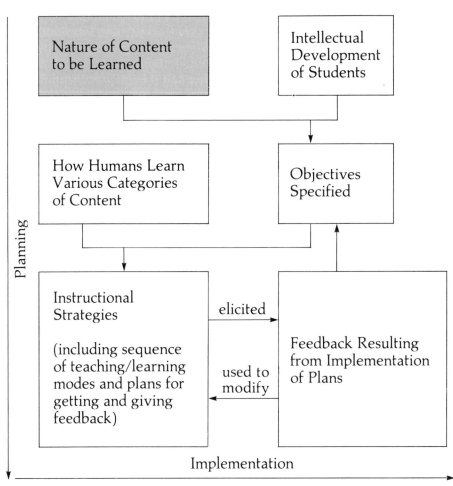

ADVANCE ORGANIZER

Few who are aware of the recent history of mathematics in the secondary-school propose that it be taught primarily for the 2 or 3 percent who will become professional mathematicians. Fewer yet of those alert to the temper of the times and the needs of contemporary society recommend treating school mathematics as an island unto itself. In fact, the goal of mathematical literacy for every citizen has been steadily gaining support among mathematics educators for over a decade. Not all who cherish this goal agree on the exact meaning of the term *mathematical literacy*. However, there are common grounds of agreement which are likely to persist and these merit the attention of all who prepare to become mathematics teachers.

Thus, it makes sense to include consideration of the implications of mathematics for everyday life. But why single out the interrelationships with science? Doesn't mathematics have some important connections with many other school subjects? We think it does, and in this chapter we sketch a few of them. However, we have chosen to concentrate on science for a number of reasons. First, mathematics is the lifeblood of modern science. The concepts and other thought models of mathematics are interwoven in every secondary-school science course which *your* students will be studying. Your students should *not* be learning that the algebra taught by the mathematics teacher differs from that used in the science class. Second, the history of mathematics and the history of science illustrate the long and fruitful union of mathematics and science. The contributions of Newton, Galileo, and others were cited in Chapter 4 as sources of teaching ideas for the mathematics class. Third, the science teacher's academic background typically includes a substantial amount of mathematics. This ought to facilitate talking to and working cooperatively with colleagues in science education. Fourth, science teachers also need *your* help in finding ways to use mathematics, rather than to abuse it, and you can depend on their expertise to help you design relevant, interesting, science-based mathematics lessons. Thus, a profitable *quid pro quo* arrangement is highly feasible. Finally, one has to start somewhere in helping students cross those artificial watertight compartments that too often isolate one school subject from another and separate school learning from out-of-school life.

Successful cooperation with science teachers can then lead you to explore similar possibilities with those who teach your students subjects such as social studies, English, industrial arts, and fine arts. Communication and cooperation among teachers can go a long way toward overcoming the feelings of isolation and frustration which too frequently engulf teachers and students alike.

We have shaded in the "Nature of Content" portion of the systems analysis schema to call attention to the mathematics/science interface so often overlooked in the literature. Many journal articles which deal with the use of science in mathematics instruction treat it as interesting, supplementary material to work in, if time allows. We recommend a more serious and deliberate approach based on a careful analysis of the nature of the two disciplines, the structure of school curricula, and applications to out-of-school life. Both the Introductory Activity and other sections of the chapter are designed to start you thinking along these lines.

10.1 INTRODUCTORY ACTIVITY

Ivory towers are lonely places these days.

You may need to consult a high-school science text in order to complete activity A.

A. Table 10.1 contains a set of science-based questions that can be used by the mathematics teacher to introduce or develop understanding of certain mathematical concepts or rules.

Table 10.1
Selected science-based questions for the mathematics classroom

Questions	The related science course or level
1. How are a doorknob, a set of stairs, a screwdriver, a baseball bat and an egg beater related?	JHS Physical Sciences and Physics
2. What shape are the compartments in a honeycomb? Would some other shape be more efficient (from the "point of view" of the bee)?	Biology
3. An impala can jump 30 to 40'; a large jack rabbit, about 25'; a kangaroo, 25 to 30'; a cricket-frog only an inch long can jump 3'. What is the record for humans in horizontal jumping? How does it compare with the data given above? (Be sure to consider the relative size of the human versus the animals cited.)	JHS Life Science
4. How could you estimate your volume? Think of at least two ways. Use the figure you obtain and your weight to calculate your specific gravity.	JHS Physical Science

1. For each question, give *at least two* reasons why the question was paired with the level or kind of *science* discipline listed next to it in Table 10.1. (If you're having trouble coming up with specifics, consult a related secondary-school text. Also ask a science major to help you, but don't be surprised if he or she pleads ignorance!)

2. Identify specific mathematics concepts/principles which could be introduced or developed by means of instructional strategies which incorporate ways to answer *each* question.

Table 10.2
Selected everyday-world questions for the mathematics classroom

Questions	Everyday world of:
1. What kind of numbers appears on soup labels, cereal boxes, the odometer of the family car, a light bulb? What is the largest and smallest number in each case?	the home
2. If all the litter found in one week in the square block area around the school were recycled, how much and how many different kinds of trash could be salvaged?	ecology and the school
3. Tyrus Cobb, a batting champion, had a lifetme batting average of .367. Compare this with the top batting average of this year's World Series players and the top batting average of our school's varsity players.	sports
4. Winning at pool is said to depend on a combination of skill and use of geometry. What geometry?	sports
5. Obtain a snapshot and an enlargement of it. Tape these on a sheet of paper. Show how lines connecting corresponding points can be used to locate a center of enlargement. Sketch a still larger photo by extending the pattern lines.	hobbies

B. Table 10.2 contains a set of questions which can be used to introduce or develop understanding of mathematical concepts or rules. However, these questions are drawn from the everyday world of potential interests of the students.

    1. Answer questions 1, 4, and 5.

    2. List the steps a class would need to take to obtain prerequisite information related to questions 2 and 3.

    3. For *each* question in Table 10.2, write additional spinoff questions which could be used to enhance the mathematics being taught.

After comparing your responses to these activities with those of other classmates, you may find that you need additional background references on certain questions. If so, check through the references provided in the final section of this chapter. However, if you're fairly clear about the potential uses of the questions in Tables 10.1 and 10.2, then you're ready to sink your teeth into the sections which follow.

10.2 MATHEMATICS THROUGH SCIENCE

You don't have to have a second field in one of the sciences in order to teach mathematics through the vehicle of science-based ideas. You do have to understand the *concept* of a mathematical model. In all of science—theoretical or applied, elementary-school level to graduate school, biological or physical—mathematical models are used to describe something going on in the world, to predict events, and to suggest solutions to complicated problems. Thus, the natural (counting) numbers are appropriate mathematical models for describing the number of planets but inadequate models for describing the size of bacteria, a virus, and a molecule. Does that use of mathematical model seem to be a far cry from the sophisticated mathematical models found in college-level physics texts and advanced biology texts? It should. However, it's exactly that simpler view of modeling, which is the key to the ideas presented in this section.

Let's start with an activity based on question 2 in Table 10.1. In a junior-high class, the teacher can show the students a picture of a cross-section of a honeycomb or, better yet, pass around (and project on the overhead) a portion of a real honeycomb. Don't assume that all the students have seen a honeycomb. Many supermarkets sell only jars of honey, minus honeycomb. Next the problem posed in the question has to be spelled out. What is meant by "efficient" in this case? (Here's a good place to involve a science colleague directly in the class, or indirectly as a resource person. Be sure to ask that science teacher about the two aspects of the bee's behavior: building the comb and storing the honey. Also seek your colleague's assistance in alerting students to the dangers of thinking about the bee's behavior in terms of human attributes and reversed cause and effect relationships. Don't feel guilty about asking for this kind of assistance since your science colleague will certainly need your help in assisting students to make proper use of mathematics in their science classes.) The "efficiency" question should eventually get restated so that quantitative techniques can be used. Students should be helped to see that there are two main issues.

1.   What other shape would "fill the plane"?

and

2.   Given a fixed surface area to use for a cell wall, which shape provides for maximum volume of the enclosed space?

Figure 10.1 depicts some common shapes that could be considered as the students try to model the situation of a comb.

Fig. 10.1   Potential "cells" for comb.

Multiple congruent cutouts of each of the shapes could be used to answer question 1. The students might be asked to "tile a floor" and, as a result, to identify those shapes which make good tiles and to determine the number of such tiles needed around a point. Be sure to include large (such as 10 cm on an edge) and small (such as 4 cm on an edge) sets of each shape so that the students don't get the idea that their answers depend on size. Six equilateral triangles fit around a point whether the triangles enclose a large or a small region. A table like Table 10.3 can be used to record data.

Table 10.3
A partially completed data table

Shape of tile	No. of sides	Fills the plane	No. of tiles around a point	Measure of one angle
Circular	—	No	—	—
Equilateral triangle	3	Yes	6	60°
Square	4	Yes	4	90°
Regular pentagon				108°
Regular hexagon				

Protractors can be made available if students have forgotten (or are just beginning to investigate) the angle measures of some of the shapes. The students can be encouraged to find patterns in the data, such as the recurring product of 360° (6 × 60°, 4 × 90°, 3 × 120°). If that pattern holds, then no wonder the regular pentagon wasn't a useful tile: 108° would have to be a factor of 360° and it isn't. Don't miss the opportunity to point out the inexactness of measurement and the differences between the cell of the comb, the cutout hexagonal shape, and the mathematical thought model, the hexagon. Moreover, be especially careful to call these generalizations "patterns," "conjectures," or "educated guesses" and *not* "statements we've proved."

The results of this part of the investigation make the construction of the comb even more puzzling. Surely it would be easier to build a three- or four-walled cell rather than a six-walled cell! Perhaps the solution of the second problem we posed originally will help here. Although the students could use the same cutout shapes and simply compute areas, we've found that a laboratory activity designed to produce a visual solution is far more productive. You can use old manila folders (or other suitable substitutes), which you will need to cut into strips 2 cm by 12 cm. Students working in groups of two or three should be asked to construct a cross-section of a comb. Some groups should construct a comb with hexagonal cells; others, a comb with square cells; and the rest, a comb with triangular cells. In *all* cases, the surface area of the wall of any *single* cell will be the same, 2 cm × 12 cm. Each group will need a supply of strips (8 to 10), tape, a scissors, and a metric ruler. (The group constructing square cells won't need a ruler. Why not?) The teacher needs to remind the class that combs do not contain double-walled cells. Furthermore, the teacher will need to precede this laboratory with a demonstration of the construction of a single cell and the subsequent attachment of an adjacent cell. As students work through the laboratory, and then compare results visually, the capacity-advantage of the hexagonal cell over either of the others should be unmistakable.

The "combs" in Fig. 10.2 were produced by groups in one class. The combs were taped to the board, and all the students observed the structural problem which had been commented on earlier by several small groups. "The triangle is a rigid figure" became a meaningful statement to all the observers. (Then, why don't the cells droop in the honeycomb? If you're not sure, check with a science colleague.)

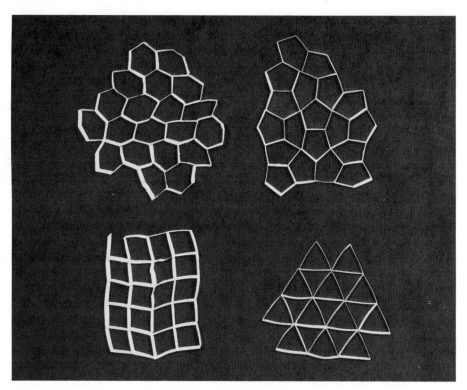

Fig. 10.2   Models of combs.

Shape, space-filling, perimeter/area/volume, and structural rigidity relationships are important to multiple fields of science. Any of the following questions can help to promote mathematical understanding in a science setting.

1.   A man coming out of a bath or shower carries a film of water which makes him 1 percent heavier. For your own weight, what weight of water would this be? What volume? (1 g of fresh water is very close to 1 cc)

2. A man is about five thousand times as heavy as a mouse. Approximately how much does a mouse weigh? Haldane (see reference in Suggestions for Further Study) says a wet mouse will carry a film of water about equal to its own weight. Why should that be true? How could we compare the *surface area* of a mouse with that of a man? (Hint: A mouse is about 4 inches long—if we ignore its tail. A man is about 68 inches tall.)

3. Why is it very difficult for a fly to escape from a puddle of water?

These are just a few of the area/volume relationships which are crucial to an understanding of certain science relationships. Many more will be found in sources cited in the final section of this chapter.

However, beyond the relationships of size and shape are the manifold properties represented by a class of mathematical models known as curves. *Pitch, frequency,* and *period* are terms associated both with properties of sound and of the sine curve. The trace of a vibrating tuning fork is depicted in Fig. 10.3. The sine curve stands out vividly.

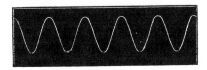

Fig. 10.3   The trace of a vibrating tuning fork.

The soda straw clarinet laboratory is designed to introduce the concepts of frequency and pitch and the mathematical relationship between them through the medium of music. The teacher must obtain a supply of soda straws, some single-edged razor blades, and a metric ruler. Measure a 3 cm length from one end of a straw and carefully make a horizontal crease at that point. Then flatten the shorter portion of the straw as shown in Fig. 10.4. To complete the construction of the reed of the clarinet, cut two congruent right triangles out of the reed about 2.5 cm from the end of the straw. Now try the clarinet.

The reed vibrates as air is blown through the straw, which acts as a wind tunnel. The sound will not be very musical, but it represents the basic note of

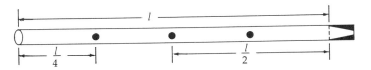

Fig. 10.4   Soda straw clarinet.

that clarinet. Each student's clarinet may sound slightly different (a good place to ask "Why?"). Be sure that each one is able to produce a sound before proceeding further. You may need to suggest that they produce a sharper crease at the end of the reed or open the trapezoidal mouth of the reed so that air will circulate into the straw. (Paper straws may have become too wet, while some plastic straws are difficult to flatten.)

The next step is to alter the clarinets so some variable sounds can be heard. The teacher must demonstrate how to cut a hole *on the top* (this is important!) of the straw and halfway down the air column (see Fig. 10.4). Then have each student "tune-up" and alternately open and close the hole with a finger to produce the basic note and the new note. The new note should sound higher; ideally, it will be pitched twice as high as the basic note. What happens when new holes are cut at the one-quarter and three-quarter marks? Let the students try these and produce the resulting sounds. The class is now ready to play their own music, and they should be encouraged to do so and to listen to the effects caused by changing the length of the air tunnel.

From here to inverse variation is an easy step. Length of air column can be related to wavelength produced by the vibrating air within that column. The *time it takes to produce a single wave* is called *the period*. If the class is studying the sine function, the teacher has a good opportunity to identify the period of $\sin x$ as 2π radians, of $\sin 2x$ as π radians, and so on. However, time, usually expressed in seconds, is used to define period in the case of sound and time is used again to define *frequency* as *the number of complete vibrations in a second*. Experiments have demonstrated that whether executed as a loud or soft sound, a pitch of middle C on a tuning fork will result in 262 vibrations per second. Moreover, a frequency of 262 will produce a pitch of middle C for all musical sounds. For example, if 262 teeth of a steam-saw cut into a log every second, you will hear middle C. If 262 blasts of air escape from a tea kettle every second, middle C is produced. Many more examples of the relationship of frequency to pitch are given in Jean's essay "Mathematics of Music" in Volume 4 of *The World of Mathematics*. The relationship of frequency, and thus pitch, to wavelength has already been characterized as an inverse variation, but what is the constant of variation? It has been empirically determined that the product of frequency and wavelength is equal to the velocity of sound. Is the amplitude of the sound curve somehow related to what we hear? Definitely, and your students who know about amplifiers can probably guess how. That's right—a louder sound will result in higher curves. A middle C on the piano will sound like middle C as long as that key is played but as the pressure on the key changes, the sound curve reflects this in a shifting amplitude.

There are many additional questions that are spinoffs from the clarinet laboratory.

1. Who in class plays a drum, a violin, or a trumpet? Where is the vibrating air column in each of these instruments? How are changes in pitch

effected? (This is an excellent place to use the expertise of the music teacher, as well as that of the physics or physical science teacher.)

2. What is the mathematical relationship of any note to a note an octave lower or higher? Compare your answers with the data from your own experiment with the clarinet. What numerical relationships are tied to notes within an octave?

3. According to the *Guinness Book of World Records* 1976 edition, the amplification for the rock group Deep Purple attained 117 decibels in London's Rainbow Theatre in 1972. Three members of the audience were said to become unconscious as a result. What is a decibel? What kind of input power is required to produce an output of 117 decibels? What is the relation of energy to perception of loudness? (See Jeans's essay for information on loudness, the threshold of pain, and the scale of sound intensity.)

The clarinet laboratory is one with long-term possibilities—for interrelating mathematics and science and for interrelating mathematics and music. This kind of triple benefit is not unusual. The honeycomb laboratory may have suggested additional excursions into geometric plane-filling patterns used by artists in Moslem tile work or those found in E. M. Escher prints. The well-known Fibonacci sequence (1, 1, 2, 3, 5, 8, . . .) can be introduced by sketching the family tree of a male bee. (Why *male*? Check with a biology colleague.) However, it can also be related to the golden rectangle, a rectangle whose sides are in the ratio of 1 to $(\sqrt{5} - 1)/2$, and to the uses of that

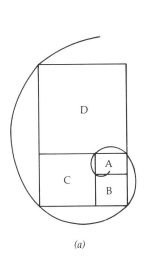

(a)   (b)

Fig. 10.5   Examples of the equiangular, or logarithmic, spiral.

rectangle in architecture and art. The spiral illustrated in a family of golden rectangles (Fig. 10.5a) is an example of an equiangular, or logarithmic, spiral which is vividly displayed in the shell of the chambered Nautilus (Fig. 10.5b). Wait a minute! The logarithmic spiral, like all other mathematical curves, is a *mathematical model*, so the shell of the chambered Nautilus exhibits a *physical model* with the characteristics of the logarithmic spiral. If your students are in a second-year algebra course, you might explore with them the extent to which this mathematical model is a good fit for the curved shape of the shell of the Nautilus. After all, the needle on a moving phonograph record describes a spiral quite different in appearance from the logarithmic spiral and, it turns out, different mathematically. However, the logarithmic spiral *is* the best fit model for the shell of the chambered Nautilus as is evident from an analysis of its structure. Before asking a science colleague to explain this, do Simulation Exercise B. You may find yourself in the position of "science teacher."

As the foregoing illustrations demonstrate, the world around us provides us with a limitless source of data which can be used to teach mathematics. Moreover, the science teacher can become a key resource person if you ask the right questions. Most mathematics majors know that high-school physics and chemistry involve considerable mathematics, but they seldom think of biology or the various junior-high science courses as an arena for mathematics/science cooperation. Further, mathematics teachers often fail to ask their science colleagues specific questions. Since many science teachers are just as vague about the background of the mathematics concepts and rules they use, the result of misguided attempts to communicate is often frustration. We've found that both parties are amazed by data such as that displayed in Table 10.4 and Table 10.5 and now each has specific topics which can be used as starting points for planning ideas, for ironing out mutual problem areas, and so on.

After examining these tables, the mathematics teacher might ask a science colleague one of the following questions.

1.   I'm teaching scientific notation and I need some very large and very small sizes or distances as illustrative examples. Would you give me some examples and some background—perhaps distance to planets and size of cells? What's Avogadro's number, by the way? Who was he?

2.   We're starting work on graphs of parabolas next week. I'd like to use some science illustrations which yield that kind of curve. What distance/time or velocity/time problems would be good to use?

The science teacher may suggest some unwieldy numbers from your point of view. Real-life measures are seldom "nice," and it is important for the mathematics teacher to introduce an awareness of this fact. However, you might explain to your science colleague that you'll use the unwieldy numbers

Table 10.4
Selected mathematics-biological science interrelationships in grades 7-9

Math grade level	Mathematics topics	Some science examples
7-9	Area and volume	Limits on cell size, water loss by leaves
7-8	Simple probability	Transmission of inherited traits
9	Mixture problems	Concentration of solutions
9	Functions	Growth curves of all types
7-9	Graphing	Population studies
7-9	Percentage	Genetics, population studies, diffusion of molecules
7-9	Rate and ratio	Enzyme action rates, ratios of elements in compounds
7-9	Metric units of measure	Centimeter, meter, gram, liter
7-9	Big and small numbers	Size of cells
7-8	Symmetry	Classification of organisms
7-9	Statistics: counting, data collecting, histograms, mean, mode	Predicting offspring types

Table 10.5
Selected mathematics-physical science interrelationships in grades 7-9

Math grade level	Mathematics topics	Some science examples
9	Function	Solubility
7-9	Geometric structures and figures	Crystal structure, shell in model of atom
7-9	Graphs	Distance-time, potential energy
7-9	Percentage (solution/mixture problems)	Percentage composition
7-9	Scientific notation	6.02×10^{23} (Avogadro's Number)
9	Slope of line	$a = \Delta v/\Delta t$
7-9	Metric units of measure	Kilogram, calorie, centimeter, liter
7-9	Variation (direct proportion; inverse proportion)	Law of lever, other simple machines
8-9	Volume	Boyle's law, Charles's law
7-9	Big and small numbers	Distance to planets
7-8	Symmetry	Crystals
7-9	Equations (solving)	Chemical equations (balancing)

in the later stages of your work but need to begin with comparatively simple numbers (few significant digits) so as to keep the students' attention on the *process*, rather than on all the details. They'll understand since they do the same thing when introducing a scientific principle. Moreover, if you're careful to share what each of you would like the students to learn from the graphs referred to in question 2, you'll find that the science teacher can't understand why mathematics students automatically divide the sheet of graph paper into four quadrants. The graph of $s = (\frac{1}{2})gt^2$ (the formula for the distance of a freely falling object where distance in feet is expressed as a function of time in seconds and g is the gravitational constant, $32'/\text{sec}^2$) doesn't make much sense when t is negative. What kind of relationships might result in the use of more than one quadrant in graphs? Temperature problems or problems involving distances above and below sea level come to mind immediately. These are just a few examples of the potential for science/mathematics cooperation. See the resources listed at the end of this chapter as you begin to add teaching ideas to each of the folders in your resource file.

10.3   MATHEMATICS AND THE EVERYDAY WORLD

When we talk about the "everyday world," we always mean the *here-and-now* world of *the students*. Recall the list of potential interests of teenagers given in Chapter 7. These areas represent starting points, with your own future students providing clues as to additional areas of interest. Although attention getting and keeping alone would be sufficient reasons to concentrate on the everyday world, there is still another reason. The vast majority of your students (97 to 98 percent) will *not* be majors in mathematics but they *will* be citizens in a world whose events are increasingly described in quantitative terms and where decisions are based on the application of mathematical models from simple odds to complex statistical systems. Thus, the everyday world of school, home, hobbies, sports, summer jobs, music, design, and even drugs makes sense to the teenager and helps give meaning to the mathematics of their present and possibly future use.

The Introductory Activity introduced you to some potential sources of everyday data which could be used to teach mathematics. How can a survey of litter on the school ground be used to teach or apply mathematics? You probably thought of the need for classifying, tabulating data, and perhaps sampling the presentation of data in graphic form. A bar graph would be an ideal way to represent data on numbers of all soda cans, beer bottles, papers, matchbooks, and miscellaneous trash. However, a histogram would *not* be an appropriate way to picture such data. Why not? Remember the continuous bars of a histogram? That's right, they must touch so that continuous data (for example, age ranges from birth to age 16) can be depicted. However, kinds of litter represent discrete data as do color of eyes, a collection of TV shows, a group of rock records, and so on. Litter suggests the area of ecology, but isn't that science? Yes, it is. Just as the physics of sound was found to necessarily relate to music, so much of the everyday world of the students will have science ties *if* we look for them.

Literature is another field with rich possibilities for teaching mathematics. Even before *The Hobbit* became popular, a creative mathematics student teacher, Constance Curley Feldt, decided to use ideas drawn from Tolkien's fantasy in order to develop applications of ratio and proportion in a ninth-grade algebra class. The students were introduced to that particular part of the adventure in which the hobbit, Bilbo Baggins, 13 dwarves, and a wizard undertake a pilgrimage to obtain the riches belonging to the dwarf Thrain. Each student was given a copy of an adaptation of Tolkien's map in color and a set of 15 questions that developed the story line and required the application of ratio and proportion principles. An artist's version of the Feldt map (see Fig. 10.6) and the first 7 questions that the students were to answer are reproduced* to illustrate how the teacher wove the story line through a sequence of mathematical exercises.

Fig. 10.6   The journey of the hobbit.

*From C. C. Feldt, "Ratios and proportions with a little help from J. R. R. Tolkien," *New York State Mathematics Teachers Journal* 27 (Winter 1977):26-27. Reprinted by permission of the editor and author.

1. To begin, if you are curious as to the height of Bilbo the hobbit and his dwarf companions, use the following information: The ratio of the height of Bilbo to that of the wizard Gandalf is 1:2 and the ratio of the height of Gandalf to that of the average dwarf is 4:3. How tall are Bilbo and the average dwarf if Gandalf is 6 feet tall?

2. Find the ratio of the height of Bilbo to that of the average dwarf.

3. The goal of our far-wandering group is to reach Lonely Mountain where the ancient dwarf king Thrain has left his riches. How many miles high is the highest peak of Lonely Mountain (as indicated by the flag)?

4. After a last good home-cooked dinner at Elrond's Last Homely House, the party must cross the Misty Mountains. Using the straight-line distance, how far is it from Elrond's front door through the mountains to the old ford?

5. But while crossing the tunnels of the Misty Mountains, our adventurous group is surprised by goblins—big goblins, great ugly-looking goblins! And Gandalf, our wizard, is nowhere to be found. That leaves 14 brave little souls in our party. If there are 6 goblins to every dwarf and 2 goblins for Bilbo, find the total number of goblins to be fought.

6. To find the height of an average goblin, go back to the information concerned with dwarves in problem (1) and note that the ratio of the height of a dwarf to that of a goblin is 6:7.

7. What is the ratio of the height of a hobbit to that of a dwarf to that of a goblin?

For further details on this successful lesson and its adaptation to a seventh-grade class, read C. C. Feldt's entire article.

Ratio and proportion could also be introduced by studying the relationship between the number of teeth on gear arrangements and the revolutions of the gears. Why not start with an example from the world of many teenagers—a three-speed bike? Why does first gear allow you to pedal with relative ease when biking up a steep hill? Turn the bike upside down and have one student turn the pedals. The number of revolutions of the pedals and the corresponding number of revolutions of the rear wheel need to be recorded for first, second, and third gear respectively. Does the ratio change if the rider goes at a faster pace? Why or why not? If you're working with a junior-high class, a simple laboratory with cardboard gears would be a good way to help them understand this application of proportion. Figure 10.7 depicts models of two different gears (30 teeth and 15 teeth), which may be copied and traced on sturdy cardboard, such as the backs of tablets. After the models have been cut out, they should be fastened to a cardboard backing by means of paper fasteners through the center of each gear. The students move the gears so

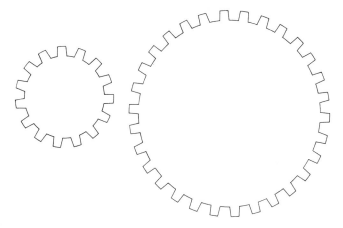

Fig. 10.7   Gear models.

that the starting points touch and then move the larger gear through a complete revolution while counting the number of revolutions of the smaller gear. A data table is completed (Table 10.6) and the students are asked to predict the number of revolutions of the smaller gear for 5, 6, and 12 revolutions of the larger gear. The students are thus guided to compute the ratio of the number of teeth of the two gears, perhaps *before* the term *ratio* is known. There are one-half as many teeth in gear B as in gear A, and it takes gear B two turns to each one of gear A. What would happen if a still smaller gear with 10 teeth were meshed with the larger gear? An extension of the laboratory would be a fine way to explore the students' conjectures; or, if time became a problem, a demonstration on the overhead projector could be used to answer the question. Spirograph gears make excellent overhead projector models for this purpose as long as the teacher first selects gears with numbers (of teeth) which are factors of one another.

Table 10.6
Data table for gear laboratory

Gear A No. of turns	Gear B No. of turns	No. turns A / No. turns B
1	2	1/2
2	4	2/4 = 1/2
3	—	—
4	—	—

However, the everyday world of the high-school student is not a complete bed of roses, and the teacher's recognition of and attention to some of the less pleasant aspects of that life can do much to restore student confidence in this adult model and direct attention to the relevance of the subject being taught. One experienced teacher of second-year algebra did just this when she contrived a set of problems which developed a real-life story line on drug use. The set of ten problems that follow were used with considerable success to demonstrate the application of several standard type problems.

1. Sammy Nerd and his friend Ed Efrom are local pushers who have just received a new batch of heroin which is 92% pure heroin. Sammy and Ed, however, are interested in earning a little extra money and want to cut the stuff to 55% heroin. How much pure quinine must they add to the 50-ounce shipment to get the correct mixture?

2. Sammy could have cut the stuff in three hours alone, while it would have taken Ed four hours by himself. They worked together for one hour, but as fate would have it, Sammy, who was also a user, died. How long did it take Ed to finish alone?

3. Knowing that he might be connected with Sammy's death, Ed quickly left the scene to sell the heroin to the closest contacts. Since the stuff was really hot, Ed got what he could for it: $50,000. If he had had two more kilograms (kilos) to sell he could have raised the price by $5,000 per kilo and gotten $112,500 instead. How many kilos did he have to sell?

4. Ed wanted to unload everything that he had just in case he got picked up. Ed had some amphetamines worth $.60 apiece and some barbiturates worth $.55 apiece. How many of each type should Ed have put into a bag if he wanted to have 14 in a bag selling for $8?

5. Ed took ten grand ($10,000) and divided it into two parts. One part he loaned to a bookie at 12% interest and the rest he invested in a numbers racket at 20% interest. He expected to receive $1,400 in interest at the end of the year. How much did he invest in each?

6. As fate would have it again, the police got onto Ed's trail. Ed heard that the police had a warrant out for his arrest and jumped into his speedboat to head upstream for his hideout which was 50 miles away. But the police knew about the hideout too, and left at the same time that Ed did for the hideout which was 60 miles downstream from the station. Their boats travel at the same rate in still water but the river's current is 7 mph. How fast was each traveling if they get to the hideout at the same time?

7. The address of Ed's hideout was a two-digit number in which the unit's digit was five less than three times the ten's digit. If the digits of the address were reversed a new number would be formed which would be 27 more than the original number. Find Ed's address.

8. The police also sent a backup car to Ed's address. After a car's brakes are applied, the distance *d* a car travels before stopping varies directly as the

square of its velocity v. If the police car's stopping distance is 32 feet when the velocity is 40 mph, what would the distance be when v is 60 mph?

9. Following the high-speed chase, the police did manage to make it just in time to arrest Ed. (Don't they always?) The ratio of Ed's age to the number of years that he was sentenced is 6:11. If his age is decreased by four and his sentence is increased by six, the resulting numbers are in a ratio of 4:9. How old will Ed be when he gets out of the slammer?

10. The grounds of the prison where Ed presently resides are rectangular, its length being three meters more than its width. A wall of uniform thickness of three meters surrounds the prison grounds. If the area of the wall's top surface is 720 square meters, what are the dimensions of the prison grounds?*

Student teachers would be well-advised to seek the advice of cooperating teachers before using potentially controversial subjects, such as drugs, in the classroom. Such subjects must be introduced with care and the data used must be real. (These problems were revised in March 1977 so that the data would reflect available information about drugs and drug-related issues.) Nothing turns teenagers off more quickly than an adult attempt at relevancy that is obviously the product of both adult ignorance of their world and too little homework on the issues.

There is really no end to the possibilities for teaching mathematics by using the everyday world. Table 10.7 lists additional types of everyday material and questions to introduce or review some related mathematics concepts/principles. By this point you should have noticed that sources of everyday materials are all around us *if* we are diligent observers. Those data tables and graphs in the local newspaper, photos of geometrically designed architecture found in last year's calendar, the Peanuts cartoon on base two addition, sports records in the school newspaper, the pine combs you found outside (Count the bracts and look at the spirals!)—these are all potential sources of teaching ideas. If they are to be on hand when you need them, they must be systematically stored in your resource file, annotated as to potential use, and able to be readily retrieved when needed. *Now* is the time to start!

10.4 CROSS-REFERENCES TO A SAMPLE CBTE EVALUATION INSTRUMENT

Which competencies are related to the emphases in this chapter? You should select 2.62, 3.11, and 3.12 immediately. If you also choose 1.1 and 3.6, you're thinking along the lines we intended. Objectives in the affective domain (1.1), especially those aimed at appreciating the role of mathematical models based

*credits: Janet McDonald
        Mathematics Teacher
        Niskayuna High School
        Niskayuna, New York

Table 10.7
Selected everyday ideas as sources of mathematics instruction

Everyday ideas	Mathematics questions
1. Statistics on deaths by car accident and/or motorcycle accident over holidays (see *Time* or *Newsweek* or check with National Safety Council).	1. Over the past three years, what are the trends in type of accident (graphs, percentage, ratio)? Predict the number of deaths per type of accident for this year, for each holiday. What factors might alter the chances of your predictions occurring?
2. Basketball freethrows (homerun record, touchdown record) at your school versus the world record.	2. What percentage increase in skill will be needed to match the world records? What have been the trends at your school? Compare the *average* record over the season of each player with the top score for a single game.
3. Obtain a completed scoring sheet from a bowling game.	3. Write a flow chart which would tell a non-bowler exactly how to compute a friend's score.
4. Obtain a city map which can be clearly projected using an opaque projector. Mark the well-known locations (school, churches, city hall, stadium, city park, etc.).	4. Have students give the coordinates of specified locations, using map indices. Have them locate their home and name its coordinates. Why are such maps usually indexed by letter and number, rather than by two numbers?
5. My size, my height, my weight—how do I compare with my classmates? A young child?	5. Estimation uses are unlimited. Make out a personal chart with your height in meters, the length of your fingernail in cm, the length of your shoe in cm, the distance from your elbow to the tips of your fingers in cm, and your weight in kg. Compare the weight of a paper clip and a penny in g. Use your table to estimate the dimensions of your desk, the height of the door, the dimensions of the hockey rink, and the weight of a softball, football, or basketball.

on an understanding of the distinction between those models and their physical-world counterparts (3.6), were repeatedly emphasized throughout the chapter. It would be easy to add 4.1, 4.3, 7.5, and many more to the list of related competencies. In actual classroom teaching, the teacher who uses the idea base in this chapter to effectively demonstrate competency 3.11, for example, will surely be simultaneously working at other competencies. So never feel as if you've missed the point if you identified a larger set of competencies than we listed. Do check on your understanding if your list did not contain those we characterized as key indicators of the focus of the chapter.

10.5 SUMMARY AND SELF-CHECK

Although constant references to students' interests were made throughout this chapter, the authors chose to shade in only the "Nature of Content" portion of the systems analysis schema. Why? We wanted to emphasize the products and processes of mathematics in their relation to other subjects and other areas of everyday life. Thus, we provided examples of simple mathematical models such as the hexagon in relation to the cells of a honeycomb. Moreover, we suggested strategies designed to *introduce* mathematical concepts and rules, as well as strategies designed to promote retention and transfer through applications in situations novel to the students. Throughout the various illustrations, the reader was reminded to consider the intellectual developmental stage of the students, to be alert to the dangers of over-generalization, and to cooperate with science colleagues in an intellectually honest presentation of the mathematics/science interface.

Now you should be able to:

1. Describe several ways in which mathematics teachers can use scientific settings in the teaching of mathematics.

2. Give *at least three* reasons to support an emphasis on the mathematics/science interface in mathematics classes.

3. Describe ways in which several areas other than science can be used to teach mathematics.

4. Cite *at least four* sources of science-related materials that have potential for teaching mathematics.

5. Cite *at least four* sources of everyday-world materials that have potential for teaching mathematics.

6. Add to your resource file ideas that employ science, other school subjects, and everyday phenomena in the teaching of mathematics.

10.6 SIMULATION/PRACTICE ACTIVITIES

A. Obtain a contemporary high-school biology text. Select *one* of the following topics from mathematics: probability, functions, graphing, rate and ratio, or symmetry; and research the ways in which that topic is referred to and utilized in the biology text. What differences (similarities) in treatment from that typically given in a mathematics class are reflected? What ways could you, the mathematics teacher, assist in transfer of these mathematics concepts/rules in the context of biology?

B. The logarithmic or equiangular spiral is a property of dead tissues rather than living tissues, according to D'Arcy Thompson. In his book *On Growth and Form* he lists several other examples of structures that display the equiangular spiral: a snail's shell, an elephant's tusk, a beaver's tooth, a cat's claws, a canary's claws, and a ram's horn.

Study any of these structures by observing the varied lines of growth in one of the structures or in a good photograph. Compare the organism at various stages of growth with the spiral shown in Fig. 10.5a. Next, look up a definition of the logarithmic spiral in a mathematics dictionary. Finally, read Thompson's explanation in Chapter 6 of his book to compare your ideas with his.

Outline a presentation based on your observations and reading which might be used with a second-year algebra class.

C. An easy way to help a junior-high class understand that the triangle is a rigid figure is to give them equal lengths of plastic straws and straight pins. Have them construct equilateral figures of 3, 4, 5, 6, 7, and 8 sides. Project sample models on the overhead. Then display the picture of a golfer shown in Fig. 10.8 and ask the students to find the triangles and to explain how these shapes help in preserving the stability needed in golf.

Fig. 10.8   The golfer.

Locate pictures of other sports that reflect the use of the triangle to promote stability. Then locate pictures of manmade structures (bridges, for example) that also utilize the triangular shape. Identify ways you would use these resources in a junior-high class as well as in a geometry class.

D. Sammy showed you the following problem from chemistry class. "I can't figure out why moles ends up on top," says Sammy.

$$\frac{\overset{1}{\cancel{11.2\ell}}}{\underset{2}{\cancel{22.4\,\ell}/\text{mole}}} = .5 \text{ moles of } H_2$$

1. How do you explain the mathematics of multiplication and division of units to Sammy?

2. Follow up your explanation by converting the speed limit on most super highways, 55 mph, to ft/sec by using identity elements such as 1 hr/3600 sec. What other identity element would you use?

3. How would you explain the mathematics teacher's usual reluctance to allow students to write labels in their work with formulas?

E. Table 10.8 below provides a way to read windchill temperatures in Fahrenheit.

1. Use a calculator and the formula $C = (5/9)(F - 32)$ to convert the data to Celsius temperatures to the *nearest degree.*

2. Design a homework assignment using the converted data which could be given during an algebra graphing unit.

Table 10.8
Windchill equivalent temperature

Temp. F°	Wind velocity (mph)								
	3	5	7	10	15	20	25	35	50
40	40	34	30	23	20	15	11	7	4
35	35	29	24	18	12	7	3	−2	−5
30	30	24	17	12	3	−1	−5	−13	−16
25	25	17	13	3	−4	−9	−15	−20	−24
20	20	12	7	−4	−13	−18	−23	−28	−33
15	15	5	−1	−11	−20	−24	−33	−37	−40
10	10	0	−7	−17	−27	−33	−38	−45	−50
0	0	−13	−18	−29	−40	−46	−53	−61	−58
−10	−10	−23	−31	−42	−53	−60	−68	−77	−65
−20	−20	−33	−43	−55	−67	−76	−85	−93	−75

F. In the problem of the honeycomb, filling the plane and Escher-type art was mentioned. Obtain a copy of E. R. Ranucci and J. L. Teeters, *Creating Escher-type Drawings* (Palo Alto: Creative Publications, 1977) for a creative treatment that both explains the mathematics underlying tesselations and aids all to become designers of tesselations. Complete *at least one* original tesselation and outline its mathematical properties.

G. The ecology activity outlined in Section 10.3 might well lead to a cooperative lesson with the social studies teacher who is working with the class on the economic and political ramifications of conservation/ecology legislation.

Identify ways in which the art teacher, the industrial arts teacher, the music teacher, the physical education teacher, the English teacher, and even the foreign-language teacher might be involved in cooperative lessons with the mathematics teacher. What kind of individual projects or class assignments could be given by the mathematics teacher and one of his or her colleagues and evaluated dually and used for two mutually supportive instructional needs?

SUGGESTIONS FOR FURTHER STUDY

Bell, M. S. *Mathematical uses and models in our everyday world* ("Studies in Mathematics," Vol. 20). Stanford, Calif.: School Mathematics Study Group, 1972.

   This remarkable little volume (149 pages) is directed toward the teacher of those who see mathematics as a mysterious game of *no* use in their world. The author dispels this notion by providing the teacher with problems—some of which ask a question, some of which invite the reader to formulate questions—all drawn from real world data. Problem materials range from data on diets and rock music to data on the speed of experimental aircraft and the characteristics of precious gems. The needed mathematics may involve simple calculations, graphing or the solution of quadratic equations. This is a prime source which should be owned by every mathematics teacher, for it merits repeated use.

Bolster, L. C., *et al. Mathematics around us.* Glenview, Ill.: Scott, Foresman and Company, 1975.

   The texts in this series contain photos, cartoons, data, and problem situations which relate mathematics to science, social studies, literature, sports, and future careers. You will find the seventh- and eighth-grade texts especially useful sources of idea material for mathematics at the secondary level.

Jacobs, H. R. *Mathematics a human endeavor.* San Francisco: W. H. Freeman, 1970.

   The subtitle says it all: "A textbook for those who think they don't like the subject." Both novice and experienced teachers have found that Jacobs knows what he's talking about. In the course of two pages, one finds questions using information about ice crystals, viruses, the planetary system, and playing-card designs. We recommend the text as a source of multiple teaching ideas for all levels of mathematics.

Mathematics and living things. Rev. ed. Stanford, Calif.: School Mathematics Study Group, 1965.

   A student text and teacher's commentary are available under this title. This valuable paperback is aimed at students who either do not like or do not understand mathematics. The approach is to begin with some interesting life science activities and to present the mathematics concepts and rules as the need arises. The science activities require only commonly available or homemade apparatus. The mathematics ideas are developed inductively for the most part by finding common patterns in the data. Main topic titles include: "Leaves and Natural Variation," "Natural Variation—Us," "Muscle Fatigue," and "Size of Cells." Our student teachers rate this set as A-1.

Mathematics through science. Rev. ed. (Parts 1, 2, and 3). Stanford, Calif.: School Mathematics Study Group, 1964.

   This three-volume set includes a student text and teacher's commentary under each of the following titles: Part 1—"Measurement and Graphing," Part 2—"Graphing, Equations and Linear Functions," and Part 3—"An Experimental Approach to Functions." The content coverage is exactly what the three titles lead one to believe and mathematics teachers will find a wealth of ideas to enrich mathematics classes from grades 6-9. The required equipment is very common and simple—rulers, string, coat hangers, marbles, washers, and the like. The texts are loaded with gems!

McWhirter, N. and R. McWhirter. *Guinness book of world records. New York: Bantam Books, 1970,* 1971, 1973, 1974, 1975, 1976, 1977.

   This paperback is regularly updated to incorporate the latest records set by both pros and amateurs. The authors have more than once illustrated ways in which such data might be used to teach mathematics. We encourage you to browse through the latest edition and be alert for ways to teach ratio, measurement, volume concepts, and so on.

Mosteller, F., *et al. Statistics by example.* Reading, Mass.: Addison-Wesley, 1973.

Statistics by example is comprised of a set of four pamphlets, each with a student text and a teacher's commentary and solution manual. The sequence of four texts: (1) "Exploring Data," (2) "Weighing Chances," (3) "Detecting Patterns," and (4) "Finding Models" incorporate mathematics skills ranging from prealgebra (1) to elementary algebra (2) and finally intermediate algebra (3 and 4). Problem materials are drawn from sports (football scores), the social sciences (vanishing women jurors), and the life sciences (the size of wildlife populations). Instructional strategy suggestions are included in the teacher's commentary. Furthermore, each illustration models many similar real-life questions to be captured from newspapers, magazines, and television reports.

Newman, J. *The world of mathematics* (Vol. 1, 2, 3, 4). New York: Simon and Schuster, 1956.

This set of four volumes includes essays on the mathematics of "gifted" birds, the kinetics of gases, crystals, mathematics in golf and gambling, and mathematics in war and art. Two essays, Jeans's and Haldane's, are particularly valuable as sources of mathematics/science ideas. We encourage you to visit these texts often. You will find multiple ways to relate mathematics to science, to technology, and to the everyday world.

Polya, G. *Mathematical methods in science* ("Studies in Mathematics," Vol. 11). Stanford, Calif.: School Mathematics Study Group, 1963.

Polya's paperback guides the reader through the struggles of Newton, Galileo, Kepler, and others. He describes the way monumental results were developed on the basis of simple mathematical concepts. The book abounds with examples of the contiguous development of mathematics and science and is highly recommended.

Schiffer, M. *Applied mathematics in the high school* ("Studies in Mathematics," Vol. 10) Stanford, Calif.: School Mathematics Study Group, 1963.

This book is based on a sequence of lectures by the author given to high-school mathematics teachers. The emphasis is on applications appropriate to elementary algebra and first-year plane/solid geometry. The author discusses and illustrates applications from science such as growth curves, optics, and refraction. It is a valuable resource for teaching ideas and a good source for science-related student projects.

Steinhaus, H. *Mathematical snapshots.* 3rd ed. New York: Oxford University Press, 1969.

"What does a mathematician do?" Steinhaus answers that question by presenting mathematical phenomena as they relate to the real world in photographs and diagrams. His illustrations include the analysis of chess strategies, the scale used for measuring shoe sizes, the patterns formed by dried mud, the properties of soap bubbles and the paths squirrels follow up a tree. This is a highly recommended source which you should put at the top of your list.

Thompson, D. W. *On growth and form.* Abridged ed. Cambridge: At The University Press, 1961.

This new version of the original 1917 Thompson text is an existence proof for the thesis that science is a large part of our everyday life. In masterful, clear, and even poetic fashion, Thompson talks of drops of rain, the spider's web, the forms of cells, and the shape of a snow crystal. Thompson's discourse on the equiangular spiral was cited earlier in this chapter. We recommend this source as one which will extend your knowledge of the mathematics/science interface and provide you with multiple teaching ideas.

CHAPTER ELEVEN

GOLD IS WHERE YOU FIND IT
RESOURCES FOR
MATHEMATICS INSTRUCTION

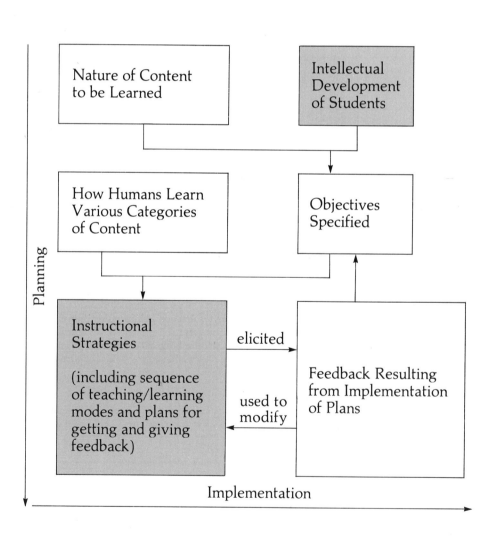

ADVANCE ORGANIZER

If you've been interacting with the text, you have no doubt realized that Chapters 1-8 comprise a cycle of instruction with each chapter focused on one or two components of the systems analysis schema. Chapter 9 was designed to help provide perspective to the teacher—a perspective which at one and the same time sheds light on the past and the present and points the way to intelligent uses of contemporary curricular materials. Chapter 10 played a complementary role in that you were alerted to the instructional potential of other school subjects and nonschool areas—areas outside the academic aspects of subject matter *per se*. These last two chapters were intended to add depth and breadth to the systematic approach to instruction which evolved in Chapters 1-8. Now you should be ready to utilize that approach in planning and teaching.

Why have we devoted an entire chapter to resources? And why so late? After all, we did recommend that you begin collecting resource ideas as early as Chapter 4. Moreover, we continually provided illustrations of a variety of references. Then what is the focus of this chapter?

In Section 11.2 you will find specific examples of learning activities not typically illustrated in secondary-school text materials. These include ideas for laboratories, demonstrations, field trips, bulletin boards, and student projects that employ homemade, easy-to-obtain, and inexpensive materials. Other sections focus on audiovisual aids, human resources, and instructional ideas particularly appropriate for use with atypical students. Finally, we have included a brief section on the all-important legal and safety aspects of teaching. The Suggestions for Further Study section of this chapter is subdivided into categories to facilitate your search for more of the kinds of ideas and information provided here.

Why did we elect to shade in the "Instructional Strategies" and the "Intellectual Development of Students" portions of the systems analysis schema? The reason for the first choice is probably obvious to you since instructional modes are key elements in strategy design. If you recall that Piagetian research demonstrates the essential role of concrete experiences in learning abstractions, then the rationale for shading in the second box becomes clear. For this same reason, we ask you to begin your interaction with the ideas in this chapter by engaging in the concrete experience which follows.

11.1 INTRODUCTORY ACTIVITY

Seeing is not the same as observing.

Go to a multipurpose store (such as Woolworth's, K. Mart, Barkers, or Korvette's). Visit the hardware, toy, notions, stationery, sewing, pet, and household departments.

A. Make a list of all inexpensive items which are potential pieces of teaching equipment. Specify the projected use of each item. For starters, think of the materials included in previous illustrations in this text.

B. Compare your list with that of two or more other classmates.

C. Keep this list handy and modify it as needed after you finish studying this chapter.

Table 11.1
Usable "Junk"

Bags and baggies (plastic)	Needles (sewing): assorted sizes
Bags: assorted sizes (paper)	Paper clips
Balloons: assorted sizes	Paper fasteners
Balls: ping pong, golf, baseball	Pins: common
Bicycle inner tube	Pipe cleaners (white and colored)
Boxes: assorted sizes (cardboard)	Plastic tape
Buttons: assorted sizes	Plastic covers off cans (coffee
Cans: assorted sizes	cans): assorted sizes
Carpenter's ruler	Poker chips
Cards: $3 \times 5, 4 \times 6, 5 \times 8$	Razor blades
Cartons: milk (½ pt., pt., ½ gal.)	Reinforcements (paper)
Clay (modeling) or putty	Sand: box
Cloth: misc. pieces and colors	Sandpaper
Clothespins (pincer type)	Scales (balance)
Coat hangers (wire)	Scissors: class set
Cotton	Scrabble game
Corks: assorted sizes	Screws: assorted
Cups (paper)	Shirt cardboards
Dowels (wooden)	Soda straws
Dress snaps	Sponges: assorted sizes
Egg cartons (colored)	Springs: assorted sizes
Elastic thread	Stopwatch
Felt scraps	String
Flashlight bulbs and receptacles	Styrofoam from florist
Foil (aluminum)	Tape: adding machine
Funnels	Thermometer
Household chemicals: sugar, food	Tongue depressors
coloring	Toy car: both battery-operated and not
Jars: baby food, both sizes,	Tubes (cardboard) from wax paper
with tops	rolls, saran wrap, etc.
Jars: pint and quart size	Thread
with screw tops	Washers (metal): assorted
Marbles	Wax paper
Mirrors: plane	Yarn (colored)
Masking tape	

11.2 INEXPENSIVE EQUIPMENT AND SUPPLIES

If you were surprised at the length of the lists compiled by you and your classmates, you are in good company. As soon as you begin to observe the teaching potential of all you encounter, the most mundane objects take on mathematical characteristics. For example, a wine bottle cork may be perceived as a truncated cone or a weight for a pendulum lab. Table 11.1 lists items like the wine bottle cork. Some may be purchased, but most may be found in the garage, the kitchen cupboard, the sewing basket, and even en route to the trash can. As you read the list in Table 11.1, try to concoct one or two ways in which each item could contribute to a lesson. Don't be overly concerned if you draw a blank on some items at first. We have provided illustrations of the uses of a few of these. After interacting with the reading, make a second attempt to identify uses of items which stumped you on the first go-around. Then be on the alert for other potentially useful pieces of "junk" and add these to your collection.

In addition to these multipurpose everyday-world materials, there are relatively inexpensive kinds of equipment which are directly related to teaching and particularly useful for the teaching of mathematics. Some of these, such as class-size sets of protractors, are requisitioned through the mathematics department and should be available for use during student teaching. (However, there may be exactly 35 protractors owned by the mathematics department, so plan ahead if you expect to use them on a particular day.) Other items, such as a geoboard, *might* be owned by the mathematics department but *can* be easily constructed with the aid of finishing nails, a

Table 11.2
Mathematics equipment and materials

Abacus	Geo-d-stix
Acetates for overhead	Georule
Angle-mirror	Hypsometer-clinometer
Calculators (class size set)	Mira (class size set)
Chalk, semipermanent	Pencils (#2) in quantity and in colors
Compasses (class size set)	Pens—marking for overhead in colors
Compasses for chalk board (use chalk on a string)	Protractors (class size set)
	Protractor—board use
Compass for overhead use (Circle Master Compass)	Protractor (transparent) for overhead use
	Rulers (metric) and meter stick
Geoboards—rectangular grid (class size set and large demonstration)	Slide rule (demonstration and class size set)
Geoboards—circular grid	Tangram set
Geoboard (transparent) for overhead use	Transit
	Weights (metric)

hammer, and squares of plywood. In Table 11.2 we list classroom equipment and materials which have been found particularly useful in the teaching of mathematics. We have included some typical items, such as rulers, but notice that they must be metric. (It is important to reinforce the adoption of the metric system in the United States by using it at every opportunity.) We've also included items, like geoboards and geo-d-stix, which may be purchased under a commercial label but which may also be homemade.

If any label seems unfamiliar, check one of the equipment sources listed in the Suggestions for Further Study section of this chapter. However, don't expect to find semipermanent chalk on sale in any typical catalogue of equipment. That's something you must concoct by using the recipe in Fig. 11.1. After chalk is treated in the way described in the recipe, it becomes very soft and can *not* be used to draw a *fine* line. However, the lines made with this chalk do not erase, although they fade over time and can easily be washed off a chalk board. The potential is enormous. For example, use the treated chalk to draw the axes, write the scale, and draw a basic curve, say $y = x^2$. Then other curves, $y = 2x^2$, $y = (1/2)x^2$, $y = x^2 + 2$, $y = x^2 - 2$, can be drawn, compared, and erased at will without disturbing the treated part of the diagram.

---

4 or 5 sticks of soft colored and white chalk.
2 jars with tight-fitting covers (Peanut butter jars are fine)
A supply of tepid tap water
3 or 4 sheets paper towels
A supply of table sugar

Half-fill one jar with water. Add sugar and stir *until* sugar will no longer dissolve. (Obtain a super-saturated solution.) Immerse chalk in solution. Take out after bubbles stop forming. Lay chalk on paper towel for a *few* minutes to blot up excess water. Store treated chalk in other jar. Be sure cover is on tight.

Solution can be used again *if* stored in tightly covered jar. Check its effectiveness after 2-3 months.

Fig. 11.1   Recipe for semipermanent chalk.

In the paragraphs which follow, selected uses of equipment from Table 11.1 and/or 11.2 will be outlined as they pertain to specific instructional modes or activities used in conjunction with one or more of these modes.

Demonstration A little string, some buttons, and two shirt cardboards are all that's needed if you want to construct a string polyhedron. Draw the same polygon on the two cardboards and cut each out carefully (see Fig. 11.2). Make puncture holes near each vertex and thread equal lengths of string

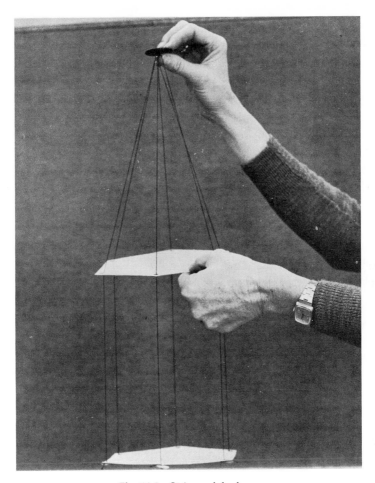

Fig. 11.2 String polyhedron.

through a large button. Thread the ends through two vertices of both layers of the polygon. Next thread each end through a small button and knot. From pyramid to prism takes one easy step. Just move one polygonal shape up. Oblique prisms can be quickly demonstrated by sliding one polygon horizontally away from the other. A twist of the wrist will send the prism into a double-napped pyramid. Once you begin to manipulate this inexpensive string model, you will discover many more uses.

What on earth can you do with plastic coffee can covers? We're sure you thought of their use as circular discs, but not very many people consider their potential as an overhead model. Draw the shape of an equilateral triangle on the cover with a flare pen. Carefully cut out the shape with a safety razor blade. Two congruent plastic shapes may be used to teach a lesson on similarity.

Manually compare one shape with the projected image of the congruent shape to help the students conclude that corresponding angles are congruent regardless of the distance of the projector from the screen. Furthermore, don't throw away the remaining part of the cover. That, too, can be used in a demonstration on the overhead. The cover becomes a stencil. You'll find it helpful to make cover stencils of the most common shapes you work with—the equilateral triangle, the square, and the regular hexagon.

A carpenter's rule can be folded to form various polygonal shapes. Form a rhombus. Have a student move the rhombus to form a rhombus of maximum area and one of minimum area. Compare the shapes. What changes? What remains the same? Form an equilateral triangle and illustrate the rigid nature of the structure. A geo-rule could be used to illustrate the same ideas and has the advantage of being brightly colored and wider than the typical carpenter's rule.

Laboratory Wax paper, pencils, and a few straightedges comprise the essential equipment for a paper-folding laboratory. Each student needs a sheet of wax paper approximately 30 cm on each edge. Instruct them to draw a straight line, l, about 5 cm from one edge and to place a point, P, about 8 cm from the same edge and halfway from the sides of the sheet (Fig. 11.3). Then as you demonstrate the folding process on the stage of the overhead projector, direct the students to match P with any point on l, fold sharply, and then open the paper. This is a good time to label the *drawn* line and the dot so that the students can keep track of the initial given line, l.

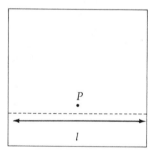

Fig. 11.3   First step in parabola construction.

Next, tell the students to match P to a different point on l and again fold and open the paper. Finally, instruct them to continue the matching, folding and opening steps and look for a pattern (Fig. 11.4). Question the students to elicit the fact that each fold is a perpendicular bisector of the line segment which joins P to its matching point. Now it is easy to develop the definition of the curve called a parabola as the locus of all points equidistant from a fixed line and a given point not on that line. The parabola has been constructed, point by point, and the folding process is the locus definition in action. The focus, the directrix, and the family of tangents have concrete meaning for the student and the laboratory activity helps to promote interaction. (Remember Piaget?)

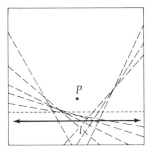

Fig. 11.4 Wax paper parabola.

Moreover, don't ignore the opportunity to investigate other physical models of the parabola and paraboloid—flashlights, floodlights, automobile headlights, the path of a projectile, the water gushing out of a fountain. A followup take-home lab to construct the ellipse and the hyperbola yield added benefits. To construct a paper-folded ellipse, draw a circle and place a point within the circle, but *not* at the center. Match the given point to points on the circle (Fig. 11.5a). A circle is also drawn as the first step in hyperbola construction but now a point is placed *outside* the circle. Again the fixed point is matched to points on the circle (Fig. 11.5b). Be sure to construct your personal model of each. You will find multiple uses for these in geometry (locus), algebra (quadratics), and analytic geometry (conics).

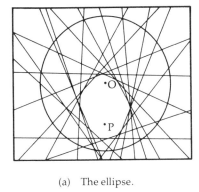

(a)   The ellipse.

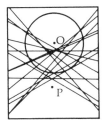

(b)   The hyperbola.

Fig. 11.5 Wax paper ellipse and hyperbola.

An interesting laboratory activity which gets at such important concepts as variation, frequency, range, and norm involves the use of sunflower seeds. Give each pair of students about 200 seeds and 14 paper cups labeled 0 through 13. Instruct them to count the number of black stripes on both sides of each seed and then place each seed in the correspondingly numbered cup. Total class results are compiled by emptying the paper cups into 14 numbered hydrometer jars (borrow these from the science department) or tall, thin, olive jars lined up on the front desk. This activity is a golden opportunity to teach students the difference between the graphing of discrete and continuous data. The sunflower seed bar graph is a concrete example on which to base developmental questions and subsequent exercises on graph paper. (Compare Sample Card in Resource File Assignment—Appendix A.)

Individual projects Straws, coat hangers, cardboard, and styrofoam chunks are the raw material from which student-designed models can be constructed. Straws, cut in equal lengths, can be fastened with straight pins to build complex geometric structures, such as the great dodecahedron. Geo-d-stix kits may be used for the same purpose. Stellated polyhedrons and a representation of a fourth dimensional cube are challenging to construct. Models of the five regular Platonic solids are an excellent starting point for a comparison of faces, edges, and vertices and an exploration of the concept of duality. (See Cundy and Rollett in the Suggestions for Further Study section for further details.)

Ask students to construct mechanical models illustrating locus problems. Some elastic, cardboard, glue, and a nail with a large head were used to construct the locus model in Fig. 11.6.

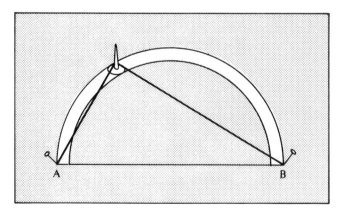

Fig. 11.6 Locus model of vertex of right angle of right triangle with fixed hypotenuse.

With thread or yarn, a needle, and cardboard, the students can design and build beautiful models using curve stitching. The simplest patterns are obtained by joining holes which are equally spaced on line segments or circles. On the reverse side of the cardboard, the thread must be taken to the next hole on the pattern; on the right side the thread is taken across to the next unoccupied hole on the far side of the design. Explaining this process to students is easier if they are told to number the holes on each side of an angle as depicted in Fig. 11.7. The same numerals appear on both arms but are reversed on one arm. Then have them connnect hole one with hole one, hole two with hole two, and so on, following the rules given earlier.

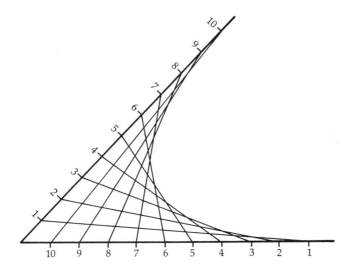

Fig. 11.7   Curve-stitched parabola.

Perhaps you're wondering if curve stitching could be used as a laboratory. The answer is an emphatic "Yes!" In fact, many of the demonstrations outlined earlier can be developed as a laboratory just as most laboratory activities have promise as a demonstration. Continue thinking of multiple uses as you read the sections which follow.

Bulletin boards A corkboard surface, a wall, a chalkboard, or even a square of poster board can be used as the backdrop for a static or moveable display. Use loops of masking tape on newspaper clippings, photos or colored cutout shapes and transform an unused section of chalkboard into display space for a demonstration. Have students bring in clippings which illustrate uses of integers and make a collage of their examples. Titles are readily constructed by "writing" in yarn with the aid of a stapler or by twisting pipe cleaners into the shape of letters. Student progress records on the "Problem of the Week" can be designed as a chart and posted on the wall. Students then record their own progress by pasting a silver star for an attempt and a gold star for success. Maps of the city can be hung from chart clips and used to initiate a lesson in distance problems, scale drawing, map reading, or even coordinate work. If you want to manipulate shapes as a basis for developing the formula for the area of a circle, a feltboard is useful. Cover one side of a sheet of plywood with felt (ask the industrial arts teacher for help). Staple it securely to the reverese side. Cut out a circular shape in a different color of felt and cut it on the lines shown in Fig. 11.8(a). Be sure that the radii cuts in each half do not extend to the edge of the disc. Then manipulate the shapes and interweave them to form Fig. 11.8(b). The felt surfaces will adhere to each other.

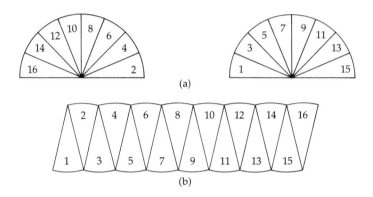

Fig. 11.8 "Area of a circle" model.

We encourage you to *use* bulletin board displays as demonstrations, as a basis for a problem-solving lesson, as a cue card for review lessons, or as an attention-getter for a unit. The topics which follow are just a few which might be the subject of a bulletin board display.

1. The Catenary—What Is It?

2. The Snowflake and Geometry

3. Magic Squares

4. Linkages—Watt's Linkage

5. Clocks and Calendars

6. The Googol and Googolplex

7. Alice in Wonderland Mathematics

8. Spirals in Nature

9. The Biggest and the Smallest

10. Music and Mathematics

11.   A Flow Chart on the Solution of an Equation

12. $(a+b)(a+b) = a^2 + 2ab + b^2$

"Move the shapes and see for yourself."

13.   Our School's Basketball Statistics

14.   The Weather and Probability

15.   Advertising and Fallacious Reasoning

Games and competitions Perhaps pleasant past experiences cause students to equate a game with fun. Whatever the reason, a well-designed and well-organized game has the potential for getting and keeping attention. When you visited the toy and games department as part of the Introductory Activity, you probably noted commercial games and kits which could be used in the mathematics classroom. Some which we have found useful are:

Forecast	Hexogram
Tuf	Chess
Numble	Checkers
Spirograph	Battleship
Dominoes	

Most of the above games are suitable for use by at most two or three students at a time. Keep them in mind when you plan diversified activities but be prepared to modify them for class-size groups. The commercial game of Battleship, for example, uses pegs on two plastic grids but with the aid of graph paper and the use of pencilled symbols to represent different kinds of ships, the entire class can play in subgroups of two. See the *School Mathematics Project Book A* for a description of the game played in quadrant I.

In similar fashion, you can modify games not originally designed for the mathematics classroom. Watch any of the popular game shows on television and take notes as to the essential features of the game—the rules, the penalties, the kind of game board, and so on. Cross-wits becomes a combination of a crossword puzzle and a "less than 20-questions" game. If your puzzle were projected on the overhead, the entire class could follow the game. The contestants might be filling in names of concepts, formulas, trigonometric identities, or citing reasons for some geometric assertions. There are no holds barred on the content of the individual items or the key question related to the individual clues. Moreover, the entire class could play the game if two teams were designated and appropriate rules for deciding the order of the contestants were outlined.

Finally, don't forget the games of childhood. Twenty questions, Who dunnit? and What's my question? are always ready to be adapted for a quick review or a probing, challenging exploration.

11.3   NONTYPICAL USES OF AUDIOVISUAL AIDS

Cassette audiotape recorders can be used to bring a wide variety of sounds into the classroom, such as the sounds of the countdown at a space center, parts of a speech by a famous mathematician, and radio commercials. Some teachers have also recorded their own mini-lectures designed to accompany a demonstration, filmstrip, or slide presentation.

Homemade charts are easy to make and valuable to have. Obtain some of those outdated pull-down maps from the social studies department. Turn them over and you have blank charts that one of your artistically inclined students can convert into custom designed visuals.

The typical spirit masters produce purple copies. It is often advantageous to produce multicolored diagrams, sketches, and flow charts. This is easily done by purchasing a few differently colored spirit carbons from an office supply store. Simply fold back the original purple carbon and replace it with another color when you are tracing the lines you wish to emphasize. Any number of different colors may be transferred to the back of a single master in this fashion. Try it. You will like it, and so will your students.

The uninterrupted and unaltered viewing of a motion picture film is not always the most effective use of this medium. Shut off the projector at key points and question students as to (1) the major points developed thus far and (2) their predictions of what will happen next. Turn off the sound and supply your own audio when the commentary is either too sophisticated or too childish for your students. Back up the film and ask students to observe a sequence a second time. Moreover, don't overlook the possibility of making a motion picture of your own design. Most schools own a good quality motion picture camera. Typically it is signed out to the athletic department on a semipermanent basis, but since it is school property, you may be able to arrange to use it.

Filmstrips and slides also lend themselves to imaginative use. Students who have been absent for several days can be put in one corner of the room with projector, filmstrip, and teacher's guide while you are busy with other students. After viewing the strip and reading the guide, the catch-up session with the teacher will be facilitated. Some filmstrips may contain only a dozen relevant frames. Cut these out and mount them in empty 35 mm slide mounts available from any photographic supply store. Homemade filmstrips are easily produced by turning a 35 mm camera on its side while photographing the scenes. Then simply instruct the developing house to "Develop only. Do not cut into separate frames." Making your own sets of instructional slides is even easier. The yearbook advisor is the likely custodian of a school-owned 35 mm camera which you can arrange to borrow. Students in the photography club can be enlisted to instruct you in its use or to supply both the camera and skilled labor to get the job done. Also, investigate some of the newer types of slide sets available from mathematics supply houses. Some can be projected on the chalkboard with room lights on so teacher and students can add labels, auxiliary segments and graph lines with chalk. These are designed to promote an inquiry approach and have proved to be quite effective.

Loop films are a fairly recent development showing great promise. These silent, repeating loops are accompanied by a detailed teacher's guide which cues the reader as to promising places to stop the motion and key questions to be asked. Again the emphasis is on an inquiry approach.

Overhead projectors have become the most common visual aid in the schools. Typically their use is limited to projecting acetate-based images on a screen, but they do have additional uses. Trace or "bake" (use a thermofax or xerox machine) a complicated diagram on acetate, project it on the chalkboard after school, go to the board, and trace the lines with a soft lead pencil. The lines will be visible to you, but not to the students, as you chalk over each in turn during a lecture and question/answer session in class the next day.

Some demonstrations are more effective if projected on a screen. We've already noted the uses to which coffee can covers might be put. Another example involves placing a straw polygon on the stage and flexing it to illustrate the lack of rigidity of non-triangular shapes. A good demonstration prior to a π laboratory in an advanced class studying probability requires the preparation of an acetate sheet with parallel lines ruled 6.1 cm apart. Then obtain a set of twenty small sticks, 4.6 cm long. (Toothpicks should serve as a good approximation.) Holding your hands high, drop the sticks onto the sheet. Have the students help count the cases where sticks cross and fail to cross the lines. If a stick does not fall on the sheet at all, drop it again. Now have groups of students repeat the demonstration a number of times, tabulate the data, and obtain the ratio of the number of sticks crossing the lines to the total number. This ratio, p, is related to π. (In this case, $\pi \approx 2(4.6)/6.1p$.) If the length of the stick is k and the distnce between the lines is a where $a > k$, the formula becomes $\pi = 2k/(a)(p)$.

Color changes in various demonstrations are also easier to see if the glass containers are set on the lighted stage of an overhead projector. In this case it is helpful to cover the projection lens with dark cloth or a cardboard mask.

A slide projector may also be used as a light source for all kinds of demonstrations. Obtain a model of a cube. Darken the room and set up the slide projector. Then adjust the distance between object and lens so as to obtain a sharp image on the wall or screen. Have the students conjecture as to the shape which will be projected as you hold the cube so that a face is parallel to the wall and then when one of the major diagonals is perpendicular to the wall. Most will be amazed at the hexagonal shape obtained in the latter case. Objects of other familiar shapes can be manipulated in like manner to introduce the students to orthographic projection drawings or even to problems dealing with various cross-sectional slices.

Television equipment is now available in almost every school and offers many unique opportunities for teachers. Mathematics-related programs can be recorded from commercial television during evenings or weekends for class viewing during regular school hours. (Care must be taken not to violate copyright laws.) It is also easy to record phenomena from field trips and laboratory work and play the tape back in slow or stop motion.

These are just a few ideas to start you thinking about seldom-used audiovisual techniques. We made no attempt to tell you how to thread, set up, or focus these devices. Each make is a bit different, and each comes with instructions for its use. The easiest way to learn the mechanics of these devices is to have someone show you how and supervise your first attempt.

11.4 FIELD TRIPS

Field trips have high potential for contributing to important goals of mathematics instruction *if* they are carefully planned and effectively executed. Among the possible goals are generating interest in the subject, collecting data for further study in class, and observing the use of mathematics in an applied area. Some examples of field trips oriented toward senior-high mathematics students are (1) visiting the drafting department of a nearby company to see and hear about the geometric and algebraic relationships being used, (2) demonstrating the use of the transit, hypsometer-clinometer, and angle mirror, (3) touring the computer center of a nearby industry or university, and (4) visiting the set design area of a college or local theater to learn about the mathematics related to design problems. Trips to a state environmental center to collect data by means of sampling techniques, a bank to learn about the mathematics involved in savings accounts, the athletic fields to obtain data on the dimensions and shapes involved, and the wooded area outside school to collect leaves for a surface area project have proved valuable for junior-high mathematics classes.

Thorough and detailed planning far in advance of the trip is essential. *You* must take the proposed trip, time it, and construct a guide sheet for students to use both before and during the trip. Safety rules, appropriate clothing, and any special student behaviors need to be specified on the guide sheet and reinforced verbally. These rules and the purposes of the trip should also be included in the parental permission slips which are to be signed and returned prior to the day of the trip. Such slips carry little legal status in a court of law but they do establish consent and often generate parental under-standing and support for your efforts. Don't forget to consult the principal's office about arrangements for students who forget to return permission slips. Every school has its own set of regulations governing field trips, and these must be followed from the earliest planning stage through the culmination of the trip.

All trips must be carefully supervised. Here's the time to enlist parental assistance and wider support of your efforts. Followup activities include written reports, laboratory work on collected data, class discussions, and letters of appreciation to those who made the trip possible. A brief oral report to the principal is in order and may well insure cooperation when future trips are planned.

11.5 HUMAN RESOURCES

Don't overlook the human resources available to you for the asking. The art teacher can demonstrate and explain the relationship of perspective in paintings to proportion in mathematics. A parent who is a carpenter can explain the mathematics used before the actual cutting of lumber needed for a bookcase. Architect's plans, scale diagrams, and a photo of the finished product can be provided by a local architect and displayed on the bulletin board. A student's older sister home for a college vacation may be willing to talk about her mathematics major requirements with interested students. A local research mathematician or college professor might be willing to consult with a talented student pursuing a mathematics fair project. The president of an amateur photography or radio club could send a member to talk about the mathematics related to each of these hobbies. The local meteorologist may be willing to explain the computation of the probabilities associated with weather forecasting. There are many laboratory technicians in nearby industries and hospitals who could talk about their work and the training required for such careers.

Then, too, don't forget your students. Many of them will have hobbies and skilled training in mathematics-related areas. Musicians, philatelists, model car- and plane-builders, orienteerers, sailing enthusiasts, and puzzle enthusiasts are but a few of the many human resources often overlooked. It is probably obvious that model car and plane hobbyists have been working with scale models, but what can be done with future sailors? Think in terms of vectors and forces!

11.6 DIFFERENTIATING INSTRUCTION

Sometimes teachers or administrators try to make the best of the differences in needs, interests, and abilities of the human resources of their students by diverse grouping arrangements. Many junior- and senior-high schools use some kind of homogeneous grouping system, for example, sections for the college-bound and noncollege-bound or classes designated as honors, average, and basic. Regardless of these attempts to subdivide students according to such criteria as IQ scores, reading test scores, grade-point average, or teacher recommendation, even a casual observer soon notices wide differences in a supposedly homogeneous group of students. For this reason, we believe that differentiation of instruction must take place in *every* class. Again we call attention to the saying: "Different strokes for different folks!" On a day-to-day basis, the teacher can work at this by (1) assigning individual projects which span a wide range of interests and skills, (2) dividing the class into several heterogeneous groups so that a variety of talents may be shared, or using more homogeneous groups, each assigned different activities, (3) initiating a "buddy" system where pairs of students work together at stated times (seek the advice of your cooperating teacher before deciding who will be paired with whom), (4) designing homework assignments with built-in choices, and (5) allowing individual students to browse through mathematics-related reading on a classroom shelf, work on a project, spend more time checking work on a calculator, and so on while the teacher and the rest of the class continue a lesson. Doesn't that kind of instruction require a lot of planning time? You bet it does! However, if the payoff for the time and energy you've expended is student learning, then the extra effort is worth it.

If you've realized that the success of such activities also depends heavily on the accurate diagnosis of student talents and weaknesses, then you've taken the first step toward differentiating instruction. Thoughtful observation of selected students followed by conferences with their teachers will help to sharpen your observation skills. At the same time you need some broad guidelines relative to atypical students. Novice teachers find it most helpful to focus on those who are often labeled the "slow learner" and the "academically talented."

The slow learner The "reluctant learner," the "underachiever," the "culturally deprived"—these are just a few of the labels used to designate students who achieve below some minimum standard. In this section we highlight the student who has poor study habits and poor learning capabilities, whatever the cause of the label. The characteristics of the slow learner include some or all of the following:

1.   Below par in both immediate and long-term memory

2.   Cognitive behaviors restricted to those of the concrete operational adolescent

3.   Poor work/study habits

4.   Reading, listening, and speaking skills far below grade level

5.   Unmotivated by long-range goals

6.   Lacks self-confidence in his or her ability to learn (due to a long history of failure in school)

Much instruction seems to fly in the face of these identifiable characteristics. Repetitive drill on arrays of meaningless (to the student) facts is bound to fail with students whose memory skills are poor. Instead, a variety of concrete, everyday activities presented in a meaningful manner helps these students to make sense out of schoolwork. They need many experiences with manipulative materials, careful directions, intermittent reinforcement for progress, and praise at each completed stage of the learning. That prescription shouldn't sound radically different from the guidelines promoted throughout the previous ten chapters of this text. The model of systematic instruction still holds, and the various illustrative examples given thus far could be adapted for slow learners.

Omit the reading in a creatively designed worksheet and audiotape the directions, being careful to speak slowly, to repeat, and to avoid complicated words. Arrange individual projects cooperatively with the industrial arts teacher and the English teacher. For example, give as much help as necessary in a model-constructing project. Help the student draw up the specifications. The completed model can be submitted jointly for industrial arts and mathematics credit. Instead of a written explanation of project design, an oral report can be audiotaped and again serve two subject areas—mathematics and English. In the junior-high π laboratory described earlier in the text, allow these students to measure more objects, encourage the use of the calculator to check divisions, and provide guidance with measuring tasks. Perhaps the class should be told to divide to the nearest whole number. Students are more willing to seek help in the measuring skills or the division skills as part of a laboratory with an immediate goal than they would be to attend to separate division and measuring lessons. Common sense is needed so that the task does not become impossible. You must decide on the modification which takes the student gradually from familiar territory to new learnings.

It should be clear that we believe that these students can learn if taught in a meaningful way with many concrete experiences. What's all-important is that the teacher demonstrate that same belief. You do slow learners no service by passing them on despite inability to meet standards. However, reasonable standards, interesting and challenging activities, and the atmosphere promoted by a thoughtful teacher will help lead the slow learner away from the all-too-familiar failure syndrome.

The academically talented These students are not always those with the highest grades, nor even those who speed through exercises. The young Galois was considered mediocre by his teachers, who advised that he be demoted. Bored and disinterested, he gave only passing attention to his studies. Subsequently, he amazed his teachers by winning the prize in the general examinations despite his barely adequate results in traditional school work. An encounter with Legendre's geometry had aroused his enthusiasm, and he began to apply his intellectual talents to geometry and then to algebra. His contribution to the contemporary view of fundamental structures, such as group and field, is considered one of the milestones in the development of abstract algebra. History is replete with similar tales of intellectual giants who in their youth were considered stupid, recalcitrant, and unmotivated. Yet even in their youth, a discerning teacher would have been able to collect clues as to their capability.

What characteristics are signs of potential academic talent? As a group, these students exhibit some or all of the following.

1. They have well-developed memory skills.

2. They display the reasoning patterns of the formal operational stage much earlier than their peers.

3. Insightful thinking, an easy grasp of the nature of a complex problem, and the utilization of original routes to problem solution are present to a high degree.

4. Reading and listening comprehension is of high quality, as are verbal skills.

5. They are capable of independent work on long-range projects of interest to them.

While repetitious drill is wasted on the slow learner for one reason, it is anathema to the talented for another. Bright students need less practice than the average student. However, they are able to go further and deeper into a problem situation than most. If the teacher were using any of the laboratory activities outlined earlier in the text, these students would be able to generalize after less experience with the concrete materials. Notice, we *didn't* say that the concrete base can be omitted. (Think back to the nested stages model.) Since the academically talented are capable of considerable self-directed learning, teachers can make excellent use of individualized projects. Several sources of ideas are included in the Suggestions for Further Study section. Also check out the mathematics-related books in the school library and direct students to one or more of these. Mathematics teaching journals are additional sources to consult—both for specific project ideas and for information about mathematics fairs and contests in your geographic area. Be sure to "publish" the results of their projects in some way. Mechanical models can be demonstrated and explained to the class, summaries of library research can be made available for other interested students to read, and several journals welcome reports of creative work by students.

In a twelfth-grade class taught by one of the authors, three gifted students had capabilities far beyond the rest of the class. By group consent, a special Friday class was designed. The three students worked in the room or the library reading and studying a special topic external to the course while the rest of the students worked with the teacher on trouble spots identified during the week. Occasional teacher conferences with the three talented students gave them a chance to report on their progress, which was impressive.

Freedom to explore an interest, to use different methods of solution, and to seek answers to atypical questions is what the talented student needs. That doesn't mean that the teacher isn't needed, or that anything goes, or that the rest of the class can be ignored. Students who are working at an individual project while the rest of the class is pursuing a lesson should have clearly defined responsibilities. A teacher may "contract" with such students both the quality and amount of work to be completed by a certain deadline. If talented students enjoy helping other students, they can become teacher aides under careful supervision. Some students find this role an exciting way of increasing their own depth in mathematics, but don't assume that all do. Diversity is paramount in the human race.

The Learning Activity Package The suggestions outlined in the previous two sections were based on the assumption that the novice teacher normally needs to find ways to accommodate atypical students within the usual instructional pattern. However, there is no doubt that instruction designed to allow students to work at varied rates and on varied materials would be a far better long-term solution to the needs of *all* students. This is the purpose of the instructional materials which we call a Learning Activity Package (LAP).

The three initials in LAP denote the major components of the learning activity package. The *L* in learning is seen as the learning of concepts, rules/principles, and other cognitive objectives plus the learning of related affective and psychomotor objectives. The *A* in activity denotes major emphasis on laboratory activities and constant attention to interaction via visual images, relevant everyday experiences, and thought-provoking questions. The *P* in package denotes a total instructional unit complete with reading material, laboratory directions, projects, instructions to work with other students, hints, feedback (including praise), test(s), answer keys, and more. The package is the student's personal instructional guide. Now the teacher is freed to deal with individuals, to offer the extra cues, challenges, and guidance needed, and to collect extensive feedback and modify individual use of the LAP based on that feedback. Students engaged in a LAP will be working at lab tables, looking at filmstrips, listening to tapes, reading and writing in personal LAP booklets, and working with two other students or with the teacher. A highly structured, well-organized, thoughtful LAP plus an observant and resourceful teacher are clearly prerequisites. A modular assignment based on this operational definition of LAP is outlined in Appendix A.

11.7 SAFETY AND LEGAL ASPECTS

Teachers are legally responsible for the welfare of students under their supervision. Attendance books have been subpoenaed for evidence in courts of law and teachers have been called upon to testify as to the accuracy of these records. Teachers have been sued for liability when injuries were sustained while the teacher stepped out in the hall for a few moments. Negligence actions have been won against teachers who allowed students to use faulty equipment, perform laboratory exercises which contained undue safety hazards, or work without using proper safety devices.

We are not lawyers and thus cannot give legal advice. However, we are experienced teachers and supervisors and can offer tips about how to avoid trouble:

1. Keep accurate attendance records for every class period. Put a written reminder in the "Routines" section of each lesson plan and put a checkmark next to the notation after you do it.

2. Present a model of correct performance while doing demonstrations. Call attention to both the techniques used and the reasons for using them.

3. Write specific safety precautions (as needed) on each laboratory guide sheet and verbally remind students of these before they begin work.

4. Check all equipment before distributing it to students and instruct them to notify you at once if anything seems to malfunction.

5. Know your state safety laws thoroughly and enforce them religiously.

6. Train the students to report *all* accidents to you immediately. You, in turn, should report these in writing according to the established policy in your school.

7. Supervise actively and thoroughly. No one expects you to be standing next to every student at the same instant, but correcting papers while youngsters work unattended is an invitation to trouble.

11.8 CROSS-REFERENCES TO A SAMPLE CBTE EVALUATION INSTRUMENT

You should be way ahead of us by this point of the text. If you earmarked category 4 of the evaluation instrument as having direct connections to the ideas presented in this chapter, you're 85 and 5/7 percent correct! That's right. Six out of the seven competencies are clear matches. Check category 4 again to identify the loner. Wait a minute! Weren't there other competencies emphasized via the work in this chapter? We think so. We identified the cluster of competencies under 2.6 as being of prime concern since we attended to human resources and differentiating instruction.

Aren't these last competencies difficult ones for a student teacher to attain? Difficult—yes. Impossible—no. However, they do depend upon prior attainment of competence in working with the entire class. Here's where a rich resource file and the implementation of plans based on the model of systematic instruction become the *sine qua non* for later individualization.

11.9 SUMMARY AND SELF-CHECK

Resources for mathematics instruction were given only incidental attention throughout the first eight chapters of this text since our primary focus was on developing and applying the model of systematic instruction. Once that major task was accomplished we attended to the nature and structure of contemporary mathematics curricula (Chapter 9). In considering the implications of recent curriculum developments we emphasized the use of national curriculum projects as excellent sources of ideas for teaching. Further attention to resources was interspersed throughout Chapter 10, where the focus was on interrelating instruction in mathematics both with other school subjects and the everyday-world concerns of youth. In the present chapter we extended the earlier work on this important concern and emphasized the use of simple and inexpensive materials in innovative ways, atypical uses of AVAs, guidelines for field trips, safety in mathematics teaching, and materials and approaches applicable to atypical students.

Now you should be able to:

1. Collect a wide variety of free and inexpensive materials and project specific uses for each item in mathematics instruction.

2. Expand your resource file to include the total range of resources needed to teach both typical and atypical students.

3. Plan and conduct mathematics learning experiences which are in accord with sound safety procedures.

4. Plan and conduct field trips in a manner consistent with both safety and educational considerations.

11.10 SIMULATION/PRACTICE ACTIVITIES

A. Obtain one of the sources of laboratory/demonstration ideas listed in the final section of this chapter.

1. Locate *two* manipulatives, *not* specifically described in this chapter, which have potential use in a junior-high class. Outline projected use. Indicate topic, specific subject matter content objectives, and teacher and/or student use.

2. Locate *two* manipulatives which would enhance a lesson in geometry, second-year algebra, or trigonometry, and outline the use of each as indicated above.

B. Each of the films in the following list has been successfully used in mathematics instruction. Locate *each* in the appropriate film catalogue. List, for each, the cost of rental or purchase, color or black and white, time, the content, and the grade level toward which the film is directed. Preview each where possible.

Film	Company
1. Adventure in Science: The Size of Things	Bailey Film Associates
2. Dance Squared	International Film Bureau
3. Donald in Mathmagic Land	Walt Disney (distributed by local film agencies)
4. Infinite Acres	Modern Learning Aids

C. The major national professional association for mathematics teachers is the National Council of Teachers of Mathematics (NCTM). Active state and regional organizations also exist throughout the country. Each of these associations can be of service to beginning teachers.

   1. Research the purposes, membership fees, publications and services of the NCTM.
      a. What special student membership and conference registration fees are offered?
      b. What types of publications are provided and at what costs? Specify the titles of *at least one* of each type.
      c. Obtain information on the location and scope of all NCTM meetings being held during this academic year.
      d. What employment/placement services are offered to members?

   2. Answer all of the above questions for one other mathematics teaching organization in your state or region.

D. In Section 11.2 we included a short list of commercially produced games. Obtain *two* such games and read and follow the instructions. (You may need to obtain one or more coplayers.) After playing the games, respond to the following questions for *each* game.

   1. What math topics might be appropriate ones for use of the game? Explain.

   2. What grade level(s) and type of students seem most appropriate for use of game? Defend your decision.

   3. Estimate the time needed to both learn and play the game. For you? For your students?

   4. What problems might be encountered if this game were used, as directed, in a classroom? What revisions or modifications might alleviate these problems?

E. Research some sources aimed at the talented and the slow learner. Prepare *at least five* cards (for your resource file) directed at enrichment for the talented student and *at least five* others directed at adapting instruction for the slow learner. These ideas must be *in addition* to those provided in this text.

F. Identify *two* locations in your geographical region which would be appropriate places for a field trip in your content area.
   For each location, outline:

   1. the instructional objectives of such a trip,

   2. the preparation the students would need prior to the trip,

   3. the followup classroom activities related to the objectives, and

   4. anticipated problems relative to this field trip.

G. In Section 11.6 we alerted you to the problem of evaluating the individual work of students under a differentiated instructional plan. Two well-known approaches to this problem are the Dalton plan and the Winnetka plan. Locate information on *each* of the above plans. Outline each approach and list some apparent advantages and disadvantages of each.

SUGGESTIONS FOR FURTHER STUDY

Catalogues of manipulatives

Creative Publications, 3977 East Bayshore Road, P.O. Box 10328, Palo Alto, Calif. 94303.

The annual catalogue includes mathematics equipment, games, posters, and teacher resource books suitable for a wide range of age and subject matter levels. See the 1978 catalogue for the "Mathematics in Nature" inserts.

Cuisenaire Co. of America, Inc., 12 Church Street, New Rochelle, N.Y. 10805.

Cuisenaire's annual catalogue contains mathematics equipment, games, and teacher resource pamphlets. The catalogue is organized into areas such as art-related materials, geometry, and problem solving.

Midwest Publications, P.O. Box 129, Troy, Mich. 48099.

Midwest publishes an annual catalogue on mathematics-related materials. Recently they supplemented this pamphlet with one called "Math Lab USA," headlining materials which are inexpensive. Manipulatives, games, charts, and teacher resource books are listed in these catalogues.

Demonstration, laboratory ideas

Buckeye, D. A. *Creative geometry experiments*. Birmingham: Midwest Publications Co., Inc., 1970.

Twenty-six experiments requiring simple, easy-to-obtain materials are described. Properties of Euclidean geometry and some simple topological properties are investigated.

Cundy, H. M., and A. P. Rollett. *Mathematical models*. 2nd ed. London: Oxford University Press, 1961.

This text is a classroom-tested guide book to the construction of simple models from linkages to stellated polyhedra. The authors provide information on needed materials and some background on the related mathematics. This is a classic which belongs in your personal library.

Farrell, M. *Geoboard geometry*. Palo Alto, Calif.: Creative Publications, 1971.

See description of contents in Chapter 9 of this text.

Gillespie, N. J. *Mira activities for junior high school geometry*. Palo Alto, Calif.: Creative Publications, 1973.

We consider many of these activities equally appropriate for senior-high geometry. Blank microscope slides can be substituted for the commercially produced mira, which is the instrument used in this series of laboratory exercises. While the activities are designed to develop further fundamental concepts of congruence, symmetry and the like, many can also be used in a unit or course on transformation geometry.

Instructional aids in mathematics. 34th Yearbook. Washington, D.C.: National Council of Teachers of Mathematics, 1973.

Included in this volume are examples of student projects, information on some uses of audiovisual equipment, and a thorough treatment of various models and mechanical devices. Unfortunately, the editors chose to consider only elementary-school uses of manipulatives. This text is an excellent reference for names and addresses of distributors and publishers and further teacher resource books and articles.

Krulik, S. *A handbook of aids for teaching junior-senior high school mathematics*. Philadelphia: W. B. Saunders Co., 1971.

The title says it. Here are included directions and diagrams of "equipment" such as Napier's Bones, "materials" such as magic squares, and games. All are easily constructed with inexpensive materials.

Laycock, M., and G. Watson. *The fabric of mathematics*. Rev. ed. Hayward, Colo.: Activity Resources Co., Inc., 1975.

This teacher resource is subdivided into major areas of mathematics (such as number, geometry, sets, and logic). For each subtopic under these major headings, a set of graded objectives is stated, activities which are heavily laboratory-oriented are outlined and sources of further teacher- and student-reference material are listed. The appendices are a valuable part of this text, for there you will find lists of manipulatives, games, visuals, and the like. Don't be put off by the pictures of third- and fourth-graders, but do check carefully the level of the objectives and the possible need to modify selected activities.

Ranucci, E. R. *Georule activities*. Palo Alto, Calif.: Creative Publications, 1971.

Teachers of junior- and senior-high students will find appropriate demonstration/laboratory activities, all of which utilize the georule.

Seymour, D. *Tangramath*. Palo Alto, Calif.: Creative Publications, 1971.

This pamphlet describes demonstration and laboratory activities based on the seven-piece tangram puzzle. Similarity, congruence, and properties of polygons are some of the topics included.

Differentiating instruction

*The items indicated by an asterisk also contain ideas, puzzles, and problems that have proved to be successful with the less-than-talented student.

*Adler, I., ed. *Readings in mathematics* (Books 1 and 2). Lexington, Mass.: Ginn and Co., 1972.

These short articles can be read with profit by teacher and a wide range of students. The topics range from deciphering a cryptogram left by Captain Kidd to the atom and its nucleus.

*Barr, S. *Second miscellany of puzzles, mathematical and otherwise*. London: Collier-Macmillan, Ltd., 1969.

This collection of puzzles can be solved with elementary arithmetic, algebra, or geometry along with a good measure of common sense. Many of these puzzles are appropriate for the typical high-school student. Look for those which demand a little more.

*Bezuska, S. *Contemporary motivated mathematics* (Books 1 and 2). Chestnut Hill, Mass.: Boston College Press, 1969.

These two booklets contain problems ranging over topics from number theory to geometry. Mathematics prerequisites, however, include algebraic skills, at most. The ideas are, thus, suitable for students of varying abilities in grades 7-10.

Burns, M. *The I hate mathematics! book*. Boston: Little, Brown, 1975.

This cartoon approach to the student who dislikes math has appeal for the slow learner and may even interest some mathematical kooks. There are laboratory and pencil-and-paper activities. The mathematics ranges from basic skills in disguise to recreational mathematics.

Charash, M., ed. *Mathematical challenges*. Washington, D.C.: National Council of Teachers of Mathematics, 1965.

This pamphlet contains a collection of problems, most of which had previously appeared in issues of the *Mathematics Student Journal*. There are problems which are suitable for the seventh-grader as well as the twelfth-grader, and solutions to all problems are included.

*Experiences in mathematical discovery. Washington, D.C.: National Council of Teachers of Mathematics, 1966, 1967, 1970, 1971.

This nine-unit series is intended as a general mathematics series for the noncollege-bound. The incorporation of a guided discovery approach makes this a prime teacher resource.

*Experiences in mathematical ideas (2 vols.). Washington, D.C.: National Council of Teachers of Mathematics, 1970.

These instructional packages include masters to be duplicated for student use, acetates and activity materials for a laboratory-oriented approach to mathematics. They are designed for use with middle-school (grades 5-8) students identified as "slow" or unmotivated.

**Kasner, E., and J. Newman. *Mathematics and the imagination*. New York: Simon and Schuster, 1940.

This recent classic is a source of a host of fascinating ideas. It can be read by the talented student or used by the teacher to initiate projects or to teach some mathematical concepts (such as, π, e, logarithm) to the "average" group of students.

Martin Gardner's sixth book of mathematical games from Scientific American. San Francisco: W. H. Freeman, 1971.

As always, Martin Gardner is able to stimulate our imagination and pique our curiosity with puzzles, paradoxes, illustrations from nature, and photos of phenomena. This is just one of the excellent source books for both the talented student and the teacher who needs ideas for the so-called "average" class.

**Ransom, W. R. *One hundred mathematical curiosities*. Portland, Me.: J. Weston Walch, 1955.

This inexpensive paperback is a treasury of riddles, paradoxes, and problems. All are mind boggling at first, although the difficulty level varies from easy to hard.

Salkind, C. T., compiler. *The MAA problem book II*. New York: Random House, 1966.

This paperback contains problems (with solutions) from the annual high-school contests sponsored by the Mathematical Association of America in 1961-1965. Solutions are included.

**Schaaf, W. L. *A bibliography of recreational mathematics*. Vol. 2. Washington, D.C.: NCTM, 1970.

Here is a good source book for the student who wants to investigate a particular topic in mathematics. The topics include puzzles, classic brain-teasers, games, uses of equipment such as linkages, science-related, sports-related, and art-related topics, and much more. The extensive bibliography should be of great help in student project work as well as teacher resource ideas.

Seymour, D., *et al. Aftermath* (Books 1-4). Palo Alto, Calif.: Creative Publications, 1971.

These books consist of collections of puzzles and problems presented by cartoon characters. Mathematical material is appropriate for seventh- through ninth-grade mathematics. They have been successfully used with slow learners as well as the average student.

The slow learner in mathematics. 35th Yearbook. Washington, D.C.: National Council of Teachers of Mathematics, 1972.

This text includes diverse activities and approaches which have been effective with the slow learner at all levels.

Games

Henderson, G. L., *et al. Let's play games in mathematics* (Vol. K-8). Skokie, Ill.: National Textbook Co., 1972.

Volumes K-8 contain collections of games and other activities cross-referenced to content topics and instructional objectives. Volumes 6 and 7 are of particular interest to the middle-school mathematics teacher, since they contain activities suitable for grades six through eight. Volume 8 is designed for use with underachievers and includes topics from addition and subtraction to ratio and proportion.

Henderson, G. L., and L. D. Glunn. *Let's play games in general mathematics*. Skokie, Ill.: National Textbook Co., 1972.

This addition to the *Let's play games . . .* series emphasizes the needs of the diverse students found in general mathematics classes. Like the other volumes in the series, this one contains games and activities suitable for small groups or the entire class. All are cross-referenced to major content areas and specific instructional objectives.

Mathematics teaching journals

Examine these for special features which contain reviews of new AVAs and books, curricular modifications for the slow and the talented learner, and ideas for demonstrations, laboratory exercises, bulletin boards, and individual projects.

The Arithmetic Teacher

A journal published by the National Council of Teachers of Mathematics (NCTM). It has articles for the teacher of mathematics (K-9). Junior-high teachers find this journal helpful— especially a tear-out section on activities.

The Mathematics Teacher

A journal published by the National Council of Teachers of Mathematics (NCTM). The emphasis in this journal is on topics and classroom tactics for teachers of grades 9-13. This journal also has a tear-out section on activities.

The Mathematics Student

This journal is published by the National Council of Teachers of Mathematics. It is directed toward students in grades 7-12.

School Science and Mathematics

Journal published by CASMT (Central Association of Science and Mathematics Teachers). It features articles on the teaching of science and mathematics on the secondary-school level.

Mathematics Teaching

Journal of the ATM (Association of Teachers of Mathematics—Great Britain). This journal is one of the best sources of ideas for creative mathematics teachers (7-12).

Other sources

Bibliography of "Free Materials for the Teaching of Mathematics" from the National Council of Teachers of Mathematics, 1906 Association Drive, Reston, Va. 22091.

How to . . . series. Washington, D.C.: National Council of Teachers of Mathematics, 1953-1970.

This series includes brief pamphlets on the use of the overhead, design of bulletin boards, conduct of field trips, and even guides to the study of mathematics.

Leffin, W. W. *Going metric: guidelines for the mathematics teacher, grades K-8*. Washington, D.C.: NCTM, 1975.

Suggested classroom activities, recommended materials, and instructions for constructing aids to teach metric concepts are outlined in this pamphlet.

Schaaf, W. L. *The high school mathematics library*. 6th ed. Washington, D.C.: NCTM, 1976.

This bibliography includes titles on metrication, computers, data processing, and geometry as well as materials for the gifted and professional texts for teachers.

Check Chapters 9 and 10 for additional references on the above areas.

CHAPTER TWELVE

CONTROL OR CHAOS IN YOUR CLASSROOM

DISCIPLINE

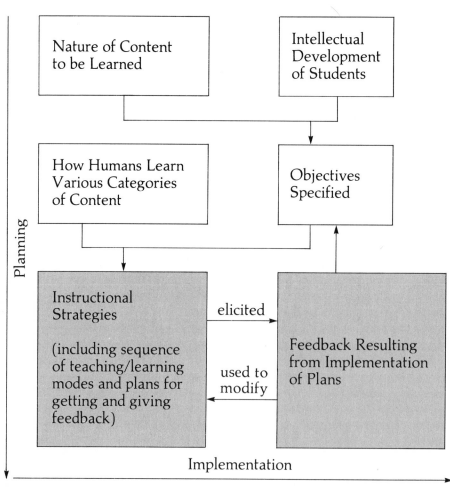

ADVANCE ORGANIZER

Some prefer to call this topic *"pupil management"*; others insist that *"instructional leadership"* is the proper term. Choose whichever title pleases you most, but what they all boil down to is *"control or chaos in your classroom."* Yes, yours. After all, you, the teacher, are supposed to be the adult in charge—the one legally responsible for the safety of students and professionally accountable for providing an environment for learning.

Not everyone is able to develop or exercise this kind of leadership. Those who cannot or will not should stay out of teaching. These are strong words, but they are based on our experience in teaching secondary school and in supervising hundreds of student teachers. But don't just take our word for it. Check out this point of view with any experienced principal. Then look up the dropout rate for teachers during their first five years of service. Also talk to college supervisors and ask them to cite the leading reason for "flunk-outs" and "withdrawals" among student teachers.

Make *no* mistake. Someone will be the leader in your class, and the choice is up to you as to whether it is you or one or more of the students. Some type of atmosphere for learning will prevail as a result of that leadership. The question here is the suitability of that climate for learning—both in the sense of promoting intellectual understanding and a positive response to reasonable rules designed to protect the rights of all. These two goals—intellectual understanding and the protection of the rights of all—are the keys to the two boxes shaded in on the opening page of this chapter. The emphasis intended by the shaded boxes, "Instructional Strategies" and "Feedback," is that a positive classroom atmosphere depends on careful planning and consistent, thoughtful response to feedback during the class period. What should and can be done about this problem?

In general, there are two approaches which work in the classroom; in the real world of the classroom *both* are needed. The first, like preventive medicine, must be employed whenever possible; the second, like surgery, is effective only in specific cases and for limited time spans. But before going into the details of either of these approaches, we should first agree on operational definitions of terms used frequently to characterize "good" versus "poor" discipline.

12.1 OPERATIONAL DEFINITIONS OF SELECTED KEY TERMS

An *absolute* prerequisite to comprehending most of the terms employed to describe classroom management is firm knowledge of the operational definitions of instructional modes as found in Chapter 1. Do you understand why this is the case? It should be fairly obvious that specific expectations for student behavior vary according to the mode being employed at the moment. Thus it will be well worth the effort required to run a quick self-check on these prerequisites prior to reading any further. Be aware of the fact that the classroom management terms defined in this section are limited to those which seem to be most useful to the novice teacher.

Appropriate noise level Immediately after a classroom observation we are frequently asked, "Was my class too noisy today?" In our judgment, the *kind* of noise should be the primary concern, but more on that later. The noise level is appropriate if *during lecture or a quiz the sounds from students approach zero (very, very closely) and during Q/A students speak one at a time with volume enough for all to hear.* Note the specific ties to the particular mode being employed at given points in the class period, and proceed to construct your own operational definitions of appropriate noise level for each of the other commonly employed modes of instruction. An obvious nonexample would be if students in a small discussion group were screaming to each other in order to communicate over the shouting level being used by another group. The inappropriate noise level of this group had been, in turn, caused by their need to hear each other over the alternately loud and quiet conversations of yet another group. Some novices fail to recognize this as a problem if all talk is relevant to the task at hand and then are surprised that one group will soon be forced to escalate its volume further to hear themselves over the other groups.

Appropriate kind of noise Like the previous operational definition, this one is tied specifically to the particular instructional mode being employed at various points in the class period. *During lecture there should be only the sound of the teacher's voice.* However, *during Q/A one-at-a-time talk on the topic by both teacher and students is desired. During laboratory work appropriate kinds of noises include the sound of footsteps, student-to-student and teacher-to-student conversation on procedures and content topics, the manipulation of equipment, and clean-up sounds.* Obvious nonexamples include talk about dates and the sounds of breaking glassware, equipment falling on the floor, or chairs being knocked over. Now consider the other instructional modes and construct your own operational definitions for the kind of noise appropriate for each.

Positive student attention signs What would be some kinds of feedback which a trained observer in your classroom would characterize as positive student attention signs? *Students facing front, being quiet, and occasionally writing in notebooks would be positive attention signs during that portion of a class period when the teacher was lecturing from the front of the room.* Similarly *many hands raised with relevant questions and answers being volunteered are positive signs during a question/ answer session.*

Negative student attention signs Quiet may be very desirable during a lecture, quiz, or while viewing a motion picture, but *silence should be interpreted as a negative attention sign during certain modes such as discussion, question/answer, and small-group laboratory work.* Similarly, *spitball throwing, manufacturing paper* *airplanes (with accompanying test flights), note passing, students eyeing the clock, reading of books hidden from the teacher's view, eating lunches, or sticking neighbors with pins can rarely be considered anything other than negative attention signs regardless of the instructional mode in use.* Now extend the list to include more of the subtle varieties and ask an experienced teacher to check it out for completeness. Be sure to include such signs as *students' gaze fixed on the spot where the teacher used to be and many requests that the teacher repeat questions.*

We have observed examples of student behavior listed in this section in many classrooms. In our experience, it is fatal for a teacher to overlook or ignore repeated instances of inappropriate level and kind of noise or negative student attention signs. Therefore, it is imperative that you be alert to this kind of student feedback from the first day you step into a classroom as a teacher. The introductory activity which follows is designed to start you thinking seriously about teacher behaviors characteristic of both effectively and ineffectively managed classrooms.

12.2 INTRODUCTORY ACTIVITY

Get it straight from the horse's mouth.

In order to perform this activity, you will need to locate three to five junior/senior-high students. Interview each of them individually to get their perceptions regarding the characteristics of good and bad teaching. Ask each student to record reactions on a form such as the one in Fig. 12.1. After collecting your data, compare it with that collected by your classmates who have interviewed different students. Look for patterns in the data, talk about these patterns with your classmates, and be prepared to discuss the matter further with your instructor.

A successful interview requires preparation as well as follow-through. One does not just walk into a secondary school and stop students in the hallway. You should talk with your instructor, who is knowledgeable about local school policies, regarding protocol. Once the arrangements for interviewing have been made, you must duplicate the form in a neat, easily readable fashion. During the interview, you may need to give operational definitions of terms which may not be clear to younger students. Encourage them to ask questions about any statement which is not clear to them. *Refuse to allow the student to use the name of any particular teacher, either on the form or in conversation with you.* Observe closely to be sure that the directions are being followed and be sure to thank the students and all school personnel who made the interviews possible.

12.3 AN OUNCE OF PREVENTION

If the data which you obtained in the introductory activity matches that which students have given us, then the notion of trying to survive in a poorly managed class probably frightens you as much as it does us. We know of no teacher who is pleased about the possibility of being in a chaotic class. It's clear that the wise teacher does all in his or her power to *prevent* such a

TO THE STUDENT

I am interested in learning to be a good teacher. I believe that student judgments about good and poor teaching can be helpful to me. Thus, I am asking you to record your honest reactions on this form. Do not sign your name. Your reactions will be kept confidential and will be added to those of other students to give me a picture of what many students think about good and poor teaching.

First, think in terms of one or two of the very best teachers YOU have had. Use the column on the left under the word BEST and place a check mark in one of the five boxes to indicate whether your best teacher(s) had a particular characteristic never, rarely, sometimes, frequently or always. Do this for each of the ten characteristics listed.

After you have finished rating all ten characteristics in terms of your best teacher(s), start again at the top of the list and place a check in the boxes in the right column which describe the worst teacher(s) YOU have had.

MY BEST
TEACHER(S)

MY WORST
TEACHER(S)

Never Rarely Sometimes Often Always

Never Rarely Sometimes Often Always

1. Made the subject matter interesting

2. Made the subject matter easy to learn

3. Kept many students participating in class

4. Had only a *few* rules for student behavior in class

5. Had *many* rules for student behavior in class

6. Enforced whatever rules he or she had for student behavior in class

7. Treated all students fairly

8. Threatened students

9. Punished only those who misbehaved

10. Punished the whole class for the misbehavior of a few students

Fig. 12.1 Student reaction form.

situation. And the first line of defense is to be found in the quality of the lesson!

Planning and execution of the lesson Getting student attention *on the lesson* is a prime consideration in the development of any lesson plan. The number and variety of ways to accomplish this are limited only by the teacher's imagination. An interesting demonstration is one of the most productive ways to accomplish this end. What makes an interest-catching demonstration? Seek one that seems to defy "common sense" or one that focuses attention on an unfamiliar (to the student) aspect of a common phenomena. If you can locate one that combines both ingredients, so much the better. Other serviceable ideas include laboratory activity (student manipulation of materials) having the same characteristics as described for demonstration, a controversial statement written on the chalkboard prior to class, or a totally restructured classroom environment that connects to the lesson (chairs and desks arranged in a circle or small clusters). The references cited in Chapters 9, 10, and 11 are rich sources of ideas for such activities.

Keeping student attention on the lesson is a more difficult task. This cannot even be approached unless the teacher *first gets attention*. Let us assume that this has been done. One obvious way to keep attention is to vary the teaching modes. If the class began with a teacher-performed demonstration coupled with question/answer, the teacher might switch to short lecture, then small-group discussion or short lecture with note taking and/or a combination of student board and seat work. Relevance to the "here-and-now" interests of students should be worked in as quickly as possible and emphasized throughout the class period. The increasing use of divergent questions as the period progresses will also help maintain attention. "Dead time" in the lesson is an invitation to trouble. When planning the lesson keep asking yourself, "Is there something constructive for every student to be doing every minute of the class period?" If the teacher cannot answer "yes," potential problems surely lie ahead. In addition, many novices beg for trouble by not informing students what they should be doing every minute of the class period. Even more make the error of not informing students *how* they are to perform the instructional tasks.

Students will work just so long without feedback on their efforts. They need to know how they are doing. Nothing is quite so satisfying as success coupled with knowledge of success. What can be more frustrating than failure or thinking you are doing a task correctly only to find out too late that you have been practicing errors? Refer back to Chapter 2 on feedback, which contains many specific suggestions on this point.

"It is a smart person who knows when to give up." Many students get lost when the teacher has failed either to control the pace or to ensure that each step in the lesson is based upon prerequisite learnings actually achieved by the students. This kind of teacher behavior breeds frustration, then a "what's the use" attitude, followed by student activities of their own design—often at odds with what the teacher had in mind. The work of Piaget and Gagné provide important guidelines for how to avoid these problems and are well worth referring to once again (see Chapters 3 and 6).

Effective and efficient handling of routines Attendance must be taken every class period. The class attendance records of a teacher are legal documents subject to examination in a court of law. The teacher must take care of this detail daily without allowing the procedure to constitute "dead time." It is important that the class session get off to a fast and interesting beginning. A seating chart *is a must* until the teacher knows at a glance who is absent and can record the same *without taking any class time* for this important routine. Useful techniques include checking attendance at the start of the period while (1) some students put answers to homework on the board and others check answers at their seats, (2) all take a quiz, or (3) the teacher conducts a five-minute question/answer review of yesterday's lesson. Sometimes a student volunteer can perform the mechanics of attendance taking, but the teacher must check the accuracy of the record at some convenient point during the period.

Collecting and passing out papers must also be reduced to a routine. Students respond readily to habit in this regard if the teacher (1) establishes a set procedure at the beginning of the year and (2) consistently uses it. For example, the teacher might hand all papers for a given row to the student sitting on the right end of that row. That student passes the papers to the left. Likewise, papers might always be collected by having students pass them to the left end of the row and then to the front left seat. Some teachers have a box near the door for routine collection of homework papers as students enter class. The main idea is to establish a procedure and make it automatic so that it does not interrupt the flow of the lesson.

The distribution and collection of working materials should be similarly handled. A key consideration here is preventing jam-ups (and consequent pushing, arguing, and so on) by planning efficient distribution patterns. Dividing materials into four sets available at the four corners of a classroom makes much sense. On the other hand, asking an entire class to clean up materials at one sink during the final three minutes of class is an invitation to chaos.

Administration of a quiz or test should begin by getting quiet and following the routines for passing out papers. Then the teacher belongs in the rear of the room, on his or her feet with eyes and ears alert. Do you see why? If a student were to cheat, what is the first thing that student would have to do? That student would have to look around to find out where the teacher is—thus attracting the teacher's attention. Students with questions should be instructed to raise their hands and the teacher should follow procedures designed to avoid disturbing nearby students when attending to individual requests for clarification. As in supervised practice, the teacher must (1) listen to a question while glancing around frequently at other students, and (2) whisper in response to a question and insist that students speak at the same level. But during a test, the teacher must (3) refuse to respond to questions seeking assistance with the content. Beginning teachers are often trapped into giving the cues they would have provided in a supervised practice situation. Students soon learn that those who ask, receive! The complaints of unfair treatment are well deserved and the teacher is on the road to a self-imposed discipline problem.

Teaching students to respond to nonverbal signals is easy and effective. Try switching the lights on and off in a class or laboratory where students are talking and/or working away at some activity. It works the first time and will work on successive occasions *if it is taught* as a signal that (1) it is time to stop work and clean up, (2) the noise has inadvertently gotten a bit too loud for the present activity, and (3) it is not used for every little disturbance. Effective instructors teach students to respond to a whole range of other nonverbal signals such as a snap of the fingers, the closing of the classroom door, or a finger touched to the lips of the teacher. Observe a real "pro" to pick up a whole repertoire of other signals to which students can be taught to respond.

Praise/reward for correct behavior Few teachers give enough attention to praising and rewarding students for behavior they wish to foster. Most attend only to inappropriate behavior and ignore that which we would like to promote—unintentional to be sure, but ineffective none the less. To be effective, the praise must be sincere, justly deserved, and not embarrassing to student recipients. Effective teachers use techniques such as (1) telling the entire group what a good job they did with today's lesson, (2) congratulating all for the fine results on yesterday's unit test, (3) calling attention to the thought-provoking quality of the question just raised by one student, and (4) asking a student's permission to read a particularly outstanding homework answer to the rest of the class. Nonverbal praise in the form of teacher smiles, hand signals, and gestures is very effective.

Other rewards often further increase the effects of praise. For example, the teacher can point out that because work progressed so effectively during the first 30 minutes today, the class can devote the final 15 minutes to an interesting educational game. Similarly, at the end of a particularly effective supervised study or practice session the teacher can announce cancellation of the previously announced written homework problems as there now appears to be no need for further practice. It is quite important to insure that students understand that the reward is a *consequence* of their good work, not a capricious act on the teacher's part.

One of the biggest mistakes we all tend to make in using praise/reward is the tendency to withhold these until student performance reaches some peak of perfection. We must learn to give encouragement for small steps in the right direction without overdoing it (and thus finding nothing left to connect with the final level of performance desired). Another all-too-common error of novices is to mistake a punishment for a reward. The typical student does not view doing problems 11-20 as much of a reward for having correctly solved problems 1-10 faster than the rest of the class!

Conveying the teacher's enthusiasm for the subject It is a truism that one can't convey what one does not have. Some prospective teachers have learned mathematics as a stagnant body of "facts" and complicated rules or principles. They can recite these flawlessly to the class (provided they rememorized the details the night before) and can work stereotype problems faster than all but their brightest students. But that is it; there isn't any more! Those who recognize the above as a self-description are in deep trouble in trying to deal with average and below-average students. "Death-warmed-over" won't work. What are these prospective teachers to do?

Self-education here and now is the *only* way out. Understanding the structure of ideas and processes for generating knowledge in mathematics can be gleaned from the sources cited and activities suggested in Chapter 4. A rich store of interesting applications to the real-world and the "here-and-now" interests of students is to be found in Chapters 10 and 11. Once the teacher becomes enthusiastic about the subject matter, then it is probable that he or she will *find* ways to demonstrate it to students. The obvious techniques include incorporating a wide variety of atypical examples, television commercials, labels from food and drug packages, newspaper articles, and other objects brought in from the outside world. Appropriate nonverbal signals such as the gleam in the eye, spring in the step, or excited tone in the voice of the teacher are hard to describe and prescribe—but these are very important. Observe an enthusiastic teacher at work and try to incorporate some of these aspects which are compatible with your own personality and classroom style.

The quiet, insistent, consistent, business-like approach Screaming simply does *not* work! In fact, students make a game out of seeing how quickly and often they can elicit this kind of response from some teachers. They will respect the calm, cool approach if the lesson is well planned and interesting. Students expect the teacher to act like an adult, not a tantrum-throwing child or an insecure teenager. This quiet approach, used consistently, has the same certain results as erosion that reduces boulders to sand grains over time.

Business-like does not mean the teacher never smiles or laughs with the students when something genuinely humorous happens. On the other hand, when we have all enjoyed our laugh, it's time to get back to work. Incidentally, it is unpardonable to laugh if it is at the *expense of a student* or group of students. No student respects a teacher who does this, but all admire a teacher who is big enough to laugh at himself or herself on occasion.

Neither this nor any other approach will work, however, if you, the teacher, present poorly prepared lessons which students perceive as dull or disjointed. A firm prerequisite to effective long-term classroom management is the planning and implementation of lessons that are both interesting and understandable. You will find that one of the most potent forces at work in the classroom is *peer pressure*. If the majority of the students are convinced that something worthwhile and interesting happens in your class, they will use very effective sanctions against the one or two whose actions tend to get out of line.

Remember that *consistency is imperative*. Words and deeds must match, and the teacher cannot allow rules to be violated for the first half of the lesson and then come on with thunder and lightning about the same breaches in conduct during the final minutes of the period. Appearances to the contrary, teenagers are really very conservative in many respects. They are creatures of habit and like to know which rules are to be enforced.

Establishing a few, simple, reasonable rules If there are too many rules, the class atmosphere is repressive and students can't remember half of them in any case. If the rules are too complicated, most students will find avoidance behavior a more appealing alternative than trying to follow the rules. If the rules seem arbitrary and codifications of the teacher's whims, it will take more time and effort to enforce them than any teacher of an academic subject can afford to devote to nonacademic matters.

Some advocate spelling out all rules the first day, but our experience suggests a different approach in order to emphasize the "no rule without a good reason" guideline. For example, on the first day of class students will enter and automatically take a seat. Suppose at the instant the bell rings, an interesting demonstration is performed and students are asked to raise their hands if they cannot see. They are then told they can keep their chosen seats as long as they can see, hear, and behave where they are today. They are requested, however, to keep the same seat for at least the first few weeks so the teacher can learn all of their names quickly. Suppose, too, that a question/answer session is woven in with the opening demonstration. Then the first time several students call out in a mixed chorus, one nonparticipating student is asked if he or she agrees with the answer given. After he or she admits confusion, the request for hand raising is made. Then students will understand that the rule is for the benefit of all. As the demonstration with question/answer continues, another mixed chorus response is likely to occur and this time *the teacher should call on someone who raised a hand and did not shout out*. Consistent teacher behavior leads quickly to desired student behavior without recourse to threats, screams, or punishment. Other simple rules, such as that no one talks while another (student or teacher) has the floor, are similarly taught as the occasion arises. Nonverbal signals, such as those described earlier, are very helpful both in establishing and maintaining the reasonable, rule-governed behavior which sets humans apart from lower animals.

12.4 THE BIG CULPRITS

The suggestions in the previous section are designed to help you eliminate the major sources of discipline problems faced by the novice. We call these sources the "big culprits" because we have seen them trap and destroy student teachers. In many situations the student teacher is deceived as to the destructive nature of these teacher behaviors because they see no apparent disastrous effects in the classrooms of experienced teachers. What novices fail to observe are the many counterbalancing moves by experienced teachers. If you would avoid the usually impossible task of regaining control once it has been lost, you must learn to recognize and avoid the big culprits.

Accepting responses of students who do not raise their hands The teacher tells the students to raise their hands and wait to be called upon prior to answering or asking questions. Shortly thereafter, someone calls out a correct answer and the teacher says "Right." What effect does this action have on the students who did raise their hands and did wait for the teacher's signal to answer? What would you do if you held your hand up while others who ignored the teacher's directions were rewarded by being recognized? The slowest student quickly realizes that the teacher's actions and words do not match. This realization leads to an increasing number of students who shout out in response to the teacher's questions. Finally, the teacher gets upset, "reams-out" the class, and proceeds further down the road to disaster by going right back to inconsistency of words and deeds. This reaction is paired with the second big culprit.

Inviting the chorus response Often there are several students calling out the *same* answer at the same time (if the teacher is lucky or everyone knows the material anyway). There are two big problems with this *unison* chorus.

First, it is inevitable that only the vocal segment of the class will continue to respond while the majority sits quietly. But will the quiet ones think along with the teacher and the minority, or will they find some other quiet activity to engage in while the show goes on?

Second, how long can this show progress before it becomes a *mixed* chorus of shouts? Observe this type of performance at your neighborhood school and keep score. Soon two or three different responses occur at once. How does the teacher pick the one to respond to with "Right," "O. K.," or "Good"? Typically the teacher only hears what is expected and ignores the other responses. Imagine the plight of the quiet and attentive student who wonders which of three answers is being identified as "Good." Now put yourself in the role of the teacher who wants to get feedback on group thinking and reinforce correct responses. Certainly, inadvertent promotion of the second big culprit is not the way to do it. Furthermore, the attendant student confusion fostered by this type of teacher behavior frequently leads to the third big culprit.

Teacher talks over student-to-student talk Inviting the chorus response not only results in student boredom but is the leading correlate of student-to-student (S-S) talk. Students ask neighbors, "What did he say?" "How can that be if this is . . .?" "I thought he told us yesterday . . . ," and so on. Typically, the teacher caught in his or her own web either remains unaware of the S-S talk or ignores it so as not to squelch the responsiveness of the class (only 4 to 10 students out of 25). When the murmur becomes a roar, our novice senses that all is not well with the class and comes down hard with recriminations, threats, or slogans which are the direct opposite of what has been reinforced by previous events. Good luck to this teacher! Experienced supervisors call this practice "pushing the self-destruct button." Some novices insist on pressing this button, and the supervisor's unhappy dilemma becomes the identification of the kindest and gentlest way to counsel them out of teaching.

Threats the teacher cannot or will not carry out "If you do not . . . immediately, I will. . . ." Is the teacher sure he or she can carry out the threat? Too often the novice hasn't the foggiest idea how he or she can make good while the students are reasonably sure that the teacher can't or won't make the ultimate decision. Usually the students know more than the novice teacher about this school's policies. After all, they have been playing the game of school for eight to eleven years in this very system while the teacher has been here only a few weeks and *doesn't even know that the principal never sees troublemakers who are sent to the office.* Those foresighted novice teachers who look up the school policy will no doubt find out that the vice-principal sees discipline cases. The student is thoroughly chastised by the vice-principal and returns to class a reformed student, right? Wrong! In more cases than we'd care to count, the student cools his or her heels in an outer office and continues the disruptive behavior for the benefit of the school secretary and visitors and peers who stroll by. When, at last, he or she confronts the vice-principal, there may be a brief heart-to-heart talk with a finale in which the student glibly promises to do better. Now who has punished whom? The student returns from the wars as a veteran with a story to tell, and you may be sure that it will be told.

There may be a time when one student must be removed from the class for the benefit of others, but the novice must determine where that student can be sent, follow up the removal, and treat this kind of action as a final step with serious consequences.

Musical chairs This variation of the popular child's party game is often played by novice teachers who are really a bit old for this game. The game begins when the teacher moves a troublesome student to another seat away from his or her friends. If the character in question is the class clown, the typical move to a front seat provides a wider audience for future antics. Who has fooled whom? Then, too, what has the teacher unwittingly done to the cooperative student who previously enjoyed the front seat? One could hardly say that this student was rewarded for good behavior by being forced to change seats. What is the teacher's next move when Charley's (the clown's) behavior deteriorates again? Too often another student is forced out of his or her seat in order to relocate Charley a second time. Frequently other students now join in the sport, the game gets more complicated, and the class slips further away from the control of the teacher.

This is *not* to say that judicious moving of one or two students is value-less. A move can be quite effective if the teacher (1) separates students who can't seem to get along together, (2) avoids creating even worse combinations (usually requires teacher homework ahead of time such as a trip to the guidance office and/or consultation with other teachers of these same students), *and* (3) makes clear to all concerned that their next seat change will have to be to somewhere *other than* in this classroom. Obviously the teacher must check the school policy thoroughly and be prepared to follow through.

Rewarding inappropriate behavior The sixth big culprit creeps in when the teacher reinforces the very behavior he or she is trying to extinguish. For example, one teacher told the class that those who worked effectively and efficiently on the worksheet during the period would have time to manipulate the abacus and begin construction of one of their own. Sam wasted time and distracted others. The teacher reprimanded him several times during the first 30 minutes of class time. Finally the teacher took away Sam's half-finished worksheet and told him to begin work on the abacus. What did this teacher's action tell those students who had worked diligently at the assigned task? What did it tell Sam?

Suppose the teacher has some students who do not wish to be in class. They are there only because (1) they had too many study halls and thus the guidance officer put them in algebra class or (2) parents insisted that they take geometry. Is it reward or punishment for these students to be expelled from a class they have tried to escape taking in the first place? There is no denying that it may occasionally be in the best interest of all concerned to remove such a student from class. However, let's neither kid ourselves that this is a punishment nor try to sell the students on this as a punitive measure. They simply won't buy it!

Inconsistent teacher behavior Yesterday the teacher came on like thunder and lightning when several students poked neighbors and disturbed other students during laboratory work. Today, the teacher ignored this same behavior without so much as a verbal reprimand. What will be the teacher's reaction to similar student behavior tomorrow? Is it any great wonder that inconsistent student behavior develops and that students come to view rules of order as capricious whims of their teacher?

Teacher's words and deeds do not match This culprit is very closely related to inconsistent teacher behavior, but it is common and important enough to deserve special emphasis. It also overlaps many other categories of culprits identified here. For example, minutes after stating that students must raise their hands and wait to be recognized, the teacher falls into the pattern of accepting shouted-out answers. Another common error occurs when the teacher announces that there will be absolute quiet during the test and then finds it necessary to make numerous oral announcements, corrections, and clarifications throughout the period. A third common invitation to break-down in morale is the practice of admonishing students to do neat and accurate work and then passing out teacher-made handouts that are almost impossible to read.

Using subject matter as punishment or threat of punishment "If this class doesn't quiet down immediately, we will have three extra homework problems for tonight." Such a statement has the same effect on student attitude toward the subject as "Because there's so much noise, we'll have a quiz right now." It is a well-known truism that students who are interested in the subject matter or "turned on" by the way the teacher presents ideas attend to the lesson and *not* to disruptive activities. Now our harassed novice has destroyed the goal toward which he or she is working. And what must students think of quizzes which are planned on the basis of noise rather than on teacher assessment of feedback on lessons? The predictable reaction of the students is undisguised grumbling, comments on the lack of fairness of the teacher, and an increased dislike for content that must be practiced for punishment, rather than need.

Pleading for personal favors One of the most difficult roles for the novice teacher is that of acting like a thoughtful, responsible leader in the classroom. During initial lessons, the beginning teacher's unwillingness to accept this responsibility is exhibited by verbal patterns such as "Would you please be quiet?" or "Keep it down, please," usually offered in a pleading, hesitant tone. The students quickly sense the teacher's failure to lead and begin taking over the reins of the classroom. Such a teacher tends to describe the situation by statements such as "*They* were very good today" or "*They* were noisy so I couldn't accomplish anything!" The students have been given the decision-making role here and have not been helped to greater self-control. Student caprice determines the days of apparent cooperation.

If this initial behavior of the teacher isn't altered fast, the observer soon is treated to the spectacle of abject pleading: "Please be quiet!" "Help me out, please!" The teacher who reaches this stage must get out of the classroom, for he or she is doing irreparable harm to the students in the process of losing his or her own human dignity.

Punishing all students for the misdeeds of a few "If the person who threw the spitball doesn't confess, everyone will stay after school for one hour." What does this treatment tell students who have been behaving as desired? They might as well get into the act, since it is likely they will be punished anyway! This also tells students that the teacher is not capable of controlling the situation. This teacher has also convicted all students as accessories to the misbehavior. Students cry "Foul play" deservedly. The teacher has gained many enemies and has helped to solidify the class in opposition to the teacher. The odds of 25 students to 1 teacher are very difficult ones for the novice to overcome.

The inappropriate teacher behaviors illustrated in these sections are those which seem to occur repeatedly and with immediate negative consequences in the classrooms of those who are unable or unwilling to accept their responsibility for ensuring a positive atmosphere. But suppose a teacher has been attending to all the suggestions and has been alert to early signs of erosion—does that teacher ever find it necessary to react to disorder? And, if so, what are some alternatives? Some of the ills for which you may need a cure and some classroom-tested cures are described in the following section.

12.5 A POUND OF CURE

The day of the hickory stick in public-school classrooms is long gone. It is true that some teachers lament its demise and a few attribute many of our present educational problems to its passing. Nonetheless, the fact of the matter is that there are very few schools today (public or private) where teachers are allowed to punish students with either fists or physical objects—the substitutes for the bygone hickory stick. Contemporary teachers must find more appropriate ways to get and maintain control over events in their classrooms.

Our own experiences as teachers and department chairpersons and that of our student teachers indicate that the approach to take is the *elimination of the need* for punishment. Careful and consistent attention to the recommendations made previously in Sections 12.3 and 12.4 have proved to be 90 to 95 percent effective. But aren't there situations where punishment is called for in spite of the best efforts of the teacher? Yes, unfortunately. Remember above all else that any alternative will work if and *only if* the teacher has been trying and continues to try to implement the positive approach. We can hold out little or no hope that anything will work for the teacher who has consistently and over a period of time ignored the "ounce of prevention" approach.

What is and is not considered punishment? *Punishment is the presentation of aversive stimuli.* Aversive to whom? The offender, of course. Anyone who hopes to use punishment effectively must remain alert to the fact that what is aversive to one student may actually be perceived as positively reinforcing to another. A loud and stern lecture on proper behavior delivered to the boy who craves attention is hardly punishment for him—particularly if it is done in the presence of his peers. Similarly, the lonely and lovesick girl sentenced to spend an hour after school with a handsome young male teacher doesn't perceive her predicament as the worst of all possible fates. Effective use of punishment, like so many other aspects of instruction, is dependent on the teacher's knowledge of individual differences among students.

Furthermore, don't make the mistake of equating punishment with the withholding of positive reinforcement or the application of "Grandma's Rule." What rule is that? Grandma might have said, "Yes, you certainly may have a portion of ice cream, as soon as you finish your meat and vegetables." Thus, *Grandma's Rule* occurs when *you make the satisfaction of someone else's wants dependent upon first achieving what you desire.* When the teacher tells the class that those who complete the set of practice problems in class will have no homework tonight while everyone else will have to complete the exercises and hand them in tomorrow, that teacher is using Grandma's Rule. Another example of a teacher using Grandma's Rule was the response of Mr. Norris to his seventh-graders. As the students entered his classroom, they began coaxing him to let them continue playing the educational game in progress at the end of yesterday's lesson. Mr. Norris's response was that if all students work effectively on the new lesson, they should have 10 to 15 minutes left for the game. Every teacher needs to master this approach because it works and because students react positively to its use when properly employed. Punishment, on the other hand, has been shown to have less predictable effects, and it often results in fostering poor attitudes toward school, subject matter, and teacher.

"O.K., but what am I to do when I *must* punish?" Let's consider some general guidelines and their application within the framework of realities set by typical public-school situations.

The first order of business is to *find out both the school and the department policy on discipline.* It is a rare school that doesn't have such policy statements written down somewhere. Seek these out and learn them before the first day you teach a class. Often the department has more specific guidelines designed to fit the general school policy. You also need to know these prior to your first day of teaching. Armed with this knowledge you should then be able to question experienced teachers regarding those all-important "unwritten" policies that everyone, including the students, knows—that is, everyone except you. It is too late to find out these things after you have embarked on a course of disciplinary action and find yourself in the embarrassing position of having to back down.

Warn only once. Then act. This is easy to carry out provided you catch "little things" which, if ignored, will rapidly grow into major problems. For example, if you notice Paul is writing definitions for his English vocabulary list during supervised practice, tell him in a whisper to put away that assignment and work on problems assigned. The next time you see him exhibit this type of misbehavior, either during the same or any subsequent class session, don't say anything to him. Simply pick up his paper, put it in your desk drawer, and refuse to discuss the matter during class. When he appears at the end of the period to plead for his paper, tell him you will deliver his paper to the English teacher with an explanation of how it got into your hands. If Paul repeats this type of behavior a third time, immediately tear up the paper, drop it in the wastebasket, and refuse to discuss the matter.

Immediacy is vital if the consequence (punishment) is to be connected in the student's mind to the cause (the act of misbehavior). Note in the illustration in the preceding paragraph that the punishment was initiated within seconds of the time the offense came to the teacher's attention. Suppose you had opted for telling Paul a second time not to do other homework in this class and you had sentenced him to stay after school to make up one period of class time. The time span between the undesired act and its consequence now becomes hours instead of seconds. Further, you are now punishing yourself by giving up after-school time needed to help slower students and to prepare materials for tomorrow's lesson. Also consider that staying after school may not be much punishment for Paul if he normally just hangs around after school with little to occupy his time. Then, too, in many school situations it is impossible to keep certain students the same day. For example, bus students must have a day's advance notice, band students cannot be kept from after-school rehearsals on Tuesdays and Thursdays, and varsity athletes are immune on game dates. By the time you get some students in after school, both you and they will barely remember the purpose! Experienced teachers learn to use staying-after-school punishment with discretion.

"Praise in public and punish in private" is an old slogan that retains quite a bit of applicability to the contemporary school scene. For example, once you have told Sue and Kathy (persistent whisperers) to see you immediately after class, go right on with the lesson. Steadfastly refuse to respond to their pleas of "We weren't doing anything." Such a conversation in the presence of other students will only serve to waste class time, divert other students' attention from the lesson, and provide an audience for whom Sue and Kathy feel compelled to provide an "act." Further, this matter is no one's business except yours and the offending students'. Experienced teachers also know that Tommy Tough Guy frequently becomes reasonable when dealt with in private and Mary Martyr climbs down from her cross when no audience of peers is around to appreciate her performance.

The punishment must fit the crime—both in kind and in severity—if it is to achieve maximum effect. An effective consequence of littering the floor with bits of paper is to arrange for the offender to spend some of his or her own spare time picking up similar debris *by hand* in several classrooms. Similarly, more than one desk carver has been cured by the experience of hand-sanding and refinishing the object back to its original condition (by all means, check school policy first on this one). Consider also the case of Spitting Sam, whom one of us caught doing his act from the front row of the balcony during an assembly program. Immediately after school he was handed a beaker, informed of the biology classes' need for 100 cc of saliva for tomorrow's laboratory experiment, and told he could leave when that need had been fulfilled. Interim drinks of water could not be allowed since the need was for saliva, not water. One *non*example of the point being made here is the practice of having students write "I must not talk out of turn during class" some horrendous number of times. Typically this serves more to develop an aversion to writing than anything else!

It is the act which must be punished, not the person. All of us must *make certain* we communicate this message to students when we administer punishment. Words alone help some, and making the kind of punishment match the kind of behavior at issue often helps even more. But it is the teacher's behavior toward the offending student *after* the punishment that can clinch the point. Every effort and means must be employed to project attitudes such as "Your mistake is over and done with, and I bear you no grudge" and "I expect you have changed and no recurrence of the problem is anticipated." A student who is left with the impression that he or she is a marked person whom the teacher will seek opportunities to pick on is very likely to become a repeated offender due to a "what's the use" attitude.

Consider three different approaches to handling collusion during a test. One approach is to tear up the offenders' papers and record zeros in the grade book. A second is to allow the offenders to complete the test believing they are undetected, score their papers, take the higher score, and award half to each paper in question—justifying this in private conversation with these students as the fairest way you could devise since you don't know who contributed which proportion of the correct information. A third approach is to do or say nothing about it this time and prepare two identical-looking but varying forms of the next test. The easiest way to do this is to vary the order of choices for each multiple-choice question. Identify the two test versions *only* by a period after the test title on one form, stack the papers so alternate forms are distributed to students sitting next to each other, proctor as usual; then collect, sort, score, and return the papers. Which of the three approaches best matches what has been pointed out about punishment thus far? In which do guilty students catch and punish themselves? If you chose the third approach, you agree with us. You should also have noticed that the third approach avoids the trap of requiring that the teacher "prove" the students cheated if irate parents phone the principal.

Consistency from day to day and from student to student is vital. You must learn to resist the tendency to overlook an infraction when perpetrated by Bright Bill and yet to come down hard on Slow Sally for the same breach of conduct. Student morale suffers badly if they even suspect their teacher is playing favorites. Likewise, tolerating loud and boisterous behavior during laboratory work one day and punishing it the next confuses students about the teacher's expectations. Developing a consistent approach takes a conscious effort on the part of most teachers, but the results are well worth the effort.

Suppose you have tried various means to deal with Terrible Tim, but nothing seems to work in his case. You have reached the end of your rope. He simply cannot be allowed to continue disrupting the lessons. Should you call for help, or would this be admitting to students, parents, and school administrators that you are incapable of handling the class? If you have done everything expected of a competent teacher, you should appeal to your cooperating teacher first and department chairperson next. If the problem persists, go for outside assistance at this point. Phone the parents, describe the evolution of the problem, tell them what you have tried to do to date, impress upon them your responsibility to others in the class, and ask for their assistance in correcting the situation. If this doesn't lead to immediate resolution of the problem, ask the guidance counselor to schedule a conference including the parent, the teacher, and the couselor. Should these measures fail, the next step is to remove the student from class and turn the matter over to the department chairperson for resolution with the vice-principal or principal. Many school systems now have access to the services of a school psychologist, who may be called into the situation at various points in the process. By all means, double-check the school policy *before* you initiate the first steps that could ultimately lead to the removal of a student from class, as this is a very serious matter.

In any case involving punishment it is critical to be sure you have first identified the right student(s) as targets for corrective action. This seems obvious, doesn't it? We call attention to it here because we have observed more than a few novices err in this regard. For example, Ms. Smith turned from writing on the board just in time to see Marge slap at Walter, who sat behind her. Marge was kept after class and lectured to on everything from the evils of distracting others to standards of ladylike behavior. Ms. Smith raved on at such a pace that Marge could not get a word in edgewise. By the time Marge was given a chance to speak, she was so angry that she refused to say anything. Her anger flared even more when, after being dismissed by Ms. Smith, she noted Wiley Walter bragging to his friends about how he had gotten away with jabbing her with a compass. What a difference it would have made if Ms. Smith had asked Marge the reason for her behavior *prior* to starting her aversive lecture!

Some techniques for catching those guilty of copying neighbors' test answers have already been described. But how can the teacher catch litterbugs and desk carvers when five different classes use the same room during a day? Let's assume Mr. Johnson finds carving on a certain desk as he is trying out a demonstration activity after school. Which of the five students who sit at that desk is the guilty one? Mr. Johnson need only take a quick tour of the area during the last minute of *each* class period for a day or so to find out. When additional embellishments are added, his question is answered. Think through other possible instances where it may not be obvious which student or students are the actual offenders and consider alternative ways of identifying the individuals at fault. This kind of thinking ahead should be excellent preparation for the moment when actual decisions will have to be made.

If you have read any books or articles on the topic of behavior modification, you should recognize that we have incorporated many of these techniques into the material presented here. We encourage the reader to use this chapter as a point of departure for further reading and especially recommend *Changing Children's Behavior* (see Suggestions for Further Study).

12.6 SO YOU'RE A TEACHER

Are you sure you want to be a teacher? What kind of person will you become? Must you stop smiling, never make mistakes, always behave in the "proper" way? In the early twentieth century citizens required their teachers to adhere to certain rules and regulations. One now-famous list given to New York City teachers admonished them against courting on weekdays, frequenting barbershops, and using curses or other foul language. Today it's a rare community that has written rules of behavior for teachers, although some still cling to unwritten rules. You should check out such unwritten rules with experienced teachers before you accept a job in the community.

But, more important than any rules are the human characteristics which encourage that positive atmosphere in which students are able to learn. We've said often that you must be able and willing to be effective and consistent. Are you willing, or are you plagued by doubts about the teacher image? Let no one tell you that students and parents will ignore the teacher who gets noisily drunk in the local tavern as they ignore the similar behavior of their favorite auto mechanic. The mechanic will often service the family car with no loss of effectiveness, but the teacher may be forced to deal with in-class misbehavior as a consequence of a similar tavern episode. How strange! Some articles in popular magazines seem to imply that teachers don't have to worry about serving as adult models, that today's students don't want their teachers to play this role. Ask a student teacher who has just given an anonymous attitudinal survey to the class about the nature of student responses to class management. Students regularly reiterate their desire to have a class environment suitable for learning and their belief that the teacher is the one to ensure that this is the case. Statements such as the following are often made: "You should have told the kids who talk to shut up," "You let the kids get away with too much noise," or "It's not fair to the ones who are trying to work when other kids clown around. You should stop them." These statements speak loud and clear. *You* are supposed to be the leader. It's your job to see that all are given an equal chance. The students look to you to exercise that leadership.

A little thought will also convince you that teenagers want adult models desperately and are today sometimes deprived of them by well-meaning, confused parents and teachers—the two classes of adults who have the most contact with teenagers. Both parents and teachers sometimes try to be friends, just like Eddie or Rose. But most teenagers have plenty of friends their own age with whom they can share secrets, exploits, and misdeeds. They occasionally want the very advice they seem to ignore—for example, the firm decision about curfew and the consequences which follow when that curfew is ignored. Tony, a sophisticated eleventh-grader who never did any homework and had a record of numerous "cuts" and failing grades, was inspired to do one unusual homework assignment by a thoughtful student teacher. The next day he told that student teacher how both his parents expressed pleasure that he had brought a schoolbook home and was doing homework. "But," said Tony, "they never told me before that they cared whether I did homework. If they'd told me how upset they were, I would have done it earlier." There's a moral in the above true story for all who work with teenagers. Attempting to reduce the barriers between student and teacher by becoming "just one of the guys" may actually create a barrier so high that no student will seek out your advice, accept your leadership, or try to attain the goals you set.

How will you accord your students the same measure of respect that you want demonstrated toward you? Will you listen thoughtfully when students are responding to a question even when they are stumbling with both ideas and words? Will you encourage students to try without fear of a verbal put-down if they make an honest error? Some teachers mistakenly believe that they are preserving the integrity of the subject matter by insisting on precision and exact answers even during the early stages of the development of a topic. In fact they discourage insight, do harm to the understanding of the content, and present an uncompromising personality.

Students have bad days, too! When Bonnie, usually a cooperative, pleasant student, suddenly erupts and spits out a defensive remark, the thinking teacher won't lash out in return, but may quietly, and for Bonnie's ears only, signal a "calm down." That teacher will probably remember to unobtrusively seek Bonnie out at the end of class—not as a punishment, but to find out if help is needed. But beware! We all have an aversion to the classic "do-gooder." A teacher's help *cannot* be forced on a student.

If students have a right to make an honest error, to try different approaches to a problem, and to expect that there should be a reason why they are required to learn something, then teachers also have a right to make an honest mistake, to admit they don't have all the answers, or to say that question never occurred to them. None of the above teacher behaviors will diminish the respect of the students for him or her if they are appropriately followed up. Ms. Steyer says she has no idea why the symbol for integration is $\int$, but admits that that's a reasonable question and says that she will find out and answer it tomorrow. Mr. Mopper was asked the same question, but he countered with, "That's a question you should all be able to answer! Since you asked, Barry, your assignment is to look up the answer and report to all of us tomorrow." Which response adds to the teacher's stature? You may be sure that Barry has learned not to ask curious relevant questions of this teacher. Notice that the word *relevant* was used in the previous sentence. Students come in all flavors, and you'll be missing the spice of life if you are never faced with one or more good-humored clowns, with their irrelevant and funny questions. It is up to you to discriminate between the good-humored clown and the so-called troublemakers. Neither one must be allowed to steal the time of other students, but your response to each differs significantly. Reread Section 12.5 and try to identify differences. In no case should you try the game of one-upmanship! Once a teacher begins to respond to a funny remark with a funny remark, the class is off in a contest of wits which has nothing to do with the lesson. Nine times out of ten the teacher is no match for the student, and even the one time the teacher wins, he or she has lost. The students have learned that the teacher is fair game and the traps will be set with repeated frequency.

Are you sure you want to be a teacher? Yes, you *can* smile and you *will* make mistakes, but you must realize that you will be seen as a model. If you are able to roll with the punches, to keep your enthusiasm for teaching your subject despite long, hard hours of work, and to find satisfaction in the long-term accomplishments of your students, then you are the kind of person who can help your students become all they are capable of becoming.

12.7 CROSS-REFERENCES TO A SAMPLE CBTE EVALUATION INSTRUMENT

You've probably already identified the competencies that match the theme of "Discipline." Category 6 stands out like the proverbial sore thumb with its references to "disruptive student behavior" and "rights and property." Notice how the specific items are inexorably connected to earlier chapter themes. A teacher who is obtaining *feedback* from a representative sample frequently or who is promoting *interaction* of students by choice of *modes* is simultaneously working at item 6.1. Similarly, if you conducted supervised practice in the manner recommended in Chapters 1 and 2, you'd alter your voice volume (7.1), be alert to the activities of many (7.2), and thereby be demonstrating the competency described in item 6.2. You can easily provide your own examples from this chapter for items 6.3 and 6.4, while item 6.5 was the theme of a major section (12.3). But doesn't "positive techniques" imply the use of praise as well? It most certainly does, and that should remind you of all the ways you can give feedback to students—ways which can serve as one positive approach to discipline.

Why, if this bugaboo of discipline is so often the downfall of beginning teachers, did we place this chapter last and these competencies among the last categories? Our conviction, supported by the experiences of the hundreds of novices we've supervised, is best described by the heading of Section 12.3: "An Ounce of Prevention." Read and believe! If you would be a teacher, put your effort into designing and implementing relevant and meaningful lessons. Then the big problems of control will not occur.

12.8 SUMMARY AND SELF-CHECK

Whether we use the words *control, discipline,* or *management,* we are labeling a set of competencies that are a necessary but not a sufficient prerequisite to instruction. In other words, if a teacher is unable to promote and maintain an environment for learning, no other instructional competencies the teacher may possess will be realized. However, the reverse is not true! The ability to stave off chaos does not automatically include the ability to instruct in an interesting and meaningful manner. The development of worthwhile lessons will require further work on *all* the components of the systems analysis model.

You would be well advised to remind yourself of the "big culprits" as you progress through your student teaching. Experienced teachers have told us that this chapter is a gold mine of practical ideas that they have found helpful. All of us at times fall into a comfortable rut of bad habits—habits that may prove to be disastrous in a new class.

At this point you should be able to:

1. Give illustrations of negative and positive student attention signs.

2. Identify instances of negative and positive student attention signs and of appropriate and inappropriate kind and amount of noise level during either a "live" or canned lesson and state reasons for your judgments.

3. Identify instances of the "big culprits" during either a "live" or canned lesson and describe the consequences which occurred when the teacher failed to attend to the situation.

4. List four to five preventive measures that should be part of each beginning teacher's behavior pattern.

5. Explain how you could apply Grandma's Rule, given a specific description of a class situation.

6. Describe remedial steps you would take to solve a specific misbehavior problem and be able to justify your decisions.

In the following exercises you are given an opportunity to test your ability to meet some of the above objectives. Be sure to check your responses with those of your classmates and your instructor. Perhaps more than any other area treated in this text, management requires more than an intellectual response. You must learn to know yourself, the attitudes you project in the classroom, and the effects of your behavior on students. You must respect both yourself and your students. It is a lifelong study for those who want to be teachers.

12.9 SIMULATION/PRACTICE ACTIVITIES

A. The first day Mr. Atkinson met his class, two students arrived a half-minute late. On the second day three students were a minute late and Mr. Atkinson waited to begin class until they were in their seats. By the end of the week he was delaying the start of class until five minutes after the bell because the students continued to trickle in throughout this time interval.

   1. What do you predict will be the situation by the end of the second week? Why?

   2. What alternatives are open to Mr. Atkinson for Monday of the second week?

   3. Which of the alternatives identified in your response to 2 are likely to lead to a worse situation? Why?

   4. Which of the alternatives identified in your response to 2 promise to lead to improvement? Why?

B. Ms. Brown complains that her class sessions begin quite promptly but that trouble starts about the middle of the period and the situation rapidly degenerates to obvious chaos by the end of the class. When asked to specify the nature of the problem, she replies that the students have no manners and such poor attention spans that even frequent switches of teaching modes get her nowhere. A check of her lesson plans reveals well-sequenced lessons, frequent change of modes, adequate plans for getting and giving feedback, many practical applications, and good relevance to the "here-and-now" interests of students. Plans for pacing seem adequate and there is an excellent match of strategies and stated objectives. On paper she looks great!

   1. Assume you are going to observe Ms. Brown in action. What would you focus on in this observation? Why?

   2. What is (or are) the most likely problem(s) you would expect to find?

   3. Cite specific suggestions you would make based upon the data you would expect to gather in 1.

C. Mr. Forgette's class proceeds smoothly as long as the modes employed are short lectures, question/answer, or teacher demonstration. Whenever he employs laboratory (activity learning), small-group discussion, or sends some students to the board, discipline disintegrates.

   1. What are possible causes of the problem?

   2. Which of the possible causes is the most probable? Why?

   3. What is the basic cure for the cause cited in 2?

D. The students in Ms. Klutz's class are busily working—that is, all but Pete and Harry. Ms. Klutz, a pretty first-year teacher, hears a sound and glances up just in time to see a note being passed from Pete to Harry. She stalks down the aisle and loudly demands the note. The rest of the students stop work and watch as Harry obsequiously hands Ms. Klutz a folded paper. Ms. Klutz threatens, "If this note is so important, maybe I should read it to the entire class!" Pete smirks and says, "Go ahead." When Ms. Klutz looks at the paper, she is horrified to find an obscene remark directed at her.

1. It is clear that Ms. Klutz has already firmly entrapped herself. Describe the best alternative(s) at this point.

2. If Ms. Klutz had it to do over, what effective approach(es) could she have used from the instant she spotted the note? Be sure to consider the limitations of each alternative approach you suggest.

E. Barbara approached Mr. Zillis after class and invited him to attend a party at her house following the Friday night pep rally and bonfire. Mr. Zillis is a young unmarried teacher with no previous commitment for that evening.

1. Describe a few potentially troublesome situations Mr. Zillis might be getting into if he accepts this invitation.

2. What further information should he seek, and from whom, prior to deciding whether to attend?

SUGGESTIONS FOR FURTHER STUDY

Home, L., and D. Tosti. *Behavior technology.* San Rafael, Calif.: Individual Learning Systems, Inc., 1971.

The two booklets in this set are composed of a series of modules each dealing with a single aspect of motivation and contingency management. Each module consists of a brief reading passage followed by a "progress check" which enables the reader to get immediate feedback on comprehension. The entire set is appropriate to the topic of discipline or pupil management and should prove useful to teachers needing help with these problems.

Krumboltz, J. D., and H. B. Krumboltz. *Changing children's behavior.* Englewood Cliffs, N. J.: Prentice-Hall, 1972.

The authors challenge the advocates of both permissive and authoritarian approaches and give numerous examples of how to encourage sought-after behavior and ways to diminish inappropriate behavior. The book is based on recent psychological research in behavior modification and is written in a clear and interesting manner. Both inexperienced and experienced teachers will find much of value in this volume.

Speeth, K., and D. Tosti. *Introductory psychology.* San Rafael, Calif.: Individual Learning Systems, Inc., 1973.

Each booklet in this series is composed of several modules. Brief reading sections in each module supply information and are followed by self-tests designed to give the reader immediate feedback on comprehension of the reading passage. Remediation exercises are spaced throughout. The modules recommended for attention in connection with discipline are "Classical Conditioning," "Operant Conditioning," "Reinforcement Schedules," and "Aversive Control and Avoidance Behaviors."

MODULAR ASSIGNMENTS

RESOURCE FILE MODULE

A. Objectives

    1. To build an organized resource file for *one* of the major units you will teach.

    2. To begin an organized resource file for each of the other *units* in the course used above.

    3. To devise an organizational scheme via 1 or 2 above that will serve as a pattern for developing resource files for each future course you teach in full-time teaching.

B. Enabling activities

    1. Study the syllabus, the text, and teacher's handbook for the chosen course and identify the major objectives for the unit to be used in A-1, and then the entire course.

    2. Consult the annotated bibliographies in this text.

    3. Look through the most promising of the sources referred to in B-2.

    4. Look for additional ideas in the university library, the public library, the professional and school libraries at the secondary school where you will student-teach, supermarkets, department stores, garages, newspapers, and popular magazines.

    5. As you find potentially useful ideas, record the bibliographic data and a summary of the material. Include Xerox copy *only* when you intend to use the material exactly as given. (N. B. A resource file is not very helpful if you must go back to the original source at a later date.)

C. Directions

    1. Obtain 8-12 manila folders. Label each with the topic of a unit in the selected course. (As your file grows, you will find it helpful to add subtopics to the outside of each folder and/or to subdivide within a folder via colored divider sheets.)

2. Return to the materials collected in accordance with B-5 and now identify projected use(s) (for example, a bulletin board display, an attention getter in a lesson on _____, an enrichment assignment in _____ for more capable Ss, a lab to introduce the rule _____, and so on). Be sure to briefly summarize details of that use, whether the lab, demo, reading, or other mode will be used to introduce, summarize, give practice in,...a concept or rule and which concept or rule will be emphasized. (Other suggested areas of resource uses are games, pictures and sketches of bulletin boards, ideas for projects, supplementary readings on a range of reading levels, transparency ideas, directions for constructing homemade equipment or models, field trip possibilities, and laboratory activities.) Be sure to note any modifications needed for the class you will have (for example, rewriting at a lower reading level). A sample card from a folder labeled "Variability/Statictics-Grade 9" is depicted in Fig. A.1.

3. The set of 8-12 folders should be sprinkled liberally with items which emphasize:

   a) Mathematics/science interrelationships, as well as those to other content areas.

   b) Content related to the Ss "here-and-now" interests.

   c) Ideas from *at least two* national and/or international curriculum projects.

   d) Ideas from *at least three* different teaching journals in your content area (as well as the interrelated mathematics/science journals).

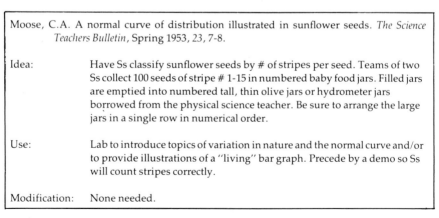

Moose, C.A. A normal curve of distribution illustrated in sunflower seeds. *The Science Teachers Bulletin*, Spring 1953, 23, 7-8.

Idea:	Have Ss classify sunflower seeds by # of stripes per seed. Teams of two Ss collect 100 seeds of stripe # 1-15 in numbered baby food jars. Filled jars are emptied into numbered tall, thin olive jars or hydrometer jars borrowed from the physical science teacher. Be sure to arrange the large jars in a single row in numerical order.
Use:	Lab to introduce topics of variation in nature and the normal curve and/or to provide illustrations of a "living" bar graph. Precede by a demo so Ss will count stripes correctly.
Modification:	None needed.

Fig. A.1 Sample card from a resource file folder.

SEQUENCE OF CONTENT MODULE

According to mathematicians, experienced teachers, and cognitive psychologists, the discipline of mathematics is a tightly hierarchical structure. Unfortunately many novice teachers view their subject matter discipline as a massive collection of facts and an arbitrary set of rules with few connections. Moreover, secondary school students are unlikely to learn anything about the structure of the subject matter unless relationships and connections are made explicit by their teacher often and in various ways. Thus, the planning of instruction consistent with the structure of the subject matter requires serious homework by the novice teacher. It is to that end that this module is directed.

A. Objectives

    1. To construct a learning hierarchy for one of the major units you will teach.

    2. To write a sample test item for each objective in the learning hierarchy.

B. Enabling activities

    1. Study both content sequences and exercises in the syllabus, text, and teacher's handbook and identify the major objectives of the unit and some of the assessment measures found in those sources.

    2. Study the learning hierarchies diagrammed in Chapter 6 of this text.

    3. Use the guidelines proposed in the "Task Analysis" section of Chapter 6 to sequence the set of objectives listed in the same section.

C. Directions

    1. Select a unit comprising two to three weeks of instruction.

    2. Write five to seven *major* cognitive instructional objectives and one sample corresponding test/quiz item per objective. Each objective and its corresponding test item should appear on a separate index card. Arrange these in a learning hierarchy by clipping or taping cards to a large sheet and adding appropriate arrows.

    3. Then analyze *each* major objective to identify prerequisites not already written. Write these additional instructional objectives and corresponding sample test/quiz items on index cards. Insert these in appropriate locations in the learning hierarchy and change arrows as needed.

    4. Finally, write four to six *future* objectives (that is, from units to be taught after the selected one) which depend upon one or more of the objectives already written. Write corresponding sample test items. Again adjust the learning hierarchy arrangement and arrows so as to correctly depict the overall sequence. Key the three types of objectives by a color code or diverse shapes. Identify the unit and grade level on the final chart.

LEARNING ACTIVITY PACKAGE (LAP) MODULE

A. Objectives: Construct a LAP with the following ingredients:

1. Specified behavioral objectives in cognitive (*must* include at least level III), affective and psychomotor (where applicable) domains,

2. Self-instructional for the most part and self-paced to some degree,

3. Streamed with plans for remediation *and* enrichment,

4. Individual and/or small group laboratory activity (the basic mode of instruction),

5. *Relevant* activities from the point of view of Ss' present interests,

6. Mathematics/science interrelationships emphasized,

7. Activities which clarify the structure of ideas of the discipline and the processes used to generate these ideas, and

8. Plan for gathering evidence on *all* specified objectives:

   a) A written test with table of specifications and key with credit distribution,

   b) Evaluation designed to assess other objectives, or to get further evidence in addition to the written test.

B. Enabling activities

1. Study Chapters 9, 10, and 11 for resource ideas.

2. Analyze LAPS designed by others (but not those that are chiefly Skinnerian programs with no emphasis on the laboratory mode).

C. Directions

1. Choose a content area that you taught which would comprise about two weeks' work. Use, as a frame of reference, the Ss you worked with during the clinical experience.

2. Write a teacher's guide to the LAP. Include:

   a) Characteristics of Ss for whom you are designing the LAP, the content area chosen, special characteristics of room.

   b) Teacher's plan for management of resources: Ss, room, furniture, special equipment.

   c) Plans for evaluation (include tests, quizzes, observational forms, keys for tests, and any other assessment means you will use).

d) Lists and descriptions, where needed, of LAP resource materials which teacher must organize and/or prepare. These might include lists of Ss' readings designed for different levels, sample equipment, models, drawings, charts, tapes, slides, transparencies, and so on (sample of the actual equipment referred to in the LAP should be appended).

e) Bibliography: List the major sources you drew on to write your LAP. You should include at least *one* national or international curriculum project and articles from at least *three* mathematics/science teaching journals among these sources. Put these in standard bibliographic form.

3. Write a LAP with cover and title as it would be designed for the students.

    a) Include instructions for Ss on *how* and *why* to do the LAP.

    b) Follow LAP samples, but modify where necessary so as to meet the objectives for this module. Note: Affective objectives may be stated in the teacher's guide only while other objectives must be included within the student's copy of the LAP.

    c) Be sure to include answer key(s) (not to tests, but for developmental questions and practice exercises), specific directions for labs, and planned ways to get and give feedback.

    4. Final preparation:

The LAP should be typed or neatly printed in a way which could actually be reproduced for a class of 30 or more students. (Note: Colored ditto masters are available and colored sheets may be used with mimeo stencils. Therefore, we suggest that you key pages or sections of pages which you would place on colored paper or type on colored masters.) It is also suggested that you arrange with several fellow students to exchange copies of the finished products. By so doing you will acquire several LAPs for future use with classes you will teach.

COMPETENCY-BASED TEACHER EDUCATION

A SAMPLE CBTE EVALUATION INSTRUMENT

STATE UNIVERSITY OF NEW YORK AT ALBANY
ALBANY MATHEMATICS SCIENCE TEACHING PROGRAM (AMST)

Student teacher _____ Subject(s) taught

_____

School in which student
teaching was done _____ Dates _____ through _____ 19____

Cooperating teacher _____ University supervisor _____

On this page, the student teacher's actual teaching assignments, the level of students in each class, out-of-class duties, rate of progress and need of supervision are described. (The school and the university attempt to plan the range of assignments so that pupils of varied learning abilities and classes at both the junior- and senior-high level can be included in the student teacher's experience. Particular attention is given to insuring experience in instructing slow learners.) Strengths unique to this teacher are highlighted. Competencies on which the teacher made progress but not in a consistent way are distinguished from those on which no or little progress was demonstrated. Data for this page is obtained from all three parties but the final responsibility for writing an overall descriptive summary is that of the college supervisor.

Also included on this page is an assessment of the long-term resource file and the two-week learning activity package developed by each student teacher. The extent to which the graduate has been able to develop materials suitable for differing individuals and situations is described.

_____
Signature of University Supervisor

CRITERION MEANS APPROPRIATE AND EFFECTIVE PERFORMANCE

Below criterion

Meets criterion

1. Selected objectives.

☐ ☐   1.1 Included those designed to effect positive attitudinal changes toward the subject matter.

☐ ☐   1.2 Included those designed to effect the attainment of intellectual skills at the levels of recall and type problems.

☐ ☐   1.3 Included those designed to effect the attainment of intellectual skills beyond the levels of recall and type problems.

☐ ☐ (SCI) 1.4 Included those designed to effect the attainment of psychomotor skills.

2. Executed the lesson plan.

☐ ☐   2.1 Used instructional strategies which matched stated objectives.

☐ ☐   2.2 Used instructional strategies consistent with the intellectual developmental stages of the student.

2.3 Demonstrated the use of instructional modes.

☐ ☐   2.31 Used more than one mode during each instructional period.

☐ ☐   2.32 Made formal oral presentations (short or long duration).

☐ ☐   2.33 Utilized sequences of developmental questions.

☐ ☐   2.34 Asked thought-provoking and/or open-ended questions.

☐ ☐   2.35 Developed productive ways of making use of student questions and answers.

☐ ☐   2.36 Initiated and sustained discussions among students.

☐ ☐   2.37 Used real objects and/or physical models in demonstrations.

☐ ☐   2.38 Utilized laboratory activities (student manipulation of materials).

☐ ☐   2.39 Provided for and supervised student practice.

☐ ☐   2.310 Designed assignments to lead to the discovery of concepts, rules or skills taught in a subsequent class session.

CRITERION MEANS APPROPRIATE AND EFFECTIVE PERFORMANCE

Below criterion | Meets criterion

☐ ☐ **2.311** Designed assignments which reinforced the concepts, rules, or skills taught in a previous class session.

2.4 Sequenced instructional modes.

☐ ☐ **2.41** Developed presentations which captured student attention at the start of lessons.

☐ ☐ **2.42** Arranged instructional modes so that student attention was maintained throughout the lessons.

☐ ☐ **2.43** Sequenced subject matter in ways meaningful to students.

☐ ☐ **2.44** Related present instruction to previous learning.

☐ ☐ **2.45** Related present instruction to future lessons.

2.5 Used feedback during the lesson.

☐ ☐ **2.51** Provided a model of correct performance of skills to be learned (to enable students to get feedback on their efforts).

☐ ☐ **2.52** Gave immediate feedback to individuals and groups.

☐ ☐ **2.53** Collected feedback frequently from a broad sampling of students.

☐ ☐ **2.54** Made immediate use of feedback to pace instruction.

☐ ☐ **2.55** Made immediate use of feedback to alter the planned sequence of instructional strategies.

2.6 Provided for individual differences.

☐ ☐ **2.61** Demonstrated regard for the worth of individual personalities of students.

☐ ☐ **2.62** Provided for applications of subject matter based on the "here-and-now" interests of individual students.

☐ ☐ **2.63** Designed individualized instructional materials based on varying objectives and learning styles.

☐ ☐ **2.64** Designed *differentiated* assignments based on student interests and levels of previous learning.

CRITERION MEANS APPROPRIATE AND EFFECTIVE PERFORMANCE

Below criterion   Meets criterion

☐ ☐   **2.65**   Used a variety of procedures such as grouping within a class and/or assignment of individual or small group projects.

☐ ☐   **2.66**   Used out-of-class time to help individual students.

☐ ☐   **2.67**   Used school resources (such as pupil personnel services, teacher, and so on) to deal with individual differences.

3. Clarified the processes and ideas of mathematics.

☐ ☐   **3.1**   Provided students with experiences in obtaining mathematical generalizations (patterns, rules, or principles).

☐ ☐   **3.2**   Provided students with experiences in using problem-solving strategies (e.g. testing and modifying conjectures, restructuring the problem).

☐ ☐   **3.3**   Distinguished between generalizations obtained from data and those obtained by deduction from assumptions.

☐ ☐   **3.4**   Provided a rationale for the terms, definitions, and algorithms of mathematics.

☐ ☐   **3.5**   Provided illustrations to show that mathematics is an invention of the human mind and an activity of human beings.

☐ ☐   **3.6**   Provided illustrations to show the distinctions between the ideal world of mathematical models (thought models) and the physical world in which the models are applied.

☐ ☐   **3.7**   Translated mathematical concepts and principles (rules) into valid forms (symbolic, schematic, or spatial).

☐ ☐   **3.8**   Called attention to content inaccuracies which appeared in verbal interaction, instructional materials, and student papers.

☐ ☐   **3.9**   Used correct and consistent mathematical language and symbolism.

☐ ☐   **3.10** Demonstrated sufficient depth and breadth of understanding in mathematics to respond to unplanned activieties, to "go beyond" text materials, *and* to spiral the teaching of concepts.

CRITERION MEANS APPROPRIATE AND EFFECTIVE PERFORMANCE

Below criterion Meets criterion

☐ ☐   **3.11** Related mathematics to other school subjects, especially science.

☐ ☐   **3.12** Provided examples of the application of mathematics to areas outside of school subjects.

4. Selected and utilized instructional resources.

☐ ☐   **4.1** Used a variety of available materials and media.

☐ ☐   **4.2** Used a variety of professional sources other than texts (e.g. professional journal articles, source books, handbooks).

☐ ☐   **4.3** Modified instructional materials to meet the special needs of the student and the limitations of the classroom.

☐ ☐   **4.4** Selected ideas from contemporary national and international curriculum projects to supplement those available in standard texts and state syllabi.

☐ ☐   **4.5** Tried out demonstrations before presenting them in class.

☐ ☐   **4.6** Demonstrated initiative in locating and using laboratory activities and demonstrations.

☐ ☐ *(SCI)* **4.7** Practiced safety precautions during laboratory and demonstration work.

5. Used tests and other evaluative measures.

☐ ☐   **5.1** Constructed tests to be consistent with emphasis of instruction.

☐ ☐   **5.2** Matched test items to instructional objectives.

☐ ☐   **5.3** Used item analysis technique to analyze the test's validity, potential weakness of instruction, and learning problems of individual students.

☐ ☐   **5.4** Constructed and used tests for diagnostic, as well as grading purposes.

☐ ☐   **5.5** Employed means other than written tests to gather evidence of learning.

☐ ☐   **5.6** Used a variety of remedial procedures based on student feedback.

CRITERION MEANS APPROPRIATE AND EFFECTIVE PERFORMANCE

Below criterion

Meets criterion

6. Managed students.

☐ ☐   **6.1** Involved nearly all students, verbally or nonverbally, in the ongoing classroom activities.

☐ ☐   **6.2** Demonstrated ability to work with single students or small groups while exhibiting awareness of the general activities.

☐ ☐   **6.3** Provided a classroom atmosphere which encouraged students to demonstrate regard for the rights and property of others.

☐ ☐   **6.4** Demonstrated the ability to control disruptive student behavior.

☐ ☐   **6.5** Used positive techniques to prevent and/or control student behavior.

7. Communicated with students.

☐ ☐   **7.1** Employed appropriate volume, modulation, pitch, and reasonable clarity of voice.

☐ ☐   **7.2** Maintained eye contact with a wide sampling of students.

☐ ☐   **7.3** Demonstrated enthusiasm for the subject via both verbal and nonverbal behavior.

☐ ☐   **7.4** Generated enthusiasm for the subject among students.

☐ ☐   **7.5** Conveyed to students the teacher's awareness of their interests and points of view.

8. Attended to the mechanics of instruction.

☐ ☐   **8.1** Demonstrated clerical efficiency with respect to the paperwork of the teacher.

☐ ☐   **8.2** Organized instructional materials in a systematic way.

☐ ☐   **8.3** Produced legible and accurate written communications (board, overhead, handouts).

☐ ☐   **8.4** Maintained a physical environment conducive to learning (such as lights, heat, ventilation).

☐ ☐   **8.5** Managed and utilized the physical organization of the room, equipment and materials.

BRIEF HISTORY OF DEVELOPMENT AND USE OF THE AMST CBTE EVALUATION INSTRUMENT

The initial draft of the AMST evaluation instrument was completed in the fall of 1971 by Dr. Walter Farmer and Dr. Margaret Farrell. Competency statements were generated from the systems analysis model used throughout this text. In the spring of 1972 this draft was submitted to a group of 12 area mathematics and science classroom teachers and chairpersons for reaction. A revised draft was then mailed to principals and department chairpersons in science and mathematics in the more than 60 secondary (7-12) public schools which had served as clinical settings for student teachers. Sixty detailed responses were received during the fall of 1972. An advisory committee consisting of an adminstrator, two mathematics/science chairpersons, six (7-12) science or mathematics teachers, and two college science/mathematics teacher education students worked through the written feedback and collaborated on needed revisions.

The revised instrument was used for the first time in the spring of 1973. At the end of that semester, a similar advisory group considered the instrument and the program on the basis of actual use. Minor revisions in sequence and language and the addition of some five competencies were incorporated into a revised instrument. Two more advisory group meetings in December 1973 and June 1974 resulted in only minor rewording of a few competency statements. Written feedback obtained each semester since June 1974 has identified no major areas of confusion.

The AMST evaluation instrument is *not* an observation tool. All student teachers and cooperating teachers are given the instrument prior to, or on the first day of, the AMST semester. However, it is used for the first time midway through the clinical phase when each of the concerned parties (college supervisor, cooperating teacher, and student teacher) independently assesses the progress of the student teacher. Consensus is achieved in a three-way meeting, which provides direction for all during the remainder of the clinical phase. Two comments should be made about the use of the instrument at this point in the student teacher's experience: (1) it serves primarily as a diagnostic tool, and (2) it is an explicit means of defining a *profile* of each student teacher's performance, thus providing baseline data for individualizing subsequent instruction with AMST trainees.

Even the "final" evaluation, as rendered by the cooperating teacher at the end of student teaching, has a distinctive diagnostic component. (The actual final evaluation which includes the finalizing of the items on the instrument and the preparation of a summary cover sheet is the responsibility of the college supervisor and occurs at the end of the semester.) This time lag is a deliberate attempt to allow student teachers to demonstrate criterion performance on a few items which do not require the student teacher to be in a classroom (see 2.63). These items are identified in an individual conference at the beginning of the postclinical phase, when the student teacher is asked to continue the self-analysis emphasized throughout the program and to make summative judgments based on available data with reference to his or her own evaluation.

But the AMST student teacher receives the instrument on the first day of the semester for another reason than that of its eventual use with reference to his or her own teaching.The evaluation instrument is a profile of the AMST program. Its objectives are the objectives of the AMST program. All instruction in each phase is designed to assist the student teacher in the eventual achievement of those objectives. Choice of clinical experiences, seminar design, and modular assignments depends on the framework of the evaluation instrument.

The instrument is novel in another way: it is not restricted to those minimal competencies which must be attained for entry into the profession. On the contrary, the competencies on the AMST instrument were meant to be descriptors of the effective teacher and it was assumed that competent novice teachers might not meet criterion on various subsets of the competencies. In many respects it is this stance which makes realistic the self-appraisal encouraged of student teachers during the AMST semester and hopefully continued by them in their teaching careers.

RELATED RESEARCH REPORTS

Farmer, W. A. , M. A. Farrell, and R. M. Clark. *A comparison of mathematics and science teachers in achieving selected teacher competencies* (a research paper presented at the annual meeting of the American Educational Research Association, March 31, 1978, Toronto, Ontario).

Farmer, W. A., M. A. Farrell, and R. M. Clark. *Test competency as related to classroom performance* (a research paper presented at the annual meeting of the New England Educational Research Association, May 5, 1977, Manchester, N. H.).

Farrell, M. A., W. A. Farmer, and R. M. Clark. *Field testing a systems analysis model for mathematics/ science teacher education* (a research paper presented at the annual meeting of the National Council of Teachers of Mathematics, 1976, Atlanta, Ga.).

FOOTNOTE REFERENCES

Ausubel, D. P. *Educational psychology: a cognitive view.* New York: Holt, Rinehart and Winston, 1968.

Ausubel, D. P. *The psychology of meaningful verbal learning: an introduction to school learning.* New York: Grune and Stratton, 1963.

Bloom, B. S., ed., M. D. Engelhardt, E. J. Furst, W. H. Hill, and D. R. Krathwohl. *Taxonomy of educational objectives. Handbook I: cognitive domain.* New York: David McKay, 1956.

Bruner, J. S. *The process of education.* Cambridge, Mass.: Harvard University Press, 1962.

Bruner, J. S. *The relevance of education.* New York: W. W. Norton, 1971.

Bruner, J. S. *Toward a theory of instruction.* New York: W. W. Norton, 1966.

Bruner, J. S., R. R. Olver, and P. M. Greenfield, eds. *Studies in cognitive growth.* New York: John Wiley and Sons, 1967.

Buckeye, D. *No read math activities* (Vol. 1,2,3). Troy, Mich.: Midwest Pub., 1975.

Dewey, J. D. *How we think.* Boston: Heath, 1910.

Elkind, D. Quantity conceptions in junior and senior high school students. *Child Development* 32 (1961): 551-560.

Farrell, M. A. *Geoboard geometry.* Palo Alto, Calif.: Creative Publications Inc., 1971.

Gagné, R. *The conditions of learning.* 2nd ed. New York: Holt, Rinehart and Winston, 1970.

Goals for the correlation of elementary science and mathematics (The Report of the Cambridge Conference). Boston: Houghton Mifflin Co., 1969.

Harrow, A. J. *A taxonomy of the psychomotor domain.* New York: David McKay, 1972.

Inhelder, B., and J. Piaget. *The growth of logical thinking from childhood to adolescence.* New York: Basic Books, 1958.

Krathwohl, D. R., B. S. Bloom, and B. B. Masia. *Taxonomy of educational objectives. Handbook II: affective domain.* New York: David McKay, 1964.

Lunzer, E. A. Problems of formal reasoning in test situations. *European Research in Cognitive Development,* (1965): 40-43.

McWhirter, N., and R. McWhirter. *Guinness book of world records*. Rev, and enl. ed. New York: Bantam Books, 1976.

National Council of Teachers of Mathematics. *Enrichment mathematics for the grades*. Washington, D.C.: NCTM, 1963.

National Council of Teachers of Mathematics. *Enrichment mathematics for high school*. Washington, D.C.: NCTM, 1963.

National Council of Teachers of Mathematics. *Historical topics for the mathematics classroom*. 31st Yearbook. Washington, D.C.: NCTM, 1969.

Nuffield Mathematics Project. *Problems—green set* (book and cards). New York: John Wiley and Sons, 1969.

Nuffield Mathematics Project. *Problems—purple set* (book and cards). New York: John Wiley and Sons, 1971.

Nuffield Mathematics Project. *Problems—red set*. (Book and cards). New York: John Wiley and Sons, 1970.

Renner, J. W., and D. G. Stafford. *Teaching science in the secondary school*. New York: Harper and Row, 1972.

Rogowski, S. *Computer clippings*. Palo Alto, Calif.: Creative Publications, 1976.

Russell, B. *Mysticism and logic*. W. W. Norton, 1929.

Sawyer, W. W. *Vision in elementary mathematics*. Baltimore: Penguin Books, 1964.

Sobel, M.A., and E. M. Maletsky. *Essentials of mathematics* (Vol. 1, 2, and 3). Boston: Ginn, 1969.

Suppes, P., D. H. Firl, E. M. Glass, J. Kaplan, C. A. Backman, and E. A. Anderson. *Mathematics one*. New York: Random House, 1974.

Thorndike, E. L. *New methods in teaching arithmetic*. Chicago: Rand McNally, 1921.

INDEX